To Catch a Virus

To Catch a Virus

SECOND EDITION

JOHN BOOSS, MD
MARIE LOUISE LANDRY, MD
WITH MARILYN J. AUGUST, PHD

WILEY

Washington, DC

Editorial Correspondence:
ASM Press, 1752 N Street, NW, Washington, DC 20036-2904, USA

Registered Offices:
John Wiley & Sons, Inc., 111 River Street, Hoboken, NJ 07030, USA

For details of our global editorial offices, customer services, and more information about Wiley products, visit us at www.wiley.com.

Wiley also publishes its books in a variety of electronic formats and by print-on-demand.

Some content that appears in standard print versions of this book may not be available in other formats.

Library of Congress Cataloging-in-Publication Data has been applied for

ISBN 9781683673736 (Paperback);
ISBN 9781683673743 (Adobe PDF);
ISBN 9781683673750 (e-Pub)

Cover image: Specter of death waiting over Panama (U. J. Keppler, 1904). Courtesy of Beinecke Rare Book and Manuscript Library, Yale University.
Cover design: Debra Naylor, Naylor Design, Inc

SKY10036151_092722

In tribute to Gueh-Djen (Edith) Hsiung, PhD,
who is remembered for her pioneering contributions
to the field of diagnostic virology, for training and inspiring
generations of diagnostic virologists with her passion for virology,
and for her social grace and generosity.

(Courtesy of Zhe Zhao.)

Contents

LIST OF ILLUSTRATIONS

Chapter 7

Chapter 10

Acknowledgments

The challenges in writing the history of the COVID-19 pandemic for this second edition have differed considerably from the challenges of writing the first edition. First have been the time spans. The viral infections and their diagnostics of the first edition (which comprise chapters 1–9 in the second edition) cover roughly a century, whereas the covered history of COVID-19 at the time of writing is about two and a half years. As a consequence, the sources used have differed considerably. Archived materials, edited books, monographs, and journals covering extended periods were examined for the viral infections in chapters 1–9. Narrative research on COVID-19 and its virus, SARS-CoV-2, has relied on contemporaneous sources. These have included prepublication reports online of reviewed and accepted journal articles, studies published online prior to peer review, and online reports from trusted news media. For example, experienced scientific and medical journalists of *The New York Times* and *The Washington Post* often have had access to emerging findings. The reader could then read the original investigation's findings through hyperlinks in online articles. Traditional print copies of newspapers, scientific and medical journals, and books continued to play an important role. Underpinning these resources has been M.L.L.'s role of directing a major university hospital's diagnostic virology laboratory.

The second difference has been the restrictions imposed by the pandemic. For the first edition, travel and meetings for interviews were unimpeded by concerns of infection. During the time of COVID-19, lockdowns, travel restrictions, and concern about contracting infection interfered with such meetings.

An experience common to both editions has been the strong and expert support of our efforts by the leadership and editorial staff at the ASM Press. For the first edition, Jeff J. Holtmeier, the Director of ASM Press at the book's inception, guided the development of the concept and provided patient encouragement. Christine Charlip, his successor, encouraged us to enlarge the audience from one that was primarily technical in orientation to one with broad interests in science and medicine. John Bell, the production editor, facilitated that shift. Artist Debra Naylor collaborated on the book's cover design.

Christine Charlip initiated the project of the second edition and provided encouragement at key points. Megan Angelini took on the role of managing developmental editor. She efficiently facilitated our work and added features, substantively adding to the value of the book. We are grateful to her and to Lindsay Williams, editorial rights specialist, for locating and gaining permission for use of figures in all chapters of the new edition.

Credit must go to laboratorians for doing the fundamental work of virological diagnosis, public health professionals for policies and practices that protect the public during the pandemic, health care workers for acute and chronic care of patients sick with COVID-19, and basic and clinical researchers for the critical work of figuring out how the disease works, how to treat it, and how to prevent it.

We are indebted to Professors Richard L. Hodinka, Irving Seidman, and Frank M. Snowden for providing in-depth expert reviews of the manuscript for the second edition as it was being developed. We are grateful to Frank Bia and Jung H. Kim for expert focused input. In a chance meeting, David Tkeshelashvili reminded J.B. of one of his core teachings. We are, of course, responsible for errors of fact, interpretation, and omission which remain.

Our thanks continue to extend to all those who contributed to the writing of the first edition, particularly Warren Andiman, Jangu Banatvala, Edward A. Beeman, Leonard N. Binn, F. Marilyn Bozeman, Irwin Braverman, Charlie Calisher, Dave Cavanagh, Gustave Davis, Walter Dowdle, Bennett L. Elisberg, Margaret M. Esiri, Durland Fish, Bagher Forghani, Harvey Friedman, D. Carleton Gajdusek, J. Robin Harris, David L. Hirschberg, Richard L. Hodinka, Robert Horne, Albert Z. Kapikian, Robert J. T. Joy, Edwin D. Kilbourne, Jung H. Kim, Diane S. Leland, W. Ian Lipkin, Dick Madeley, Kenneth McIntosh, Michael B. A. Oldstone, Stanley Plotkin, Morris Pollard, Philip K. Russell, Karen-Beth G. Scholthof, Gregory Tignor, and Alex Tselis.

Historians and curators are the guardians of the traces of our past, and several provided indispensable help for the first edition. At Yale they included Toby Appel, Melissa Grafe, Frank Snowden, and Susan Wheeler. Elsewhere, Steven Greenberg, Col. Richard C. V. Gunn, Sally Smith Hughes, and Sarah Wilmot provided invaluable guidance. Translations of scientific papers in French and German were essential. Mary Ann Booss and Brigitte Griffith offered expert advice in translating scientific publications in French, while Carolin I. Dohle provided expert translations of scientific publications in German.

A number of individuals helped to secure especially difficult-to-find figures or provided critical help in developing the first edition's manuscript. Zhe Zhao, Dr. Hsiung's grandniece, was kind enough to provide the portrait used with the dedication in both editions. Special thanks are extended to Paul Theerman and Ginny A. Roth at the National Library of Medicine for providing many high-resolution images from the library collection. Others included Joyce Almeida, Debbie Beauvais, Claire Booss, Robert B. Daroff, John and Donna Jean Donaldson, Will Fleeson, Emma Gilgunn-Jones, Tina Henle, Albert Z. Kapikian, David Keegan, Edwin Paul Lennette, Rich McManus, Venita Paul, Thomas Ruska, and Irving Seidman.

It seemed particularly important to gain an understanding of the first diagnostic virology and rickettsiology lab established anywhere. It was established in January 1941 at the Walter Reed Army Medical Center. In investigating the establishment and operation of that laboratory for the first edition, we had exceptional support from committed and knowledgeable personnel. We extended special thanks to Michael P. Fiedler, research librarian, Andrew H. Rogalsky, archivist, and Leonard N. Binn, whose career in virology at Walter Reed spanned over five decades. The first diagnostic virology and rickettsial lab in the U.S. civilian sector was at the State of California Public Health Labs. We are grateful to Bagher Forghani for generously and graciously providing a full understanding of that very important lab and its leaders.

For the first edition, M.J.A. was continually cheered by her dear, now late father, Ralph August, by friends near and far away, and by her Let's Look at Art docent colleagues at the San Jose Museum of Art. Through the second edition and the pandemic, she has also been inspired by her friends and colleagues at Art Yard BKLYN.

M.L.L. was grateful to be able to contribute to the unprecedented virology laboratory efforts necessitated by the COVID-19 pandemic, and to share some of those experiences in this book. Through two years of recurring testing challenges, she was buoyed as always by the support, wise counsel, and optimism of her husband, children, and grandchildren.

The first edition could not have been written without the support of J.B.'s family. Months turned into years, and stacks of books and reprints turned into multiple file cases. Patience and encouragement were constant. That has remained the case for the second edition. J.B.'s wife, Mary Ann, has offered daily insights from her public health background, assessing developments in COVID-19; Christine

Rankin, J.B.'s daughter, provided computer programs guidance and a keen reading of a draft, making very constructive suggestions; Dave, J.B.'s son, provided much-needed computer support and drew his attention to topics which hadn't yet been identified. J.B. is humbled and grateful for their support.

JOHN BOOSS, MD
MARIE LOUISE LANDRY, MD
with MARILYN J. AUGUST, PHD

Foreword

The first edition of *To Catch a Virus*, by John Booss and Marilyn J. August, was published in 2013. It provided an outstanding and entertaining historical account of diagnostic virology, from the initial demonstration that yellow fever was caused by a filterable agent to the development of molecular techniques that provide rapid and precise identification of many human pathogens.

Much has happened in the infectious disease world since 2013, most notably the COVID-19 pandemic, caused by a novel virus called SARS-CoV-2 (severe acute respiratory syndrome coronavirus 2). This pandemic occurred one century after the "Spanish" influenza virus that stretched from 1918 to 1920 and was the most lethal pandemic of the 20th century. Although that pandemic appeared at the time to be caused by a respiratory pathogen, it was not until the 1930s that influenza viruses were isolated and identified, initially in animals and subsequently in humans. Vaccines against human influenza took another decade to be developed and licensed by the U.S. Food and Drug Administration, and the first antiviral drug for influenza A (amantadine) was not approved until 1966.

Booss and Marie Landry in the new edition elegantly describe both COVID-19 disease surveillance and the molecular techniques for detection of SARS-CoV-2 that have allowed much more rapid development and implementation of diagnostic, vaccination, and therapeutic efforts, all within 2 years from the first outbreak in Wuhan, China, in late 2019. International investigators used next-generation techniques to rapidly determine the genomic sequence of the original virus, recognize new variants, and design safe and effective vaccines and treatments. Booss and Landry identify many scientific and public health heroes of the pandemic (Li Wenliang, Zhang Yongzhen, Sharon Peacock, Barney Graham, Jason McLellan, Katalin Karikó, Drew Weissman, Tony Fauci, and others) and present interesting biosketches of them. The spread of SARS-CoV-2 across the globe and the herculean efforts of such scientists to develop diagnostic, preventative, and therapeutic modalities, often in the face of difficult political impediments, are beautifully described.

No one knows when the next viral pandemic will occur or what virus will emerge to cause it. We now know that there are hundreds, possibly thousands, of viruses capable of causing human disease, and are better prepared to detect novel viruses, quickly identify them, and develop both vaccines and treatments for them. The progress made in the last decade is monumental, and Booss and Landry have described the process in gripping fashion.

Booss and Landry have again dedicated this book to their late mentor, Dr. Gueh-Djen (Edith) Hsiung, a major pioneer in the field of diagnostic virology. Dr. Hsiung would be proud to see the advances that have been made since her death in 2006, and the grandly successful efforts of her mentees in describing this progress in the current edition.

MARTIN S. HIRSCH, MD
Professor of Medicine, Harvard University
Division of Infectious Diseases, Massachusetts General Hospital
Editor-in-Chief Emeritus, The Journal of Infectious Diseases (2003–2022)

Preface

We are pleased with the reception of the first edition of *To Catch a Virus*, suggesting that there is a place in the literature for a book describing the history of how viruses are captured and identified.

With a nod to *To Catch a Thief*, Alfred Hitchcock's 1955 classic mystery film, the first edition served as a chronicle of discovery and diagnosis. It was a history of diagnostic virology from the initial diagnosis of a human viral illness at the turn of the 20th century by Walter Reed and the Yellow Fever Commission to the emergence of molecular methods of diagnosis and the human immunodeficiency virus (HIV/AIDS) epidemic more than a century later. We covered the first diagnostic virology lab at the Walter Reed Army Medical Center at the start of World War II. Diagnostic virology would emerge to sit astride the confluence of dynamic developments in science, public health struggles with epidemics and emerging diseases, and the intensive medical care of individual patients.

Since the publication of the first edition in 2013, several epidemic viral diseases have threatened human life. These included Ebola in 2014, Zika in 2015, and COVID-19 in 2019/2020. These came close on the heels of H1N1 influenza in 2009 and Middle East respiratory syndrome (MERS) in 2012.

Given the critical importance of the work to diagnose and treat these emerging viruses, Christine Charlip, Director of ASM Press, expressed an interest in a second edition of *To Catch a Virus*. Dr. John Booss contacted Dr. Marie Landry, Director of the Clinical Virology Laboratory at the Yale New Haven Hospital, to ask whether she would be interested in assuming the lead. Dr. Landry, who had

had a fundamental role in the first edition, although not credited, agreed to take on the task. Very soon thereafter, the COVID-19/SARS-CoV-2 (severe acute respiratory syndrome coronavirus 2) pandemic burst upon us, and Dr. Landry's diagnostic virology laboratory work became even more demanding. She agreed to continue with the second edition if Dr. Booss would oversee the project. Dr. Marilyn J. August, the coauthor of the first edition, was interested in supporting the new project and agreed to assist.

This edition, like the first, is dedicated to Dr. Gueh-Djen (Edith) Hsiung, a pioneering diagnostic virologist. She imbued dozens of trainees and hundreds of students with a passion for virology. She was a mentor to each of us.

The structure of the book remains the same as the previous edition, with most chapters built around specific viral diseases, such as yellow fever in the first chapter, to demonstrate the development of a technology. In the second chapter, polio, rabies, and influenza are described to show the use of animals and embryonated chicken eggs to isolate and identify viruses. Smallpox is described in the third chapter to demonstrate that the body's immune system, like the brain, has memory. Immunological memory provides protection from reinfection and allows the measurement of antibodies to identify a virus. Jennerian vaccination, inducing immunity, was the basis of the remarkable smallpox global eradication project. Some viral infections, like smallpox virus, rabies virus, and the herpes family of viruses, leave footprints called inclusion bodies. The detective work to recognize those footprints, the development of Rudolf Ludwig Virchow's concepts of cellular pathology, and the beginnings of electron microscopy (EM) are unraveled in chapter 4.

Chapter 5 details the events leading up to the development of tissue culture and the recognition of viral cytopathology (CPE) which would win the Nobel Prize for John Enders, Thomas Weller, and Frederick Robbins. President F. D. Roosevelt's polio and his "March of Dimes" would drive the nation's interest and funding.. Chapter 6 describes the virtual torrent of viruses captured and identified by the development of tissue culture for virus isolation and identification. At the National Institutes of Health in Bethesda, MD, the Laboratory of Infectious Diseases was a hotbed of viral discovery and disease investigation. Led by Robert J. Huebner, investigators such as Wallace Rowe, Robert Chanock, and Albert Kapikian made discovery after discovery linking viruses to illness or, conversely, showing them to be nonpathogenic passengers. The role of diagnostic virology labs at the state level, at the Communicable Diseases Center, and in university hospitals exploded from the latter 1950s onward. These laboratories defined individual patients' illnesses, often after the acute phase had passed. They also alerted the country to the appearance of epidemics such as influenza.

The next three chapters trace developments which brought diagnostic virology into active patient management. Chapter 7 describes the clinical application of EM and of fluorescent-antibody staining. EM came into its own with several advances, notably negative staining developed by Sydney Brenner and Robert Horne, allowing the detailed description of virus architecture. Viral gastroenteritis is used as the exemplar of disease in which EM played a defining role. Fluorescent-antibody staining, developed though the imagination and tenacity of Albert Hewlett Coons, allowed "taillights" to be put on molecules. That is, the technique would allow the identification of an offending virus. This methodology was applied by Phillip S. Gardner and Joyce McQuillin in pioneering efforts at Newcastle-upon-Tyne in the United Kingdom. They aimed to provide clinicians with a "rapid viral diagnosis," particularly of acute respiratory disease in infants and children, within 24 hours.

Chapter 8 describes the evolution of our understanding of viral hepatitis and how innovative immunological techniques, developed by both serendipity and ingenuity, led to the identification of some of the culprits. In the case of radio-immunoassay (RIA), Rosalyn Yalow and Solomon Berson originally developed the highly sensitive and accurate technique to measure human insulin and other endocrine molecules. When applied to viral diseases, it allowed the screening of the blood supply for hepatitis viruses. Less complicated and nonradioactive assays, such as the enzyme-linked immunosorbent assay (ELISA) and the enzyme immunoassay (EIA), soon followed. They transformed many aspects of biology, including the diagnostic process in virology.

Chapter 9 examines the molecular revolution in diagnosing and managing viral diseases. It tracks HIV/AIDS, and the application of molecular methods for discovery and control. The syndrome first appeared in 1981 as a virtually inevitable death sentence. That characterization was transformed by the use of highly active antiretroviral therapy, including a protease inhibitor, in 1996. HIV/AIDS current status is as a managed chronic disease in those individuals who have access to molecular viral diagnostic assays and a wide spectrum of specifically targeted antiretroviral drugs.

The molecular foundation for these developments started with the demonstration of DNA as the basis of heredity by Oswald Avery in the 1940s. The demonstration of the double helix by James D. Watson and Francis Crick, using X-ray crystallographic data of Rosalind Franklin, allowed the cracking of the genetic code and the molecular biological revolution which followed. Watson and Crick, with Maurice Wilkins, would win the Nobel Prize. Franklin, whose data had been crucial, was not included. Another key development was the demonstration of the enzyme reverse transcriptase, which transmits genetic information from RNA to DNA and facilitated the discovery of HIV. In another crucial development, exquisitely sensitive detection and quantitation assays were produced. They are based on nucleic acid amplification principles developed for PCR by Kary Mullis. PCR and other molecular techniques such as nucleic acid sequencing allow the measurement of the amount of HIV in plasma, i.e., "viral load," and the determination of mutations of the virus, facilitating management of antiviral therapy.

Molecular diagnostics have been streamlined so that the individual steps of nucleic acid extraction, amplification, measurement, and detection are done in closed systems in "real time." Hence, the highly trained diagnostic virology specialists of the tissue culture era are being replaced by less specialized staff. Commercially available tests have reduced the complexity of laboratory procedures with the benefits of automation and high throughput. For high-volume testing, robotic pipetting instruments are employed, with workflow guided by computers. In the opinion of a number of experienced diagnostic virologists, the diagnostic virology lab, as we have known it, is becoming a thing of the past. Many of its diagnostic functions are being consolidated to "infectious disease" molecular laboratories that include nonviral pathogens or are being transferred to "point-of-care" locations such as clinics, medical offices and public health settings, and with COVID-19 even to home testing. The process of viral discovery makes use of high-throughput nucleic acid sequencing and information comparisons using bioinformatics. These are fundamentally important to identifying new viruses or old viruses in "new clothes" that emerge to attack, frighten, and baffle.

In the final chapter, new for this edition, we tell of the astonishing COVID-19 pandemic, which within months changed life as we had known it. Also astonishing have been the advances in next-generation genetic sequencing. This technique allowed the identification of and the virus (which was to be named SARS-CoV-2), its epidemiologic spread, and the emergence of variants, and was essential to the design of structure-based vaccines. The emergence and application of new vaccine platforms, notably messenger RNA (mRNA) and virus vector based, allowed production of highly effective vaccines in record time.

We introduce our COVID-19 coverage with a selected history of events in the pandemic, including biographies of two heroic individuals. Discussion of the molecular virology and variants follows. We include the biography of a remarkable woman who has organized and led rapid sequencing efforts in the United Kingdom. The chapter moves on to vaccines, with biographies of four highly innovative scientists who were fundamental to their development. Diagnostic tests, including molecular assays, rapid antigen tests, and antibody assays, as well as sample collection are discussed. All of the virologically related topics are presented with an historical context.

We then discuss the worrisome, and biologically fascinating, clinical dilemmas that have arisen from the pandemic, such as long COVID. Treatments including dexamethasone, monoclonal antibodies, and the advent of successful oral medications top off the clinical section. Our description of the cultural impacts is bolstered by a biographical description of a physician-scientist who has been a remarkable public servant. We conclude with thoughts on the future, observing that diagnostic virology has assumed a crucial role in international relations.

As readers will recognize, staying abreast of the rapidly moving pandemic has been an ever-challenging task. For example, reports of the Omicron variant exploded into the world's consciousness at the time of Thanksgiving in the United States—November 2021—less than a month before the manuscript was due to our editor. Fortunately our editor, Megan Angelini, has facilitated updating the text as events merited until the final work on the galley proofs in July 2022.

Throughout the book we have sought illustrations from the general social context to illustrate perceptions of viral infections. Several other types of figures have also been chosen to support the text. In addition to photographic portraits of key historical figures, diagrams of diagnostic procedures and micrographs of virus-infected cells have been selected as examples of the kinds of work that diagnostic virologists have performed.

This edition, similar to the first, is directed to a diverse audience. There have always been fascination, curiosity, and fear of viral epidemics that threaten the lives of individuals and the fabric of society. This was true for the yellow fever outbreak in 1793 in Philadelphia, as it was in the 1980s when AIDS first made its mysterious entrance, and through the COVID-19 pandemic. We hope that virologists, microbiologists, epidemiologists, and clinicians will appreciate the broad canvas that we have attempted to paint. We hope, too, that the general public will appreciate having clinical, public health, and virological information woven together. Finally, if we help students of various disciplines to see the impact of this little spool of nucleic acid, able to outwit highly evolved humankind and its cultures, we will have served an important mission.

JOHN BOOSS, MD
MARIE LOUISE LANDRY, MD
with MARILYN J. AUGUST, PHD

About the Authors

JOHN BOOSS is Professor Emeritus of Neurology and Laboratory Medicine at the Yale University School of Medicine. For twelve years he was the National Program Director of Neurology for the U.S. Department of Veterans Affairs. Following residency in neurology, he trained in experimental virology with G.-D. Hsiung, to whom this book is dedicated. He subsequently trained with E. F. Wheelock in viral immunology. Dr. Booss has studied the modulation of immune function by murine cytomegalovirus and experimental models of viral infection of the brain. He studied T cells in the brains of persons with multiple sclerosis with Margaret M. Esiri at Oxford. With Professor Esiri he coauthored *Viral Encephalitis in Humans* (ASM Press, 2003). He examined the local host response to xenogenic brain cells transplantation with Claude Jacque at the Hôpital Salpêtrière in Paris. Dr. Booss' clinical interests focused on encephalitis, multiple sclerosis, and the neurology of HIV/AIDS. In retirement he has explored the impact of epidemics on Native American history. Most recently he examined the role of smallpox on the turning point of the French and Indian War (in press, *New York History*, Winter 2022–2023).

MARIE LOUISE LANDRY is Professor of Laboratory Medicine and Medicine (Infectious Diseases) at Yale University School of Medicine. She trained in internal medicine, infectious diseases, and clinical virology at Yale, and after completing a postdoctoral fellowship with G.-D. Hsiung, she changed her primary focus to virology. She is founder and Director of the Clinical Virology Laboratory at Yale New Haven Hospital and also helped found and is former Director of the Veterans Affairs National Virology Reference Laboratory at VA Connecticut. Since 1980, she has been the primary lecturer in the medical virology course for Yale medical students. She served as Councilor, Secretary-Treasurer, and then President of the Pan American Society for Clinical Virology, the main professional society for diagnostic virologists in the Americas, and received the Diagnostic Virology Award in 2005. Since 2003, spanning editions 9 through 13, she has been the Virology Volume Editor for the *Manual of Clinical Microbiology*, ASM Press, and since 2020, for *ClinMicroNow*. She has over 200 publications in clinical and diagnostic virology.

1 Fear or Terror on Every Countenance

YELLOW FEVER

> *The production of yellow fever by the injection of blood-serum that had previously been through a filter capable of removing all test bacteria is, we think, a matter of extreme interest and importance.*
>
> Reed and Carroll, 1902 (1)

INTRODUCTION

In 1793, within two decades of the writing of the Constitution of the United States and the Declaration of Independence, Philadelphia experienced an outbreak of yellow fever which shredded the fabric of civil society. While the Declaration of Independence and the Constitution have stood as blueprints for the philosophical and practical bases of representative government, the understanding of yellow fever at that time was still mired in the miasma of pre-germ theory speculation.

The first case of yellow fever in the Philadelphia 1793 outbreak was recognized in August by Benjamin Rush as the "bilious remitting yellow fever" (2). As the outbreak grew, there was no consensus on its origin. Rush attributed it to "putrid coffee" which "had emitted its noxious effluvia" after being dumped on a dock. The College of Physicians was "of the opinion that this disease was imported to Philadelphia by some of the vessels which were in the port after the middle of July." The role of the mosquito as a vector for disease was not to be recognized until decades later. In the 1793 Philadelphia outbreak, "Fear or Terror was set on every

To Catch a Virus, Second Edition. Authored by John Booss and Marie Louise Landry.
© 2023 American Society for Microbiology. DOI: 10.1128/9781683673828.ch01

countenance." The effect on families was devastating. In reporting the horror of the desertion of sick wives by husbands, the desertion of sick husbands by wives, and the departure of parents from sick children, Mathew Carey, another contemporary observer, noted that those actions ". . . seemed to indicate a total dissolution of the bonds of society in the nearest and dearest connexions." He commented on "the extraordinary panic and the great law of self-preservation. . ." (3). Rush reported on the exodus, "The streets and roads leading from the city were crowded with families flying in every direction for safety in the country" (2). J. H. Powell, the modern-day chronicler of the 1793 Philadelphia epidemic, noted that business languished and public administration virtually halted. With widespread sickness, over 40,000 deaths, and diminished population, the economy of the city collapsed. It was not until November 1793 that the city began to rebound, ". . .a time of recovery—of moral, psychological, intellectual reconstruction" (4).

Rush, who remained in the city, worked relentlessly, at times seeing upwards of 150 people in a day. At the end of his 1794 account of the epidemic, Rush tells of the effect on himself in a "Narrative of the state of the Author's body and mind" (2). Following the death of his sister, he wrote, ". . .my short and imperfect sleep was disturbed by distressing or frightful dreams. The scenes of these were derived altogether from sickrooms and graveyards." This courageous, indefatigable physician embodied the paradox of latter 18th-century Philadelphia, which was the site of advanced social-governmental thinking but backward in scientific-medical thinking. Beyond his medical pursuits, Rush was an advanced social thinker, a delegate to the Continental Congress, and a signer of the Declaration of Independence. He promoted improved conditions for mental patients and prisoners, promoted education, and promoted the abolition of slavery (5). Yet Rush also reflected the confusion and ignorance of infectious diseases before the advent of laboratory methods. Ascribing yellow fever to the effluvia of putrefying coffee, he treated infected individuals with powerful purging and bloodletting and considered all diseases derived from one cause, comparing the "multiplication of diseases" to polytheism (2). Unrecognized at this time was the association of microbes with infectious diseases, which would come in the next century, along with the recognition that specific insect species could be vectors for disease transmission. Elsewhere, too, outbreaks of yellow fever were seen as striking suddenly and "in an unaccountable fashion." A chronicler of epidemics of colonial America, John Duffy quoted from an outbreak in Charleston ". . . 'the Distemper raged, and the destroying Angel slaughtered so furiously with his Avenging Sword of Pestilence'. . ." (6). Thus, the metaphors of divine punishment, of an angry God, were the means of understanding the ravages of infection. The people were reduced to struggling with the effects of the epidemics: "'nothing was done but carrying medicines, digging graves, (and) carting the dead '. . ." (6).

The understanding of infectious diseases was to change dramatically in the next century, with the work of Louis Pasteur and Robert Koch establishing the germ theory. Just about a century after the 1793 yellow fever outbreak, the first understanding of viruses as filterable agents requiring living cells for propagation was established separately in the 1890s by Dmitri Ivanowski (7) and Martinus Beijerinck (8). Shortly thereafter, yellow fever was the first human virus shown to be a filterable agent (9). With the Philadelphia epidemic of 1793 as a dramatic backdrop, the details follow of how germ theory was proven and how the concepts of viral diseases, including yellow fever, were experimentally determined.

GERM THEORY

Seeing with one's own eyes is important for understanding the causation of infectious diseases. The microscopic or submicroscopic size of microbes was the root cause of centuries of misunderstanding of infectious diseases. For millennia, diseases were conceived as the work of demonic spirits, the wrath of God, or the miasmic emanations of decaying matter (10). These "invisible" microbes spawned massive epidemics and fear (Fig. 1). The reigning theoretical concept of disease causation was that of humoralism, of an imbalance of the four humors: blood, phlegm, black bile, and yellow bile. Interventions such as bleeding and purging were designed to restore the balance of the humors. The concept originated with Hippocrates and Galen and held sway for centuries (11). It did not account for microbes as the cause of infectious illness.

That is not to say that there weren't glimmers of recognition of transmissible infectious agents. Girolamo Fracastoro (Fracastorius), whose poem about the shepherd Syphilis named that disease, wrote of its contagiousness in the 16th century (12). In his 1546 work *On Contagion*, he described germs as transmitters of disease (13), according to Garrison the first scientific statement on the nature of contagion (11). However, it was with the development of the first crucial piece of laboratory equipment, the microscope (14), that the particulate microbial nature of infectious diseases was visualized. With improved magnifying lenses introduced by Antony van Leeuwenhoek and Robert Hooke in the 17th century, it was finally possible to describe the microscopic world (15). van Leeuwenhoek called bacteria "animalcules" (Fig. 2).

In the 19th century, Louis Pasteur laid to rest the magical thinking implicit in an unseen world when he disproved the theory of spontaneous generation. This advance relied on a second crucial innovation: artificial growth medium in which microbes could visibly multiply. Pasteur's swan-necked flask contained a growth-supportive fluid, which showed turbidity when exposed to the atmosphere and remained clear and uninfected when unexposed. Further, Pasteur's studies with silkworms established the crucial concept that specific pathological conditions were associated with

FIGURE 1 *Specter of death waiting over Panama (U. J. Keppler, 1904). Yellow fever, which had been termed "the American Plague," struck Philadelphia in 1793. It later threatened the construction of the Panama Canal, as shown in this cover illustration for Puck, a political satire and humor magazine. (Courtesy of Beinecke Rare Book and Manuscript Library, Yale University.)*

specific causes—a concept we now take for granted (16). After years of experimentation with the silkworm diseases pébrine and flacherie, Pasteur demonstrated their causation and means of prevention by eliminating the offending microbes.

Robert Koch, the genius who laid bare the specific causes of infectious diseases, refined the tools for laboratory diagnosis of infection (Fig. 3). He markedly facilitated the viewing of microbes through a microscope with the development of a sub-stage condenser, a lens that concentrates light from the source through the object studied. The visualization of microbes was further enhanced through the application of histological stains to differentiate the organelles from other structures in

FIGURE 2 *van Leeuwenhoek exhibiting his microscopes for Catherine of England (painting by Pierre Brissaud). Leeuwenhoek first described bacteria viewed through his early microscopes as "animalcules." (Courtesy of the Abbott Historical Archives, Leeuenhoek exhibiting his microscopes for Catherine of England, 1939.)*

specimens (17). With his development of photomicroscopic methods, Koch was able to share his observations, and it was a revelation when his first figures were published.

Koch worked to improve another crucial laboratory tool, solid culture medium. Following the observation of the growth of bacteria on sliced potatoes and with the advice of Fannie Hesse, the wife of a physician working in his lab, Koch incorporated agar into nutrient broth and so created solid medium as a means of selectively growing bacteria (14). To this day, the growth and isolation of pure bacterial cultures on agar medium remain the standard of practice for microbiological research and diagnosis.

In his work on tuberculosis, Koch approached the vexing question of etiology. Koch's postulates had their precedent in the work of Jacob Henle (18). While these tenets have evolved over time, Koch's postulates are usually understood to include the following:

1. Regular isolation of an organism from the diseased organs and absence from healthy organs
2. Growth of the organism in pure culture
3. Re-creation of disease on transmission to a susceptible host
4. Reisolation of the offending organism from the experimental host

FIGURE 3 *Robert Koch, about 1908. Koch developed the methodology that allowed the emergence of bacteriology as a science. In addition, he isolated the bacteria of tuberculosis and cholera, age-old scourges of humankind. (Courtesy of the National Library of Medicine.)*

Once it was possible to demonstrate the growth of microbes on artificial media and to see these microbes microscopically, the determination of the agents of specific infectious diseases simply exploded. These developments heralded the "golden age" of microbiology between 1877 and 1906 (Table 1). Koch revealed

TABLE 1 Milestones in the golden age of bacteriology[a]

Disease	Bacterium	Year(s) isolated
Anthrax	*Bacillus anthracis*	1877
Suppuration	*Staphylococcus, Streptococcus*	1878, 1881
Tuberculosis	*Mycobacterium tuberculosis*	1882
Cholera	*Vibrio cholerae*	1883
Pneumonia	*Streptococcus pneumoniae*	1886
Meningitis	*Neisseria meningitidis*	1887
Gangrene	*Clostridium perfringens*	1892
Plague	*Yersinia pestis*	1904
Syphilis	*Treponema pallidum*	1902
Whooping cough	*Bordetella pertussis*	1906

[a]Modified from T. D. Brock (17), p. 290. Nomenclature retained.

the age-old scourges of civilization, "Asiatic" cholera and consumption (tuberculosis), to be associated with specific organisms. Likewise, staphylococcus and streptococcus were identified as the causes of wound suppuration, limb amputation, and frequently death during wartime and after surgical procedures. A follow-on to the recognition of lethal bacterial wound contamination was Lister's development of techniques for antiseptic surgery (19). The golden age was thus characterized by advances not only in laboratory techniques and demonstration of specific microbial causes of infections but also in the means to prevent those infections.

BIRTH OF VIROLOGY, "FILTERABLE VIRUSES"

Notably absent from the list of infectious diseases were the scourges that we now know as viral diseases. Diseases with telltale skin lesions such as disfiguring smallpox, measles, and the dramatic yellowing of malignant bilious fever or yellow fever did not succumb to isolation attempts on artificial media, nor did rage (rabies), long known for its transmission through the bite of a rabid dog. Fear of these diseases sparked scientists of the day to investigate their etiology and control. Edward Jenner's revolutionary inoculation of vesicular material from cowpox lesions blunted or prevented dreaded smallpox in recipients. Likewise, Pasteur developed attenuated rabies virus material that he used to inoculate young Joseph Meister, bitten by a vicious, rabid dog. Meister's miraculous survival unleashed a popular demand for vaccination against this frightful disease (16).

Epidemiological studies also fostered an understanding of viral disease even before viruses were actually defined and understood in the laboratory. Such was the case when Peter Ludwig Panum followed the transmission of measles to the Faroe Islands, a group of islands in the North Atlantic Ocean. The consequences of a ship's carpenter incubating measles and delivering it to the remote Faroe Islands allowed Panum to define the transmission of this viral disease decades before the isolation and demonstration of the virus (20). In like manner, the mode of transmission of yellow fever was demonstrated before its characterization as a virus (21). That characterization depended on the defining tool of the virology laboratory; oddly enough, it was a filter customarily used to exclude bacteria. So important was that instrument that viruses were long known as "filterable viruses."

Like Pasteur's studies of silkworm diseases, the first recognition of viral diseases resulted from commercial urgency: a threat to the tobacco crop by mosaic disease. In 1885, Adolf Mayer demonstrated that infection could be transmitted by sap from diseased to healthy plants (22). In 1892, Ivanowski reported to the Academy of Science of St. Petersburg on the mosaic disease of the tobacco plant, "According to my experiments the filtered extract introduced into healthy plants produces the symptoms of the disease just as surely as does the unfiltered sap" (7). Independently, Martinus Beijerinck reported in 1898 that the tobacco disease could be transmitted through a porcelain candle filter and that tests for bacteria were negative (Fig. 4). He characterized the infection as a *"contagium vivum fluidum,"* or contagious living liquid, and suggested that one form could be ". . . a contagium that exists only in living tissues . . ." (8). Hence, two laboratory requirements defined viruses: passage through a porcelain filter and the need for living cells on which to grow.

For the first demonstration of a filterable agent in mammals, again the drive was commercial. Foot-and-mouth disease impaired cattle breeding and reduced milk production, bringing severe economic hardship to Prussian agriculture (23). Friedrich Loeffler, a collaborator of Koch and the discoverer of the diphtheria bacterium and toxin, was appointed in 1897 by the Prussian Ministry of Cultural Affairs to study foot-and-mouth disease. While readily transmissible experimentally, the agent was not visible microscopically. In collaboration with Paul Frosch, Loeffler demonstrated that foot-and-mouth disease could be transmitted to cows by bacterium-free lymph which had been passed through a Berkefeld filter (24). Experiments designed to detect the presence of a toxin resulted in retention of disease-producing activity, even at extremely high dilutions. This observation raised the possibility that the effect was produced by a germ which could multiply. The hope was that a vaccine could be developed. Unfortunately, as discussed by H.-P. Schmiedebach and more recently by A. Kenubih, a vaccine with persistent immunity-inducing capacity did not result, and foot-and-mouth disease remains a scourge to this day (23, 25).

FIGURE 4 *Martinus Beijerinck in his laboratory, May 1921. Beijerinck, like Ivanowski, demonstrated that tobacco mosaic disease could be transmitted by sap which had passed through bacteriological filters. He also demonstrated the need for living cells to replicate the disease-causing factor, which he called* contagium vivum fluidum.

The work of Loeffler and Frosch stimulated work on yellow fever, a terrifying human viral disease which had been the subject of intense debate in the 19th century between the contagionists and the anticontagionists (26). Yellow fever was at once a political and commercial battle as well as a scientific matter. The politically liberal anticontagionists were against quarantines and the bureaucratic apparatus that supported quarantines. The French Academy of Medicine weighed in on the issue in 1828 against yellow fever quarantines. Thus, the matter of disease transmission had political overtones as well as commercial consequences. Walter Reed, as director of the U.S. Army Yellow Fever Commission, would definitively demonstrate the mode of yellow fever transmission. William Welch, the dean of American pathology, pointed out to Reed the possible relevance of the work of Loeffler and Frosch on foot-and-mouth disease to the etiology of yellow fever (9).

WALTER REED AND THE YELLOW FEVER COMMISSION

By 1900, several pieces of the puzzle were in place that would help to explain the spread of yellow fever. Of principal interest was the mode of transmission; the precedent had been set in 1897 with malaria when mosquitoes were identified as vectors. Walter Reed acknowledged ". . . the splendid work of Ross, Bignami, and others with regard to the propagation of malarial fever..." (9). Reed also acknowledged the work of J. C. Nott in 1848 in suggesting ". . . that the spread of yellow fever could not be assumed by the assumption of a diffusible miasm in the atmosphere but required the presence of an intermediate host. " Unfortunately, Nott, a prominent southern U.S. physician of the 19th century, also advanced racialist thinking with justification for slavery (27). The specific mosquito, then called *Stegomyia fasciata*, later called *Culex fasciatus* and finally *Aedes aegypti* (28), had been identified by Carlos Finlay in 1886. However, Finlay had failed to convince his colleagues that this mosquito was responsible for disease spread. One of the principal reasons for that failure was ignorance of an extrinsic incubation period, a time during which the virus matures in the mosquito.

Reed recognized the careful work of Henry Rose Carter (29) (Fig. 5) in two small towns in Mississippi in 1898, "'demonstrating the interval between the infecting and secondary cases of yellow fever.'" Reed was gracious in declaring, "To Dr. Carlos J. Finlay, of Havana, must be given, however, full credit for the theory of the propagation of yellow fever by means of the mosquito" (9).

Sill missing was a crucial piece of the puzzle: isolation of an agreed-upon etiological agent of yellow fever. Although this was a time of exciting discoveries in medical bacteriology, credit in disproving putative bacterial causes must go to George Miller Sternberg, a pioneer American bacteriologist. An author of early American textbooks of bacteriology in the 1890s, he spent most of his career in

FIGURE 5 *Henry Rose Carter, 1909. As a member of the Marine Hospital Service, he was able to deduce a delay between primary and secondary cases of yellow fever. This extrinsic incubation period implied the need for another, nonhuman, host, later shown to be the mosquito. He was assigned to the Panama Canal Zone in 1904 to work on yellow fever. (Courtesy of Historical Collections & Services, Claude Moore Health Sciences Library, University of Virginia.)*

the U.S. Army and was largely self-taught in bacteriology. In 1890, he published *Report on the Etiology and Prevention of Yellow Fever*, in which he thoroughly disposed of the several candidate bacteria as the cause of yellow fever (30). In 1897, Sternberg appointed Reed and James Carroll to investigate yet another candidate, *Bacillus icteroides* (Sanarelli) along with his own candidate, Bacillus X. Reed and Carroll, who were joined by Aristedes Agramonte in Cuba in 1898, demonstrated that *Bacillus icteroides* "bore no relation to the disease" (31).

In 1900, by-then Surgeon General Sternberg appointed Reed, Carroll, Agramonte, and Jesse W. Lazear to a board of army medical officers to investigate yellow fever in Cuba (31). The board first met 25 June 1900 (32). Astonishingly, within 4 months the board was able to report at the Annual Meeting of the American Public Health Association in October 1900 that *Culex fasciatus* served as the intermediate host for yellow fever (21). In clearing the field of bacterial contenders and in appointing the U.S. Army Yellow Fever Commission in 1900, Sternberg can be credited as the catalyst of these findings on the transmission and etiology of yellow fever (Fig. 6).

Without a bacterial agent identified in culture, the need remained to do studies in human subjects. The use of experimental animals was to come later (33–35). Finlay had already used human subjects in his earlier studies (36). Still, the investigators recognized the ethical implications of the studies in humans and offered themselves first. In James Carroll's words, "Then arose the question of the tremendous responsibility involved in the use of human beings for experimental purposes. It was concluded that the results, if positive, would be sufficient justification of the undertaking. It was suggested that we subject ourselves to the same risk, and this suggestion was accepted by Dr. Reed and Dr. Lazear" (31). Carroll became the first experimental subject, accepting the risks ahead of other volunteers.

The circumstances to further study yellow fever were propitious. Following the Spanish-American War, yellow fever appeared yet again in Cuba, placing the populace and American troops at risk. Soon after arrival in Cuba, Reed and his Commission colleague, Agramonte (Fig. 7), visited an army barracks at Pinar del Rio where an outbreak was occurring. Observations made on that visit "... did not tend to strengthen one's belief in the theory of the propagation of yellow fever by fomites" (9). A curious story was told of that visit. Only one of nine prisoners, well guarded in jail, had come down with yellow fever. Speculation was raised that an insect such as a mosquito had bitten the one prisoner. That speculation was buttressed by the observations of Carter of the interval between infecting and secondary cases (37, 38). It was decided to test Finlay's theory of mosquito transmission of yellow fever.

In Reed's words, "... the search for the specific agent of yellow fever while not abandoned, should be given secondary consideration, until we had first definitely

FIGURE 6 *George Miller Sternberg. Known as America's first bacteriologist, he produced the first textbook of bacteriology in the United States. He was Surgeon General of the Army from 1893 to 1902, during which time he appointed the Yellow Fever Commission. (Courtesy of the Historical Collections & Services, Claude Moore Health Services Library, University of Virginia.)*

learned something about the way or ways in which the disease was propagated from the sick to the well" (9). In preliminary experiments by Lazear, mosquito eggs were supplied by Finlay, and mosquitoes were raised in the laboratory, allowed to feed on yellow fever patients, and allowed to bite human subjects. First among the subjects was Carroll, who fell ill and almost perished (38). Lazear, apparently bitten by a stray mosquito in 1900, was a victim of their research efforts: he contracted yellow fever and died. The results of the experiments showed that 2 of 11

FIGURE 7 *The Yellow Fever Commission consisted of (upper left) Walter Reed, who led the Commission; (lower left) James Carroll, who performed the filtration experiment; (upper right) Aristides Agramonte; and (lower right) Jesse W. Lazear, who became infected in the course of the experiments and died. In a remarkably brief period of time at the turn of the 20th century, the Commission under Reed demonstrated that the disease was transmitted by mosquitoes and that it could be transmitted by filtered blood and thus was caused by a virus. (Courtesy of the Historical Collections & Services, Claude Moore Health Sciences Library, University of Virginia, except for the image of Walter Reed, courtesy of The National Library of Medicine.)*

experimentally infected subjects developed yellow fever. It was concluded that *"The mosquito acts as the intermediate host for the parasite of yellow fever, and it is highly probable that the disease is only propagated through the bite of this insect"* (italics in the original) (21).

There followed the construction of two small buildings in an open field to compare the transmission of yellow fever by fomites with transmission by the bites of infected mosquitoes or inoculation of infected blood. The "Infected Mosquito Building" was well ventilated and divided into two compartments by a screen. The "Infected Clothing and Bedding Building" was purposely not well ventilated so as to retain any noxious effects of bed clothing, pajamas, and other items from previously infected cases. After some early discouraging results, John R. Kissinger, a soldier who Reed praised for having volunteered "solely in the interest of humanity and the cause of science" and who would accept no payment, came down with experimental yellow fever from the bites of infected mosquitoes (9). In these experiments, six of seven "non-immunes" bitten by infected mosquitoes in the Infected Mosquito Building became ill with yellow fever (39). None of the seven subjects in the Infected Clothing and Bedding Building exposed to fomites from cases of yellow fever became ill, nor did subjects become ill who had remained behind the screen, not bitten by mosquitoes.

The clarity of the design of comparison groups and the results were decisive: 85.71% infected by mosquitoes versus 0% by fomites. In the definitive publication in *JAMA*, "The Etiology of Yellow Fever: an Additional Note," Reed, Carroll, and Agramonte ended with several major conclusions. In addition to confirming that "*C. fasciatus* serves as the intermediate host," they determined that 12 days or more was required after contamination for the mosquito to transmit the infection. Thus, they determined experimentally what Carter had observed epidemiologically. They found that yellow fever could be transmitted by blood subcutaneously inoculated when taken from a patient on the first 2 days of the illness. They concluded that yellow fever resulting from a mosquito bite "confers immunity" against attempted reinfection with infected blood (39).

In memory of Lazear, the experimental station established by Reed, where the crucial studies were conducted demonstrating the transmission of yellow fever by mosquitoes and not by fomites, was christened Camp Lazear. Ironically, although Carroll recovered from acute yellow fever infection, he tragically died 7 years later of myocarditis attributed to that attack of yellow fever.

An important piece of the puzzle still remained to fall in place. Walter Reed and his colleagues' final conclusion of their *JAMA* report was that "... the specific cause of this disease remains to be discovered" (39). Having turned away from that goal in their transmission studies, Carroll returned to the project. Initially confronted with local objections to further experimentation, Carroll resumed his studies in September 1901 in Cuba on the nature of the infecting agent (31). In the crucial experiment, six individuals were exposed to the bites of infected mosquitoes (1). Four did not develop yellow fever, but two did. Blood was taken from patients I and II for further transmission study, but due to an accident to the vacuum pump, the blood from patient I could not be used. The blood from patient II was divided into three aliquots of partially defibrinated and diluted serum. The first aliquot, a positive control, was left untreated and successfully transmitted yellow fever to patient III. The second aliquot was heated to 55°C for 10 minutes and failed to transmit disease to patients IV, V, and VI. Based on previous work on heat stability with toxins, Reed and Carroll argued against a toxin. The third aliquot was "slowly filtered through a new Berkefeld laboratory-filter" and the filtrate was inoculated into patients VII, VIII, and IX. Patients VII and VIII developed "unmistakable" attacks of yellow fever; patient IX remained well. The scientific data were presented at the annual meeting of the Society of American Bacteriologists, 31 December 1901 and 1 January 1902. Thus, clinical virology can be said to have started in the first years of the 20th century.

Presciently, they noted that the most effective means of controlling the spread of yellow fever was through destruction of mosquito breeding areas and prevention of mosquitoes biting the sick (40). This strategy was employed with extraordinary success by William Crawford Gorgas of the U.S. Army, also present in Havana at that time. A brilliant story unto himself, Gorgas cleared Havana of yellow fever. An interesting connection can be noted here between Gorgas and J. C. Nott, mentioned above, who had suggested that yellow fever transmission required an intermediate host. Nott, coincidentally, was the doctor who delivered the infant Gorgas, whose success in the story of yellow fever eradication was based on Nott's theory. A few years later, Gorgas also cleared the Canal Zone of yellow fever and malaria, allowing the successful construction of the Panama Canal (41–43). Thus, the yellow fever-mosquito story was intimately wound into America's expanding international role (44).

It was clear from the classic studies of the Yellow Fever Commission that transmission experiments had to be performed in human subjects. However, that presented significant ethical issues, not only for potentially lethal viral infections such as yellow fever but also for permanently disabling anterior poliomyelitis.

Although Carroll reported in 1904 that others had also shown that the agent of yellow fever was filterable (31), attempts to identify a bacterial cause continued (32). It was not until a successful experimental animal host, the rhesus monkey, was demonstrated in 1928 and then the successful use of intracerebral inoculation of white mice (34, 35) that large-scale studies of the yellow fever virus could be undertaken and the bacterial candidates dismissed.

REFERENCES

1. **Reed W, Carroll J.** 1902. The etiology of yellow fever: a supplemental note. *Am Med* **3:**301–305.
2. **Rush B.** 1794. *An Account of the Bilious Remitting Yellow Fever as It Appeared in the City of Philadelphia in the Year 1793.* Thomas Dobson, Philadelphia, PA.
3. **Carey M.** 1959. Yellow fever, p 114–118. *In* Major RH (ed), *Classic Descriptions of Disease with Biographical Sketches of the Authors,* 3rd ed. Charles C Thomas, Springfield, IL.
4. **Powell JH.** 1949. *Bring Out Your Dead: the Great Plague of Yellow Fever in Philadelphia in 1793.* University of Pennsylvania Press, Philadelphia, PA. http://dx.doi.org/10.9783/9780812291179.
5. **Shryock RH.** 1935. *Benjamin Rush. Dictionary of American Biography base set. American Council of Learned Societies,* 1928–1936. American Council of Learned Societies, Washington, DC. http://pbs .org/wnet/redgold/printable/p_rush.html. Accessed 16 June 2010.
6. **Duffy J.** 1953. *Epidemics in Colonial America.* Louisiana State University Press, Baton Rouge, LA.
7. **Ivanowski D.** 1942. Concerning the mosaic disease of the tobacco plant, 1892, p 25–30. *In* Johnson J (trans. ed.), *Phytopathological Classics, No. 7.* American Phytopathological Society, St Paul, MN.
8. **Beijerinck MW.** 1942. Concerning a contagium vivum fluidum as cause of the spot disease of tobacco leaves, 1898, with 2 plates, p 33–52. *In* Johnson J (trans. ed.), *Phytopathological Classics, No. 7.* American Phytopathological Society, St Paul, MN.
9. **Reed W.** 1901. The propagation of yellow fever; observations based on recent researches. *Med Rec* **60:**201–209.
10. **Winslow C-EA.** 1944. *The Conquest of Epidemic Disease: a Chapter in the History of Ideas.* Princeton University Press, Princeton, NJ.
11. **Garrison FH.** 1960. *An Introduction to the History of Medicine,* 4th ed. W B Saunders Co, Philadelphia, PA.
12. **Clendening L.** 1960. *Source Book of Medical History.* Dover Publications, New York, NY.
13. **Wright W.** 1930. *Hieronymi Fracastorii: de Contagione et Contagiosis Morbis et Eorum Curatione, Libri III.* G P Putnam's Sons, New York, NY.
14. **Bradbury S.** 1967. *The Evolution of the Microscope.* Pergamon, Oxford, United Kingdom.
15. **Gest H.** 2007. Fresh views of 17th century discoveries by Hooke and van Leeuwenhoek. *Microbe* **2:**483–488 http://dx.doi.org/10.1128/microbe.2.483.1.
16. **Dubos RJ.** 1950. *Louis Pasteur: Free Lance of Science.* Little, Brown, Boston, MA.
17. **Brock TD.** 1999. *Robert Koch: a Life in Medicine and Bacteriology.* American Society for Microbiology, Washington, DC.
18. **Carter KC.** 1985. Koch's postulates in relation to the work of Jacob Henle and Edwin Klebs. *Med Hist* **29:**353–374 http://dx.doi.org/10.1017/S0025727300044689.
19. **Lister J.** 1867. On a new method to treat compound fracture, abscess, etc. *Lancet* **89:**326–329 http:// dx.doi.org/10.1016/S0140-6736(02)51192-2.
20 **Panum PL.** 1939. Observations made during the epidemic of measles on the Faroe Islands in the year 1846, p 802–886. *In* Kelly EC (ed), *Medical Classics.* Williams & Wilkins, Baltimore, MD.
21. **Reed W, Carroll J, Agramonte A, Lazear JW.** 1900. The etiology of yellow fever. *Public Health Pap Rep* **26:**37–53.
22. **Mayer A.** 1942. Concerning the mosaic disease of tobacco, 1886, p 8–24. *In* Johnson J (trans. ed.), *Phytopathological Classics, No. 7.* American Phytopathological Society, St Paul, MN.
23. **Schmiedebach H-P.** 1999. The Prussian State and microbiological research—Friedrich Loeffler and his approach to the "invisible" virus, p 9–23. *In* Calisher CH, Horzinek MC (ed), *100 Years of Virology: the Birth and Growth of a Discipline.* Springer-Verlag, Vienna, Austria.
24. **Loeffler F, Frosch P.** 1898. Berichte der Commission zur Erforschung der Maulund Klauenseuche beidem Institut fur Infectionskrankheiten in Berlin. *Dtsch Med Wochenschr* **24:**80–83, 97–100 http:// dx.doi.org/10.1055/s-0029-1204235.
25. **Kenubih A.** 2021. Foot and mouth disease vaccine development and challenges in inducing long-lasting immunity: trends and current perspectives. *Vet Med (Auckl)* **12:**205–215.

26. **Ackerknecht EH.** 1948. Anticontagionism between 1821 and 1867. *Bull Hist Med* **22:**562–593.
27. **Horsmann R.** Josiah C. Nott, *on* Encyclopedia of Alabama. http://encyclopediaofalabama.org/article/m-4475. Accessed 16 April 2022.
28. **Hemmeter JC.** 1927. *Master Minds in Medicine.* Medical Life Press, New York, NY.
29. **Carter HR.** 1898. Communication with a town infected with yellow fever. *Med Newsl (Lond)* **72:**546–547.
30. **Sternberg GM.** 1890. *Report on the Etiology and Prevention of Yellow Fever.* US Government Printing Office, Washington, DC.
31. **Carroll, J.** 1904. A brief review of the etiology of yellow fever. *NY Med J and Philadelphia Med J* **79:**241–245, 302–310.
32. **Agramonte A.** 1928. A review of research in yellow fever. *Ann Intern Med* **2:**138–154 http://dx.doi.org/10.7326/0003-4819-2-2-138.
33. **Stokes A, Bauer JH, Hudson NP.** 1928. Experimental transmission of yellow fever to laboratory animals. *Am J Trop Med* **8:**103–164 http://dx.doi.org/10.4269/ajtmh.1928.s1-8.103.
34. **Theiler M.** 1930. Susceptibility of white mice to the virus of yellow fever. *Science* **71:**367 http://dx.doi.org/10.1126/science.71.1840.367.a.
35. **Theiler M.** 1930. Studies on the action of yellow fever virus in mice. *Ann Trop Med Parasitol* **24:**249–272 http://dx.doi.org/10.1080/00034983.1930.11684639.
36. **Finlay C.** 1886. Yellow fever: its transmission by means of the culex mosquito. *Am J Med Sci* **92:**395–409 http://dx.doi.org/10.1097/00000441-188610000-00006.
37. **Carter HR.** 1900. A note on the interval between infecting and secondary cases of yellow fever from the records of the yellow fever at Orwood and Taylor, Miss., in 1898. *New Orleans Med Surg J* **52:**617–636.
38. **Kelly HA.** 1923. *Walter Reed and Yellow Fever.* Norman Remington, Baltimore, MD.
39. **Reed W, Carroll J, Agramonte A.** 1901. The etiology of yellow fever: an additional note. *JAMA* **36:**431–440 http://dx.doi.org/10.1001/jama.1901.52470070017001f.
40. **Reed W, Carroll J.** 1901. The prevention of yellow fever. *Public Health Pap Rep* **27:**113–129.
41. **Gibson JM.** 1950. *Physician to the World: the Life of General William C. Gorgas.* The University of Alabama Press, Tuscaloosa, AL.
42. **Gorgas WC.** 1909. Sanitation of the tropics with special reference to malaria and yellow fever. *JAMA* **52:**1075–1077 http://dx.doi.org/10.1001/jama.1909.25420400001001.
43. **McCullough D.** 1977. *The Path between the Seas: the Creation of the Panama Canal, 1870–1914.* Simon & Schuster, New York, NY.
44. **Morgan HW.** 1965. *America's Road to Empire: the War with Spain and Overseas Expansion.* John Wiley, New York, NY.

2 Of Mice and Men

ANIMAL MODELS OF VIRAL INFECTION

Our results with ferrets, so far as they have gone, are consistent with the view that epidemic influenza in man is caused primarily by a virus infection.

Smith, Andrewes, and Laidlaw, 1933 (1)

INTRODUCTION

Work with human volunteers for the study of yellow fever had created a backlash in Cuba, and Walter Reed had recognized that it was impossible to proceed. Unlike bacterial agents, which could often be grown on artificial media, viral agents required living cells in which to replicate. In viral diseases of plants, such as tobacco mosaic disease, the natural host was readily available and appropriate. The same was true of viral zoonoses, such as foot-and-mouth disease. For human diseases, such as rabies and polio, nonhuman hosts were necessary to isolate and characterize the etiological agent. Hence, efforts were made to isolate filterable viruses in a range of animal hosts. In addition, the embryonated egg was adapted from studies of embryology and physiology for the study of filterable viruses. In the case of rabies, a zoonosis that spreads from animals to humans with catastrophic results, the development of animal models might be expected to be productive. In contrast, polio and human influenza were not known to have animal hosts, nor was an animal host known for yellow fever. Yet, isolation and study of the agents for each of these diseases were achieved in experimental hosts. First for discussion is rabies.

To Catch a Virus, Second Edition. Authored by John Booss and Marie Louise Landry.
© 2023 American Society for Microbiology. DOI: 10.1128/9781683673828.ch02

RABIES: DOGS AND RABBITS

Extensive experimental work on what would turn out to be the second human disease shown to be caused by a filterable virus, rabies, had been under way for virtually a century. The medical historian Lise Wilkinson credits Georg Gottfried Zinke's 1804 publication as being ". . . the first description of experiments specifically intended to follow the transmission of the unknown agent of rabies" (2). Zinke reported successful transmission of rabies to several species, including dogs, cats, rabbits, and fowl (3). While certain characteristics, such as incubation time and clinical characteristics, differed from those in later reports, Zinke's published work was an important landmark. Studies of diseases in animals in the 18th and 19th centuries involved agents of agricultural import—cattle plague (rinderpest), foot-and-mouth disease, bovine pleuropneumonia, and glanders—and established the basis for comparative medicine, the study of human diseases in animals (4). Rabies straddled human and animal medicine.

Rabies is horrific in all aspects: in the savage bites by crazed wolves or dogs to implant infection, in the anxiety and fear in anticipation of whether the disease will develop, in the torturing expression of the acute disease, and in the knowledge that once expressed, rabies is an essentially fatal disease. The fear induced by the

FIGURE 1 *Mad Dog. The fear of rabid dogs has been portrayed throughout history. This caricature by T. L. Busby was published in London in 1826. (Courtesy Yale University, Harvey Cushing/ John Whitney Medical Library.)*

approach of a rabid dog was used to good effect in Harper Lee's *To Kill a Mockingbird*: "... motivated by an invisible force that was inclining him toward us. We could see him shiver like a horse shedding flies; his jaw opened and shut; he was alist, but he was being pulled gradually toward us" (5) (Fig. 1). Accounts of terrifying attacks by rabid wolves in country villages have played a role in the development of treatment. It has often been cited that one of the motivations for Pasteur's study of rabies was a childhood memory of a rabid wolf attack on the village of Villers-Farlay and the town of Arbois (6) (Fig. 2). It was reported that Pasteur was 8 when he listened in horror to the screams of victims who had come to a blacksmith's shop for "treatment"—cauterization. In 1886, soon after Pasteur's publicity spread worldwide concerning postexposure prophylaxis treatment of rabies, 19 Russians came to Paris seeking treatment. In the cases of two victims of a rabid wolf's attack, one had a lip and cheek bitten off, and the other's face had been ripped off. Of the 19, 3 died during treatment and 16 ultimately returned to Russia (7). In 1955, the WHO, reporting on the use of hyperimmune serum and vaccine, described the effects of a rabid wolf's attack on an Iranian village (8). Twenty-nine persons were bitten, 18 with severe head wounds, including a 6-year-old boy in whom a skull bone was crushed and the dura mater covering of the brain slashed. Remarkably, 25 of 29 wounded survived, including the young boy, who received intensive treatment with hyperimmune serum and vaccine.

Rabies in humans may start after an incubation period of weeks or months with itching at the site of the bite and growing apprehension. Thereafter, aerophobia and hydrophobia may develop. The description of hydrophobia by D. A. Warrell et al. serves well: "The patient picks up a cup to drink but, even before the liquid has reached his lips, his arm begins to shake, he takes a rapid succession of inspiratory gasps, his neck muscles are seen to contract forcibly, and the spasm ends with throwing away his cup and falling back with neck extended" (9). Death ensues after periods of marked agitation and fearfulness alternating with lucidity. It is no wonder that rabies attracted attention out of proportion to its incidence.

In studies of rabies transmission to animals, several experimental aspects were considered and refined, many by Pasteur and his colleagues on the road to producing a postexposure vaccine. One was the type of inoculum, infected central nervous system (CNS) tissue. Other crucial considerations were the animal species used and the route of inoculation.

In 1821, F. Magendie reported the transmission of rabies from a sick person to a dog by injecting saliva beneath the skin (10). Magendie reported that the dog became rabid in about a month. This appears to be the first acknowledged transmission of rabies from a human to an experimental animal (2). Use of the dog as the test animal, while logical as the original species tested, is dangerous

FIGURE 2 *Louis Pasteur. Pasteur was one of the principal founders of germ theory. He disproved the theory of spontaneous generation; linked agricultural, animal, and human diseases to specific infections; and developed vaccinations, including that to prevent rabies. (Courtesy of the National Library of Medicine.)*

because the "furious" type of rabies is expressed. The indiscriminate and aggressive biting presented very significant risks to experimenters. In contrast, a more tractable model, expressing "dumb" or paralytic rabies, would have much to recommend it. Another distinct weakness of the model as first employed was the long and variable period of incubation before disease expression. Pasteur and his colleagues would later address that issue, modifying the nature of the inoculum and route of delivery in a different host.

Great progress in the development of an experimental rabbit model was achieved by M. Galtier, who held the chair of pathology in the veterinary school at Lyon, France (4). In 1879, he reported that rabies in the dog could be transmitted to the rabbit and could be serially transmitted from rabbit to rabbit (11). Significantly, he found that the predominant symptoms in the rabbit were paralysis and convulsions, hence of much less risk to the experimenter than the furious biting behavior of the infected dog. Galtier reported that once the disease declared itself, the rabbit was able to live for 3 or 4 days, thereby facilitating further observations and transmission studies as well as an accurate demonstration of positive transmission. Of great importance as a model, the rabbit revealed a shorter incubation period than other species: the peak onset of illness in 25 cases occurred between 13 and 23 days. Thus, the rabbit model had been well characterized in Galtier's hands by 1879, before Pasteur and his associates launched their epic work. From an examination of Pasteur's laboratory notebooks, Gerald L. Geison confirmed that Pasteur's work on rabies began at the end of the following year, 10 December 1880, to be precise (6). Pasteur's laboratory notebooks, numbering over 100, had become available to scholars at the Bibliothèque nationale (National Library of France) in the 1970s, so Galtier's priority in establishing and refining the rabbit model could be verified (6). Patrice Debré, a biographer of Pasteur, suggested that Galtier had been considered for the Nobel Prize in 1908, but he died before the award decision, precluding the honor (7).

Galtier had also attempted, without success, transmission of rabies by CNS tissue. However, Henri Duboué, a Paris physician, found otherwise. According to Debré, Duboué wrote to Pasteur in January 1881 that the "'morbid agent'" traveled to the medulla oblongata from the site of the bite (7). Vallery-Radot's biography of Pasteur confirmed the theory of Duboué but indicated that it was unsupported by experimental evidence (12). It remained for Pasteur and his colleagues, notably Emile Roux, to exploit the CNS aspects of the model. The goal was to shorten the period of incubation and to achieve full success in experimental transmission of rabies in a convenient animal model (12).

The motivation to select a particular experimental problem by a scientist is usually complex and may reflect the less tangible facts of the scientist's life. The distinguished Pasteur biographer René J. Dubos, an eminent 20th-century biologist, noted that "There is something odd in the selection by Pasteur of rabies as the next subject for his experimental studies" (13). Perhaps Pasteur's horrifying experience as a young boy witnessing the effects of a rabid wolf's attack was a motivation. More pressing than the relatively low prevalence of rabies was the nature of the disease itself. As well summarized by Geison, ". . . rabies embodied the ultimate in agony and degradation, stripping its victims of their sanity and reducing them to quivering convulsive shadows of their former selves" (6).

Pasteur refined the animal model. Vallery-Radot noted that "Evidently the saliva was not a sure agent for experiments, and if more information was to be obtained, some other means had to be found of obtaining it" (12). With the growing realization of a seat of infection in the brain, Pasteur and his collaborators—"actually it was Emile Roux" as noted by Geison—developed an intracranial method of inoculation (8). The skull was trephinated, and infected cerebral matter from the rabid animal was deposited under the dura mater on the surface of the brain. This method achieved success: in the words of Vallery-Radot, ". . . rabies was contracted surely and swiftly" (12), and according to Geison, these collaborators ". . . had at last developed a uniformly successful method of transmitting the disease from animal-to-animal" (6). The next step was development of a brief and reproducible incubation period. Subdural transmission in rabbits resulted in a progressively shortened, "fixed" incubation period, as brief as 7 days (12).

What followed in the Pasteur story of rabies postexposure prophylaxis is the stuff of legend. Briefly told, a method of attenuating the infectivity of the rabid rabbit CNS by desiccation was devised, and the material could protect dogs from virulent infection. Circumstances led to the historic postexposure treatment in 1885 of Joseph Meister, a 9-year-old boy brought to Pasteur from Alsace (Fig. 3). Because the boy was suffering severe bites from a rabid dog, his case was thought likely to be fatal (2). The dog was killed and its stomach was found to contain hay, straw, and wood fragments, apparently leaving "no doubt that the dog was rabid" (14). Such was the state of diagnostic virology! After a series of inoculations of progressively more virulent rabies spinal cord tissue, Meister survived. The floodgates of postexposure treatment opened, with hundreds arriving from all parts of the globe. Geison, in his review of Pasteur's laboratory notebooks, observed "some remarkable discrepancies between the public and private versions of this celebrated story" (6). According to the notebooks, Joseph Meister was not the first person in whom the antirabies treatment was attempted. Of importance, however, is the train of events leading to the demonstration of rabies as a filterable virus.

In 1884, "... Pasteur did concede that he and his collaborators had still not managed to isolate and cultivate a rabies microbe in artificial media" (6). While rabies virus was incapable of growth on artificial media, one of the criteria that defined a viral agent, rabies was transmissible to susceptible animal hosts. The stage was set for the demonstration that rabies was a filterable, "ultravisible" agent. That task would prove to be a technical challenge. However, in 1903, Paul Remlinger, the Director of the Imperial Institute of Bacteriology at Constantinople, reported "Le passage du virus rabique à travers les filtres" (15). His success depended on the techniques of filtration and inoculation and on the number of experimental animals

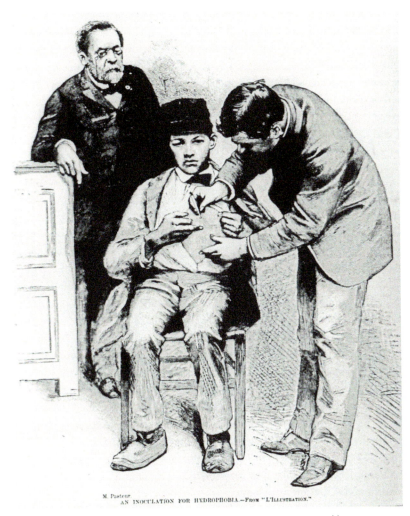

M. Pasteur
AN INOCULATION FOR HYDROPHOBIA.—From "L'ILLUSTRATION."

FIGURE 3 *Rabies prevention. The recognition that the Pasteurian treatment could prevent rabies after the bite of a rabid animal resulted in dramatic public acceptance. In this print from 1885, Pasteur is depicted observing the inoculation of a boy to prevent hydrophobia (rabies in humans). (Courtesy of the National Library of Medicine.)*

tested. As summarized by a contemporary, John McFadyean, a veterinarian, the most porous of the Berkefeld filters would allow passage of the virus, whereas more finely porous filters would hold the agent back. The filtrates were deposited under the dura, and several rabbits were inoculated with each filtrate (16). The success in filtration was confirmed by others (16). Remlinger was gracious, attributing to Pasteur the idea of rabies as ultramicroscopic (15). It was a fitting tribute to Pasteur.

Coincidentally, in 1903, A. Negri described inclusions in the neurons of the CNS of rabid animals and humans (17). While the inclusions were misinterpreted as parasites, they became in the following years "an extremely important method of diagnosis" (18). Thus, 1903 was a seminal year in the study of rabies. Using the animal model developed by Galtier and refined by Pasteur and his colleagues, Remlinger demonstrated the filterable, that is, viral, nature of the agent of rabies. Negri developed a histopathological hallmark for rabies which, when positive, was a "rapid viral diagnostic technique."

POLIO: MONKEYS

Thought to exist for centuries as a sporadic disease, polio was a rare crippler of infants and children. John Paul, a student of polio, used a figure of an ancient Egyptian stele showing a young priest with a withered leg as evidence of such early observations in his 1971 volume, *A History of Poliomyelitis* (19). As Paul described, "Under such endemic conditions paralytic poliomyelitis could have remained under cover for centuries in populations in which infant mortality was high." Yet the individual experience of the paralytic form of the disease, from sporadic to epidemic times, has been consistent. Walter Scott was born in 1771 in an area characterized as ". . . a cramped dimly lit alleyway with poor sanitation and little fresh air" (20). He described the onset of polio at 18 months: "One night, I have often been told, I showed great reluctance to be caught and put to bed, and after being chased about the room, was apprehended and consigned to my dormitory with some difficulty. It was the last time I was to show such personal agility. In the morning I was discovered to be affected with the fever which often accompanies the cutting of large teeth. It held me three days. On the fourth, when they went to bathe me as usual, they discovered that I had lost the power of my right leg" (19). The contagiousness of polio was not recognized in the centuries when polio was understood to be a sporadic disease, and Scott's fever was attributed to the cutting of teeth.

The evolution of polio from an endemic, sporadic disease to an epidemic pattern occurred toward the end of the 19th century. In the United States, for example, Charles Solomon Caverly of the Vermont State Board of Health reported in November 1894 that "Early in the summer just passed, physicians in certain parts of Rutland County, Vermont, noticed that an acute nervous system disease, which was almost invariably attended with some paralysis, was epidemic" (21). Caverly investigated this "'new disease" that was affecting the children and initially found 123 cases. He reported that "The territory covered is mainly the narrowest part of the Otter Creek Valley in Rutland County, bounded on the east by the Green Mountain range and on the west by the Taconic range. . . " Of 110 cases in which paralysis occurred, 50 persons recovered fully and 10 died, "leaving fifty who are apparently permanently disabled." It is of note that a microbe was suspected: "... the microbic or infectious nature of the disease is generally believed in by us..." However, Caverly reported no evidence of contagiousness, since it "affected almost invariably but a single member of a household." He published a fuller report of his investigations in 1896 (22). It is remarkable that in his first report just months after the epidemic, Caverly noted the apparent absence of communicability. Presciently, however, Caverly suspected the microbial nature of the disease, which was to be shown by Karl Landsteiner and Erwin Popper 14 years later (23). Thus, with reports by others, the stage was set for proving the contagious and filterable virus nature of this affliction.

Communicability was demonstrated by Ivar Wickman 11 years later in Sweden, showing the role of nonparalytic cases in sustaining the epidemic (24). John Paul emphasized Wickman's contribution: "Considering the importance of the contributions of Ivar Wickman, I do not believe that his work is fully appreciated today" (19). Later, Wickman failed to be appointed the Chair of Pediatrics at Stockholm, a position formerly held by his mentor in polio studies, Karl Oscar Medin. Perhaps in despair, though the reason was never ascertained, Wickman took his own life at the age of 42. John Paul emphasizes Wickman's contributions as identifying the contagious nature of the disease, describing the importance of nonparalytic forms, estimating the incubation period as 3 to 4 days, and classifying different forms of the disease (24). Polio transmission to monkeys was soon to be shown by others. While crucial in demonstrating the viral nature of the disease, studies with monkeys may have deflected attention from key pathogenic aspects of the human disease.

The first decade of the 20th century proved to be decisive in the demonstration of polio as caused by a contagious, filterable virus. Unfortunately, those proofs were dependent on the emergence of epidemics of polio in Europe and North America. Wickman's extensive studies of the epidemic in Sweden in 1905 were published as journal reports and as a monograph which was translated into English in 1913 (24). While the infectious nature of the disease had been suspected, no bacterial cause had been successfully demonstrated. Small laboratory animals also had not yielded an agent. However, in a meeting in December 1908 in Vienna, Austria, Karl Landsteiner, acknowledging the assistance of collaborator Erwin Popper, reported transmission of polio to monkeys, demonstrating for the first time a virus as the etiological agent of polio (Fig. 4). Two monkeys inoculated by the intraperitoneal route with spinal cord from a fatal human case demonstrated the symptoms and microscopic lesions of polio (23). A fuller presentation of Landsteiner and Popper's findings was published the following year (25). Clinical findings of paralysis were observed in one of the monkeys, and both showed the characteristic histopathological findings of polio in the spinal cord exactly like those observed

FIGURE 4 *Karl Landsteiner. Landsteiner was awarded the Nobel Prize in 1930 for the discovery of human blood groups. Another significant contribution was the demonstration that polio could be transmitted to monkeys by using spinal cord tissue from children who had died from the illness. (Courtesy of the Lasker Foundation.)*

in human cases. In contrast, rabbits, guinea pigs, and mice, also inoculated by the intraperitoneal route, were negative. Thus, the importance of the monkey for isolation of poliovirus was established.

Simon Flexner and Paul A. Lewis at the Rockefeller Institute in New York, NY, quickly picked up the observations of Landsteiner and Popper and demonstrated serial passage of polio in monkeys using intracerebral inoculation (26) of spinal cord material. The importance of serial passage was to eliminate the possibility of a nonreplicating toxic factor, an experimental principle that had been established with the first demonstration of a viral disease in animals, foot-and-mouth disease by F. Loeffler and P. Frosch (27). Success in passage of polio to monkeys benefited from serendipity. Landsteiner and Popper noted that the monkeys they had used had been left over from previous studies of syphilis (25). As pointed out by John Paul, Landsteiner and Popper used Old World monkeys, not the "relatively unsusceptible" New World monkeys (19). It remained to be shown that the experimental disease could be transmitted by filtered material. Landsteiner, teamed with Constantin Levaditi of the Metchnikoff Laboratory at the Pasteur Institute, reported on 28 November 1909 the successful transmission of polio to a monkey after intracerebral inoculation of an emulsion which had been passed through a Berkefeld V filter (28). Likewise, on 18 December 1909, Flexner and Lewis at the Rockefeller Institute reported transmission of polio to a monkey by intracerebral inoculation of spinal cord material passed through a Berkefeld filter (29). Thus, within 1 year, December 1908 to November and December 1909, clinical and pathological experimental replication of poliovirus in monkeys had been achieved, with serial passage and confirmed demonstration of the filterability of the infectious agent.

ARTHROPOD-BORNE DISEASES, YELLOW FEVER, AND EPIDEMIC ENCEPHALITIDES: MONKEYS AND MICE

The experiments of Walter Reed et al. and the Yellow Fever Commission in Cuba following the Spanish-American War were conducted with human volunteers. Reed and Carroll's studies on the filterable agent (30), while confirmed by others, could not be carried further because of the ethical objections to use of human volunteers in potentially lethal experiments. Hence, while extraordinary progress was made in control of the yellow fever due to the work of William Crawford Gorgas in controlling the mosquito vector (31), little progress was made in either laboratory diagnosis or seroepidemiological studies. The field remained hampered for demonstration of potential etiological agents until an animal susceptible to infection by yellow fever virus was identified. Thus, when Hideyo Noguchi inoculated guinea pigs with material from yellow fever patients in Guayaquil, Ecuador, and *Leptospira icteroides* was isolated, it was thought to be the etiological agent of yellow fever (32). Noguchi's investigations were done under the auspices of a commission established by the Rockefeller Foundation.

Because of their remarkable contributions, the origins of the Rockefeller Foundation and its International Health Commission should be mentioned. The International Health Commission was formed in 1913, and its first task was control of yellow fever. It had been feared that the opening of the Panama Canal would result in the spread of yellow fever to the Orient because of "radical changes in trade relations" (33). Hence, the International Health Commission set as its goal to "give aid in the eradication of this disease in those areas where the infection is epidemic." A first target for elimination of the disease was Guayaquil. An antimosquito campaign started at the end of 1918 resulted in no further cases of yellow fever after June 1919. That accomplishment was accompanied by Noguchi's report of a leptospira as the cause of yellow fever. As A. J. Warren commented, "Dr. Noguchi's error is easily understandable." Clinically, yellow fever and the illness caused by the spirochete closely approximate each other and may coexist in a population (33). It remained for another Rockefeller commission to finally disentangle the etiology of yellow fever.

In 1925, the West Africa Yellow Fever Commission was established (33). It was in Nigeria that Adrian Stokes, Johannes H. Bauer, and N. Paul Hudson used *Macacus rhesus* monkeys to confirm the filterable viral etiology of yellow fever (34). In contrast to the findings of Noguchi, no spirochetes were demonstrated. However, they did demonstrate the transmission of yellow fever from human to monkey and from monkey to monkey. While another species, *Macacus sinicus*, was shown to be only moderately susceptible, *Macacus rhesus* replicated the clinical disease and pathology as found in humans. Thus, the agent fulfilled the requirement to pass through Berkefeld V and N grade filters. In efforts to defend his theory of the leptospira etiology, Noguchi traveled to West Africa. Tragically, he did not live to resolve this conundrum; he died in 1928 of yellow fever at age 51 before learning the answer.

Other types of animals in these studies, including guinea pigs, were not found to be susceptible. The investigators credited the director of the Commission, Henry Beeuwkes, with obtaining animals for experimentation from areas of the world away from West Africa. A tragic note was found at the head of the report that the lead author, Adrian Stokes, "fell victim to yellow fever." It was suspected that the infection was acquired in the laboratory. Yellow fever surely took its revenge on those seeking to unlock its secrets, including Jesse W. Lazear, of the Commission led by Walter Reed, and Hideyo Noguchi.

While the report by Stokes et al. on the susceptibility of rhesus monkeys to yellow fever ushered in a laboratory model for study, monkeys were expensive and complex to manage. In 1930, Max Theiler, the investigator who was later to receive the Nobel Prize for developing the yellow fever vaccine, 17D, demonstrated the susceptibility of white mice after intracerebral inoculation (35–37). This model allowed the development of the mouse protection assay, which, in turn, facilitated large-scale epidemiological studies of viral prevalence (38).

Isolation in monkeys, and particularly in mice, was exceptionally productive in the 1930s for filterable viruses that caused epidemics of encephalitis. These viruses were transmitted by mosquitoes and came to be called "arthropod-borne" (39), later shortened to "arbovirus." The first of these agents was St. Louis encephalitis virus, named after the city in which an epidemic occurred in 1933. Virus was isolated in monkeys by R. S. Muckenfuss et al. (40) and in mice by L. T. Webster and G. L. Fite (41). Another epidemic encephalitis, which had blanketed many countries in Asia and put millions of people at risk, was noted in Japan as early as the 1870s. It was originally called Japanese B encephalitis to distinguish it from encephalitis lethargica, the first recognized epidemic encephalitis. A number of laboratories reported isolation of Japanese B virus in the 1930s, including that of R. Kawamura, which isolated the virus in monkeys and mice (42). Subsequently, several other epidemic arboviral encephalitides were recognized in the 1930s, and their etiological filterable viruses were also isolated in the same hosts (43).

INFLUENZA: FERRETS

The influenza pandemic of 1918–1919 is one of the greatest natural disasters of recorded history. First recorded at an Army Camp in Kansas in 1918 toward the conclusion of World War I, influenza killed more people worldwide than did that conflict. The exact number will never be known because of incomplete global health records, inaccurate records of death, and censorship associated with the war. One analysis estimated that between 24.7 and 39.3 million lives were lost to the pandemic worldwide (44). The war facilitated the pandemic in concentrating and moving troops and disrupting the civilian populations. Reciprocally, conduct of the conflict was hampered on both sides by disabled troops (Fig. 5). Rail travel and transoceanic steamships took the pandemic to all but a few remote locations. In other influenza epidemics, it has been the very young and the elderly who have suffered the brunt of the mortality. However, the unique feature of the 1918–1919 pandemic of influenza was the predilection for killing previously healthy young adults. It now seems likely that such mortality resulted from an overexuberant host response, with the body's own defenses turning against the body in an attempt to ward off infection.

The effects on society and on individuals are poignantly described in Katherine Anne Porter's story *Pale Horse, Pale Rider* (45). The medical historian Alfred W. Crosby said of Porter's story, "It is the most accurate depiction of American society in the fall of 1918 in literature" (46). Porter was herself desperately ill and lost her lover, an Army lieutenant, to influenza. In the fictionalized account in *Pale Horse, Pale Rider*, the lover of the protagonist, Miranda, describes the effects of the disease on the city: "'It's as bad as anything can be,' said Adam, 'all the theaters and nearly all the shops and restaurants are closed, and the streets have been full of funerals all day and ambulances all night…'" Miranda is taken to the hospital while Adam is out getting them ice cream and hot coffee. She never sees him again, for he was not allowed to visit her. Tragically, she learned of Adam's death in an army camp hospital upon her

FIGURE 5 *"Coughs and Sneezes Spread Diseases," a slogan and poster campaign that was designed to cut down on the spread of respiratory diseases in the United Kingdom in World War II. Disease was portrayed as hampering the war effort at home. No doubt lurking in the minds of the Ministry of Health officials were the devastating effects of the 1919 influenza pandemic. (Courtesy of Yale University, Harvey Cushing/John Whitney Medical Library.)*

recovery. Crosby reports that Porter said of the effect of the pandemic on her life, "It just simply divided my life, cut across it like that." (46). It was true, commented Crosby, for many of her generation. He dedicated his account of the pandemic, *America's Forgotten Pandemic*, "To Katherine Anne Porter, who survived" (46).

Although the pandemic of 1918–1919 was shattering on a global scale, "The first pandemic of influenza in the bacteriological era," according to Richard Shope, "was that of 1889–1890" (47) (Fig. 6). During that epidemic, R. Pfeiffer isolated the bacillus *Haemophilus influenzae* (48). Pfeiffer held that the bacillus was present in cases associated with influenza but not in other individuals. John R. Mote reviewed the literature disputing this view but nonetheless assented that "... Pfeiffer's bacillus was for years considered by many workers to be the cause of epidemic influenza" (49). As Patrick Playfair Laidlaw described in his Linacre lecture, *H. influenzae* "... had held an almost undisputed position since the end of the pandemic 1899–1890" (50).

The accepted ethical norms of disease investigation in humans seemed to be suspended when the influenza pandemic of 1918–1919 came toward the end of World War I. Historically, wartime overthrew many usual constraints on the study of illness. As the science writer Gina Kolata put it, "But in 1918, such ethical arguments were rarely considered. Instead, the justification for a risky study with human beings was that it was better to subject a few to a great danger in order to save the many" (51). Negative studies with sailors in Boston, MA, and San Francisco, CA, were recounted by Kolata: intense exposure of presumed susceptible subjects to infected persons and their mucus, blood, exhalations, and coughs failed to induce the disease. While other groups of investigators appeared to have some success, the results were not consistent enough to reach a conclusion. As expressed by Laidlaw, one of the investigators who successfully transmitted influenza to ferrets, "One can never be sure that the experimental subject is susceptible and therefore negative results can be discounted; while positive results, though perhaps more significant, are always open to the suspicion that infection was picked up in some accidental manner" (50).

Attempts to transfer the illness to animals that had proven susceptible to other human viral pathogens were inconclusive (49). Monkeys, rabbits, guinea pigs, and mice were among those tested. Mote's compilation serves to emphasize the importance of at least three experimental variables: the material transmitted, blood versus sputum or throat washings; the site of transmission, intraperitoneal versus intranasal; and the susceptibility of the host. No clear etiological agent emerged from the 1918–1919 influenza pandemic. It was not until the study of swine influenza in two epizootics in 1928 and 1929 in Iowa by Richard Shope and Paul Lewis of the Department of Animal Pathology of the Rockefeller Institute at Princeton, NJ (52–54), that a new story began to emerge.

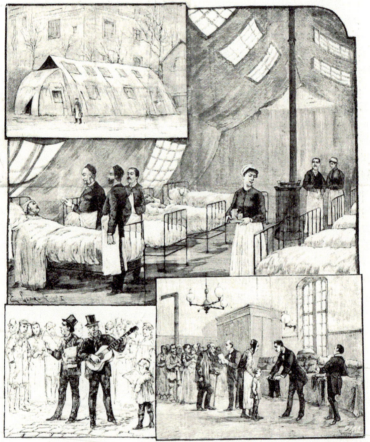

FIGURE 6 *L'influenza à Paris. This cover is from a Parisian weekly in 1890 during the influenza pandemic of 1889–1890. Originating in Russia and spreading westward, influenza became known as "Russian flu." The cover depicts four scenes relating to the epidemic in Paris (clockwise from top left): a tent set up in a hospital courtyard, the interior of tent ward for the sick, the distribution of clothes to families of victims, and two men singing a new song, "L'influenza, tout l'monde l'a!" ("Influenza, Everyone Has It!"). (Courtesy of the National Library of Medicine.)*

Swine influenza had been recognized during 1918–1919 by J. S. Koen of the U.S. Bureau of Animal Industry. According to Richard Shope's review in 1958, Koen recognized in swine similarities in prevalence and symptoms to human influenza, but he was criticized for calling it "flu" (47). The objections were economic and arose from a fear that the swine-flu connection would turn away the public from the consumption of pork. The epizootic in pigs was massive, with millions becoming sick and thousands dying. Recognizing the annual recurrence and enormity of the epizootics with such devastating economic consequences, Shope and Lewis began in earnest their investigations in 1928.

What they uncovered, reported in 1931 in a series of three articles in the *Journal of Experimental Medicine*, was quite remarkable (52–54). They discovered two agents working synergistically to produce the disease, with neither producing severe disease on its own. The first organism, very much like the Pfeiffer bacillus, was named *Haemophilus influenzae suis* and failed to induce experimental disease by itself. The second organism, a newly recognized filterable virus, induced a milder disease in experimental pigs than seen on pig farms. The more severe disease in swine resulted when the two agents were administered simultaneously. It later emerged in the report by Wilson Smith, C. H. Andrewes, and Patrick Playfair Laidlaw that the human and swine influenza viruses were antigenically closely related (1). As Shope put it, ". . . despite the failure of human investigators of the 1918 influenza pandemic to discover the cause of the outbreak, Mother Nature, using swine as her experimental animals, had done so" (47). Shope reported that he and Laidlaw independently reached the conclusion of the ". . . likelihood that swine had indeed acquired their infection from man in 1918. "

Prior to his work with Smith and Andrewes in which the virus of human influenza was isolated in ferrets, Laidlaw had collaborated with S. W. Dunkin on the experimental study of canine distemper in ferrets (55–57). The work was conducted at the National Institute for Medical Research Farm Laboratories, Mill Hill, London, United Kingdom, where thorough procedures were employed to prevent exogenous infection of experimental animals. The reasons to use the ferret were that it could "be confined in a small space with ease and comfort" and that keepers claimed that ferrets were very susceptible to dog distemper, which could wipe out an entire breeding colony. Hence, special buildings, cages, personnel practices, and experimental procedures to study ferrets were well established at Mill Hill by the 1920s. In developing the ferret model under such controlled conditions, Laidlaw and Dunkin successfully showed experimental transmission of canine distemper virus to the ferret with overt expression of clinical and pathological features of disease. They demonstrated the disease to be caused by a filterable virus, not by the bacterium *Bacillus bronchisepticus* (*Bordetella bronchiseptica*), which they characterized as a secondary invader (57). When influenza appeared again in London in 1933, "The ferret obviously was the animal to test for susceptibility to influenza" (58). Laidlaw's team was the right group to perform the studies to isolate a filterable agent.

Isolation of human influenza virus in ferrets, reported in the 8 July 1933 issue of *The Lancet* by Smith, Andrewes, and Laidlaw, was a signal event in the history of human influenza (1). The article described work that was fastidious in the care to exclude exogenous infection and elegant in the clarity of its results. It also demonstrated the experimental serendipity of susceptibility to various viral infections by different species of animals, so-called "species specificity." Although the possibility of influenza being a viral disease had been raised in about 1914 (50), ambiguous results had been obtained in humans, and unsuccessful attempts were made in other species. Smith et al. commented that "The filtrates, proved to be bacteriologically sterile, were used in attempts to infect many different species. All such attempts were entirely unsuccessful until the ferret was used " (1).

The report in 1933, which the investigators termed "a preliminary communication," detailed a number of critical parameters for experimentation. These included the source and nature of the inoculum, throat washings from people sick with influenza. Experimental manipulation and important clinical observations in ferrets included intranasal instillation, the biphasic clinical course, the nasal histopathology in infected ferrets, and serial passage. Finally, the characteristics of the agent were documented, including filterability, the absence of bacterial growth, and the neutralization of the virus by serum taken from people who had recovered from clinical influenza and from ferrets that had recovered from experimental infection. Smith and colleagues also studied the relationship to the virus of swine influenza received from Richard Shope; they found "a close antigenic relationship" (1). However, "[i]n striking contrast to swine influenza," there was no synergistic role for *H. influenzae suis* in the production of experimental disease in ferrets.

The Mill Hill investigators' report on the successful use of the ferret as an animal model received prompt confirmation from investigators on other continents. For example, T. Francis, at the Rockefeller Institute in New York City, working with sputum obtained from patients in a 1934 influenza epidemic in Puerto Rico, transmitted the disease to ferrets (59). F. M. Burnet reported the experimental transmission of influenza to ferrets from a 1935 epidemic in Melbourne, Australia (60).

The successful isolation of human influenza virus in the ferret in several laboratories triggered the exploration of other biological systems. Andrewes et al. reported the successful transmission of ferret-passed virus to mice (61), which was also reported in the following month by Francis (59). Reports of the successful cultivation of the virus in minced chicken embryo soon emerged (62, 63). Smith concluded that the egg membrane technique was "unsuitable for the study of this virus" (63). However, further studies would show that this biological system, the embryonated egg of chickens, was remarkably productive for the understanding of human influenza infection.

There are parallels between the yellow fever story and the study of the cause of human influenza. Noguchi claimed to have isolated a bacterial cause of yellow fever, *Leptospira icteroides*, in guinea pigs. The bacterium, while causing symptoms in guinea pigs similar to those of yellow fever, turned out not to be the cause of yellow fever, and the guinea pig turned out not to be a susceptible host of yellow fever. As noted above, Pfeiffer in 1893 reported the isolation of a Gram-negative bacterium as the cause of human influenza (48). *Haemophilus influenzae*, or Pfeiffer's bacillus, as it came to be known, was for some time thought to be the cause of influenza (49). Richard Shope and Paul Lewis found *Haemophilus influenzae suis* to be associated with swine influenza (52–54). It was to be shown that the ferret was the model of choice to study human influenza and that a synergistic bacterial infection was not present (1).

There was also a tragic link to three of the diseases discussed in this chapter. Paul A. Lewis was one of the first to transmit poliovirus serially in monkeys (26) and to identify it as a filterable agent in his work with Flexner (29). He had collaborated with Shope on his studies of swine influenza (64), only to later succumb to yellow fever while studying it in Brazil.

EMBRYONATED EGGS

"The recognition of the potentialities of the method for virus research was almost entirely due to Ernest William Goodpasture and his collaborators," wrote W. I. B. Beveridge and F. M. Burnet in their 1946 monograph on the use of the chicken embryo for the cultivation of viruses and rickettsiae (65). Goodpasture was a Tennessee-born pathologist who early in his career had trained with William Welch, who was one of the founders of American pathology at Johns Hopkins (64). As the medical writer Greer Williams told the story, working at Vanderbilt University, Goodpasture had put an M.D. pathology trainee, Eugene Woodruff, to work on investigating the pathogenesis of fowlpox (64). His wife, Alice Woodruff, a Ph.D. physiologist, joined the effort somewhat later in an attempt to culture the virus. After their unsuccessful attempts in tissue culture, Goodpasture suggested to Alice Woodruff that they try embryonated chicken eggs for the study of fowlpox.

As later reviewed by Goodpasture, there was a very long history of the study of embryonated eggs in experimental biology (66). He noted that the first reported creation of a "window" in the egg to observe the embryo was by L. Beguelin in 1749. He also noted that the first published study of the use of embryonated eggs in infection was in 1905 by C. Levaditi working with the spirillum of fowl. The laurel for the first application of embryonated eggs to the study of viral infection goes to Peyton Rous and James B. Murphy, who in 1911 reported the successful transplantation of a transmissible sarcoma of fowl (67). The tumor was transmitted to the chick not only by finely divided tumor but also by "a filtrate free of the tumor cells," resulting in tumors in the membranes of the embryo.

Starting with the technique of E. R. Clark for operating on chicken embryos (68), Goodpasture's group developed and adapted the technique for virological studies. The methods to deposit the virus in a sterile fashion on the chorioallantoic membrane or in the allantoic or amniotic cavities were carefully described (69). The first report by the Goodpasture group on the use of the embryonated egg was that of Alice Woodruff and Goodpasture for infection with fowlpox virus (70). Goodpasture's principal interest was in the pathogenesis of viral infections; however, with others he also developed the embryonated egg for vaccine production (69). That work laid the foundation for the development of influenza vaccine.

The next major development was the report by Macfarlane Burnet in 1935 of the growth of influenza virus in embryonated eggs (71). Serial passage of influenza in eggs was required before egg membrane lesions became obvious. Hence, it did not appear that the technique would prove of use for primary virus isolation. Presciently, Burnet suggested that the method could serve as a means of antigen preparation for immunization.

The necessary step to bring the technique into primary diagnostic work resulted from a chance observation. George Hirst, working at the Rockefeller Foundation in New York, reported that "When the allantoic fluid from chick embryos previously infected with strains of influenza A virus was being removed, it was noted that the red cells of the infected chick, coming from ruptured vessels, agglutinated in the allantoic fluid" (72). This observation was confirmed by L. McClelland and R. Hare at the University of Toronto (73). Further expansion of this observation facilitated the development of diagnostic tests for influenza, both primary isolation and antibody development. For virus isolation, clinical specimens were inoculated into eggs, the amniotic fluids were harvested, and a dilution of fluid was mixed with red blood cells. Agglutination indicated likely influenza virus infection, which was proven by the inhibition of agglutination by virus-specific antisera. Henceforth, use of the ferret, with its attendant costs, housing, and susceptibility to exogenous infection, was superseded by use of embryonated eggs (74). The embryonated egg proved to be a boon to virological studies in general and to influenza in particular. As a measure of the importance of the technique, Nobel Laureate Macfarlane Burnet simplified his description of many years of research to say, "From 1935–1955, one can summarize my life as learning about influenza virus in chick embryos" (58).

The use of experimental animals and that of embryonated eggs were the principal means of virus isolation prior to the development of cell culture in the 1950s.

Animals were crucial to the isolation and characterization of herpes viruses, among many others. In parallel, development of serological techniques to measure virus-specific antibodies was to become the major means of viral diagnosis for several decades. That story is told in the next chapter.

REFERENCES

1. **Smith W, Andrewes CH, Laidlaw PP.** 1933. A virus obtained from influenza patients. *Lancet* **ii:**66–69 http://dx.doi.org/10.1016/S0140-6736(00)78541-2.
2. **Wilkinson L.** 2002. History, p 1–22. *In* Jackson AC, Wunner WH (ed), *Rabies.* Academic Press, Amsterdam, The Netherlands.
3. **Zinke GG.** 1804. *Neue Ansichten der Hundswuth, ihrer Ursachen und Folgen, nebst einer sichern Behandlungsart der von tollen Tieren gebissenen Menschen.* C B Gabler, Jena, Germany.
4. **Wilkinson L.** 1992. *Animals and Disease: an Introduction to the History of Comparative Medicine.* Cambridge University Press, New York, NY.
5. **Lee H.** 1960. *To Kill a Mockingbird.* Warner Books, New York, NY.
6. **Geison GL.** 1995. *The Private Science of Louis Pasteur.* Princeton University Press, Princeton, NJ.
7. **Debré P.** 1998. *Louis Pasteur.* (Forster E, translator.) Johns Hopkins University Press, Baltimore, MD.
8. **Baltazard M, Bahmanyar M, Chodssi M, Sabeti A, Gajdusek C, Rouzbehi E.** 1955. Essai pratique du sérum antirabique chez les mordus par loups enragés. *Bull World Health Organ* **13:**747–772.
9. **Warrell DA, Davidson NM, Pope HM, Bailie WE, Lawrie JH, Ormerod LD, Kertesz A, Lewis P.** 1976. Pathophysiologic studies in human rabies. *Am J Med* **60:**180–190 http://dx.doi.org/10.1016/0002-9343(76)90427-7.
10. **Magendie F.** 1821. Expérience sur la rage. *J Physiol Exp Pathol* **1:**41–47.
11. **Galtier M.** 1879. Études sur la rage. *C R Seances Soc Biol* **89:**444–446.
12. **Vallery-Radot R.**1900. *The Life of Pasteur.* (Devonshire RL, translated from "La Vie de Pasteur," 1900, Hachette, Paris.) Garden City Publishing Co, New York, NY.
13. **Dubos RJ.** 1950. *Louis Pasteur, Free Lance of Science.* Little, Brown, Boston, MA.
14. **Lechevalier H, Solotorovsky M.** 1974. *Three Centuries of Microbiology.* Dover, New York, NY.
15. **Remlinger D.** 1903. Le passage du virus rabique à travers les filtres. *Ann Inst Pasteur (Paris)* **17:**834–849.
16. **McFadyean J.** 1908. The ultravisible viruses. *J Comp Pathol Ther* **21:**58–68, 168–175, 232–246 http://dx.doi.org/10.1016/S0368-1742(08)80008-2.
17. **Negri A.** 1903. Beitrag zum Stadium der Aetiologie der Tolwuth. *Hyg Infektionskr* **43:**507–528.
18. **Hiss PH Jr, Zinsser H.** 1919. *A Text-Book of Bacteriology. A Practical Treatise for Students and Practitioners of Medicine,* 4th ed. D Appleton, New York, NY.
19. **Paul JR.** 1971. *A History of Poliomyelitis.* Yale University Press, New Haven, CT.
20. **Edinburgh University Library.** Accessed 14 September 2008. The Walter Scott digital archive. http://www.walterscott.lib.ed.ac.uk/.
21. **Caverly CS.** 1894. Preliminary report of an epidemic of paralytic disease, occurring in Vermont, in the summer of 1894. *Yale Med J* **1:**1–5.
22. **Caverly CS.** 1896. Notes of an epidemic of acute anterior poliomyelitis. *JAMA* **26:**1–5.
23. **Lott G, Bartel J.** 1908. Offizielles Protokoll der k.k. Gesellschaft der Aerzte in Wien. *Wien Klin Wochenschr* **52:**1829–1831.
24. **Wickman I.** 1913. *Acute Poliomyelitis (Heine-Medin's Disease).* (Maloney WJAM, translator.) The Journal of Nervous and Mental Disease Publishing Company, New York, NY.
25. **Landsteiner K, Popper E.** 1909. Übertragung der Poliomyelitis acuta auf Affen. *Z Immunitatsforsch* **2:**377–390.

26. **Flexner S, Lewis PA.** 1909. The transmission of acute poliomyelitis to monkeys. *JAMA* **53:**1639 http://dx.doi.org/10.1001/jama.1909.92550200027002g.

27. **Loeffler F, Frosch P.** 1898. Berichte der Kommission zur Erforschung der Maulund Klauenseuche bei dem Institut fur Infectionskrankheiten in Berlin. *Dtsch Med Wochenschr* **24:**80–83, 97–100 http://dx.doi.org/10.1055/s-0029-1204235.

28. **Landsteiner K, Levaditi C.** 1909. La transmission de la paralysie infantile aux singes. *C R Seances Soc Biol* **67:**592–594.

29. **Flexner S, Lewis PA.** 1909. The nature of the virus of epidemic poliomyelitis. *JAMA* **53:**2095 http://dx.doi.org/10.1001/jama.1909.925502500010011.

30. **Reed W, Carroll J.** 1902. The etiology of yellow fever: a supplemental note. *Am Med* **3:**301–305.

31. **Gorgas WC.** 1909. Sanitation of the tropics with special reference to malaria and yellow fever. *JAMA* **52:**1075–1077 http://dx.doi.org/10.1001/jama.1909.25420400001001.

32. **Noguchi H.** 1919. Etiology of yellow fever. II. Transmission experiments on yellow fever. *J Exp Med* **29:**565–584 http://dx.doi.org/10.1084/jem.29.6.565.

33. **Warren AJ.** 1951. Landmarks in the conquest of yellow fever, p 1–38. *In* Strode GK (ed), *Yellow Fever.* McGraw-Hill, New York, NY.

34. **Stokes A, Bauer JH, Hudson NP.** 1928. Experimental transmission of yellow fever to laboratory animals. *Am J Trop Med* **8:**103–164 http://dx.doi.org/10.4269/ajtmh.1928.s1-8.103.

35. **Theiler M.** 1951. The virus, p 39–136. *In* Strode GK (ed), *Yellow Fever.* McGraw-Hill, New York, NY.

36. **Theiler M.** 1930. Susceptibility of white mice to the virus of yellow fever. *Science* **71:**367 http://dx.doi.org/10.1126/science.71.1840.367.a.

37. **Theiler M, Smith HH.** 1937. The use of yellow fever virus modified by *in vitro* cultivation for human immunization. *J Exp Med* **65:**787–800 http://dx.doi.org/10.1084/jem.65.6.787.

38. **Sawyer WA, Lloyd W.** 1931. The use of mice in tests of immunity against yellow fever. *J Exp Med* **54:**533–555 http://dx.doi.org/10.1084/jem.54.4.533.

39. **Hammon WM.** 1943. The epidemic encephalitides of North America. *Med Clin North Am* **1943:**632–650 http://dx.doi.org/10.1016/S0025-7125(16)36287-3.

40. **Muckenfuss RS, Armstrong C, McCordock HA.** 1933. Encephalitis: studies on experimental transmission. *Public Health Rep* **48:**1341–1343 http://dx.doi.org/10.2307/4580968.

41. **Webster LT, Fite GL.** 1933. A virus encountered in the study of material from cases of encephalitis in the St. Louis and Kansas City epidemics of 1933. *Science* **78:**463–465 http://dx.doi.org/10.1126/science.78.2029.463.

42. **Kawamura R, Kodama M, Ito T, Yasaki T, Kobayakawa Y.** 1936. Epidemic encephalitis in Japan. The causative agent compared with that in the St. Louis epidemic. *Arch Pathol (Chic)* **22:**510–523.

43. **Booss J, Tselis A.** 2014. A history of viral infections of the central nervous system. *In Neurovirology. The Handbook of Clinical Neurology.* Elsevier, Amsterdam, The Netherlands. http://dx.doi.org/10.1016/B978-0-444-53488-0.00001-8.

44. **Patterson KD, Pyle GF.** 1991. The geography and mortality of the 1918 influenza pandemic. *Bull Hist Med* **65:**4–21.

45. **Porter KA.** 1972. Pale horse, pale rider, p 269–317. *In* Porter KA, *The Collected Stories of Katherine Anne Porter.* Harcourt, New York, NY.

46. **Crosby AW.** 2003. *America's Forgotten Pandemic: the Influenza of 1918,* 2nd ed. Cambridge University Press, New York, NY. http://dx.doi.org/10.1017/CBO9780511586576.

47. **Shope RE.** 1958. Influenza: history, epidemiology, and speculation. *Public Health Rep* **73:**165–178 http://dx.doi.org/10.2307/4590072.

48. **Pfeiffer R.** 1893. Die Aetiologie der Influenza. *Z Hyg* **13:**357–386 http://dx.doi.org/10.1007/BF02284284.

49. **Mote JR.** 1940. Human and swine influenzas, p 429–516. *In* Zinsser H (ed), *Virus and Rickettsial Diseases with Special Consideration of Their Public Health Significance.* Harvard University Press, Cambridge, MA.

50. **Laidlaw PP.** 1935. Epidemic influenza: a virus disease. *Lancet* **225:**1118–1124 http://dx.doi.org/10.1016/S0140-6736(01)19376-1.
51. **Kolata G.** 1999. *Flu: the Story of the Great Influenza Pandemic of 1918 and the Search for the Virus that Caused It.* Farrar, Straus and Giroux, New York, NY.
52. **Lewis PA, Shope RE.** 1931. Swine influenza. II. A hemophilic bacillus from the respiratory tract of infected swine. *J Exp Med* **54:**361–371 http://dx.doi.org/10.1084/jem.54.3.361.
53. **Shope RE.** 1931. Swine influenza. I. Experimental transmission and pathology. *J Exp Med* **54:**349–359 http://dx.doi.org/10.1084/jem.54.3.349.
54. **Shope RE.** 1931. Swine influenza. III. Filtration experiments and etiology. *J Exp Med* **54:**373–385 http://dx.doi.org/10.1084/jem.54.3.373.
55. **Dunkin GW, Laidlaw PP.** 1926. Studies in dog-distemper. I. Dog-distemper in the ferret. *J Comp Pathol Ther* **39:**201–212 http://dx.doi.org/10.1016/S0368-1742(26)80020-7.
56. **Dunkin GW, Laidlaw PP.** 1926. Studies in dog-distemper. II. Experimental distemper in the dog. *J Comp Pathol Ther* **39:**213–221 http://dx.doi.org/10.1016/S0368-1742(26)80021-9.
57. **Dunkin GW, Laidlaw PP.** 1926. Studies in dog-distemper. III. The nature of the virus. *J Comp Pathol Ther* **39:**222–230 http://dx.doi.org/10.1016/S0368-1742(26)80021-9.
58. **Burnet M.** 1969. *Changing Patterns: an Atypical Autobiography.* American Elsevier, New York, NY.
59. **Francis T Jr.** 1934. Transmission of influenza by a filterable virus. *Science* **80:**457–459 http://dx.doi.org/10.1126/science.80.2081.457.b.
60. **Burnet FM.** 1935. Influenza virus isolated from an Australian epidemic. *Med J Aust* **2:**651–653 http://dx.doi.org/10.5694/j.1326-5377.1935.tb43332.x.
61. **Andrewes CH, Laidlaw PP, Smith W.** 1934. The susceptibility of mice to the viruses of human and swine influenza. *Lancet* **224:**859–882 http://dx.doi.org/10.1016/S0140-6736(00)74657-5.
62. **Francis T Jr, Magill TP.** 1935. Cultivation of human influenza virus in an artificial medium. *Science* **82:**353–354 http://dx.doi.org/10.1126/science.82.2128.353.
63. **Smith W.** 1935. Cultivation of the virus of influenza. *Br J Exp Pathol* **16:**508–512.
64. **Williams G.** 1959. *Virus Hunters.* Alfred A. Knopf, New York, NY.
65. **Beveridge WIB, Burnet FM.** 1946. *The Cultivation of Viruses and Rickettsiae in the Chick Embryo.* His Majesty's Stationery Office, London, United Kingdom.
66. **Goodpasture EW.** 1938. Some uses of the chick embryo for the study of infection and immunity. *Am J Hyg* **28:**111–129.
67. **Rous P, Murphy JB.** 1911. Tumor implications in the developing embryo. *JAMA* **56:**741–742.
68. **Clark ER.** 1920. Technique of operating on chick embryos. *Science* **51:**371–373 http://dx.doi.org/10.1126/science.51.1319.371.
69. **Goodpasture EW, Buddingh GJ.** 1935. The preparation of anti-smallpox vaccine by culture of the virus in the chorio-allantoic membrane of chick embryos, and its use in human immunization. *Am J Hyg* **21:**319–360.
70. **Woodruff AM, Goodpasture EW.** 1931. The susceptibility of the chorio-allantoic membrane of chick embryos to infection with the fowlpox virus. *Am J Pathol* **7:**209–222, 5.
71. **Burnet FM.** 1935. Propagation of the virus of epidemic influenza on the developing egg. *Med J Aust* **2:**687–689 http://dx.doi.org/10.5694/j.1326-5377.1935.tb43360.x.
72. **Hirst GK.** 1941. The agglutination of red cells by allantoic fluid of chick embryos infected with influenza virus. *Science* **94:**22–23 http://dx.doi.org/10.1126/science.94.2427.22.
73. **McClelland L, Hare R.** 1941. The adsorption of influenza virus by red cells and a new in vitro method of measuring antibodies for influenza virus. *Can J Public Health* **32:**530–538.
74. **Hirst GK.** 1948. Influenza, p 93–121. *In* Francis T Jr (ed), *Diagnostic Procedures for Virus and Rickettsial Diseases.* American Public Health Association, New York, NY.

3 Filling the Churchyard with Corpses
SMALLPOX AND THE IMMUNE RESPONSE

> *In the final analysis, only an antibody response in the host*
> *constitutes definitive evidence*
> *of infection with a specific virus.*
>
> Principles of Internal Medicine, 1962 (1)

INTRODUCTION

The host is well defended against virus infections, for the immune system has several weapons at its disposal. Among these weapons are antibodies which develop in response to and are specifically targeted against the infecting agent. Antibodies perform a number of host defense functions, including combining with the virus, thereby neutralizing it, and attaching to infected cells to promote their destruction. Each of these activities prevents further multiplication and spread of the virus. Antibodies develop from days to weeks after infection, leaving evidence of specific viral infection. For decades the measurement of antibodies was the principal means by which viral diagnostic labs established the identity of a viral infection. Serological measurement of antibodies was less cumbersome and costly than the isolation of viruses in animal hosts. In addition to its highly accurate specificity, another remarkable feature of the immune system is memory, the basis by which the host recognizes the appearance of a previously encountered virus (2). Immunological memory is also the basis of vaccination. If a host can be exposed by vaccination to a virus or

To Catch a Virus, Second Edition. Authored by John Booss and Marie Louise Landry.
© 2023 American Society for Microbiology. DOI: 10.1128/9781683673828.ch03

its components in a less harmful fashion than natural infection, immunity will be induced to protect against future infection.

The emergence of immunological concepts in the latter half of the 19th century and at the start of the 20th century and the development of assays to measure antibodies against invading pathogens transformed therapeutic approaches and management of infectious diseases. Like virology, the science of immunology developed in the wake of advances in bacteriology. Crucial to the development of immunology was the understanding of mechanisms of defense. These were first understood in the protection against reinfection for certain well-recognized diseases such as smallpox.

The early story of the immune response includes smallpox vaccination and the phagocytosis-versus-humoral immunity debate with the joint award of the Nobel Prize to Elie Metchnikoff and Paul Ehrlich in 1908 for their seminal work (3). Early assays including complement fixation, neutralization, and hemagglutination were fundamental to early viral diagnostic labs and made clinical laboratory diagnosis more accessible. With these key discoveries, Koch's postulates gained new meaning as applied by T. M. Rivers for viral diseases (4).

PROTECTION: THE CASE OF SMALLPOX

The role of recovery from a first episode of an infectious disease in providing protection against another attack has been known for centuries. Characteristic traces of smallpox, pockmarks on the skin, were recognized as evidence of past infection and hallmarks of immunity to reinfection. Acute smallpox infection had to be distinguished from other viral exanthema such as measles. Rhazes, a Persian physician of the 10th century in Baghdad, wrote *A Treatise on the Smallpox and Measles*, which is regarded as a landmark in clinical description (5). Rhazes reported the symptoms that may precede the skin eruptions: ". . . a continued fever, pain in the back, itching in the nose, and terrors in sleep." The authority of this treatise in Europe extended into the 17th century.

Two millennia before Rhazes, Ramses V died in 1157 BCE of what was assumed to be smallpox, and his body was mummified. Donald R. Hopkins recounts a remarkable episode in which he was granted permission by Egyptian president Anwar el Sadat to examine the front upper half of the unwrapped mummy in 1979 (6). After describing the rash of elevated "pustules," Hopkins concluded that the appearance and distribution were "similar to smallpox rashes I have seen in more recent victims." Hopkins, a physician who participated in the WHO Smallpox Global Eradication Programme, also wrote the classic treatise on the history of smallpox, *Princes and Paupers*. It was first published in 1983 and republished in 2002 with a new introduction as *The Greatest Killer*. While the exact origins of smallpox cannot be documented with certainty, it has been assumed that it arose millennia ago when aggregations of populations in towns and small cities emerged that supported epidemic spread.

The terror and dread occasioned by smallpox were well captured by the British historian Thomas Babington Macaulay in his *History of England*: ". . . the smallpox was always present, filling the churchyards with corpses, tormenting with constant fears all whom it had not yet stricken, leaving on those whose lives it spared the hideous traces of its power, turning the babe into a changeling at which the mother shuddered, and making the eyes and cheeks of the betrothed maiden objects of horror to the lover" (quoted in reference 7). Macaulay's characterization of "all whom it had not yet stricken" is an implicit statement of the immunological protection of those "whose lives it spared."

In contrast to lands where "the smallpox was always present," populations in which smallpox had not previously existed were devastated on its first appearance. Nowhere, perhaps, is this more stunning and graphically illustrated than with its introduction into the western hemisphere—so stunning, in fact, that the American historian William McNeill began his seminal *Plagues and Peoples* asking how Hernando Cortez, with fewer than 600 soldiers, was able to conquer the Aztec empire of Mexico with its millions of inhabitants and later how Pizarro was able to conquer the Inca empire in South America (8). The scourge of smallpox nearly wiped out the previously unexposed Aztec and Inca populations, allowing Cortez easy victory over his diminished opposition. McNeill implicitly notes the role of immunological protection in that the Spaniards had the advantage of previous exposure to smallpox, whereas the Amerindians had not. He writes that smallpox was ". . .a disease that killed only Indians and left Spaniards unharmed." McNeill comments, "the lopsided impact of infectious disease upon Amerindian populations therefore offered a key to understanding the ease of the Spanish conquest of America—not only militarily but culturally as well." Elsewhere on the North American continent, Francis Parkman wrote of a smallpox outbreak among the Huron Indians north of Lake Ontario in 1636: "Terror was universal . . . its ravages were appalling. . .No house was left unvisited. . ..Everywhere was heard the wail of the sick and dying children" (quoted in reference 9). European history was also significantly influenced by smallpox: McNeill documents that smallpox altered the course of British political history. Hopkins termed the killing of kings, queens, an emperor, and a tsar in or around the 18th century ". . .a regicidal rampage without parallel. . ." (6). Thus, whole populations were devastated by smallpox in the New World, and leadership as well as the common people was destroyed in the Old World.

When the opportunity presented itself to offer protection against smallpox by inoculation with smallpox material, a practice known as variolation, it was accepted in some European nations more successfully than in others. The procedure had been known elsewhere, including nasal insufflation, in which scabs from a mild case of smallpox were blown into the nostril. It had been practiced in ancient China as "planting of flowers," and inoculation had been known in India "since before the Christian era" (6). The campaign to bring inoculation against smallpox to England was waged by Lady Mary Wortley Montagu (Fig. 1), who first encountered it while in Turkey with her husband, the British ambassador. Lady Montagu

FIGURE 1 *Lady Mary Wortley Montagu. Lady Montagu, an aristocrat of considerable intellectual sophistication and beauty, brought the practice of variolation against smallpox to England from Turkey. It consisted of inoculating smallpox material and preceded Jenner's discovery of vaccination, the inoculation of cowpox to prevent smallpox. Lady Montagu is shown in a Turkish embellished costume with a jeweled turban in an illustration from* The Letters of Horace Walpole. *(Courtesy of the James Smith Noel Collection, Louisiana State University, Shreveport, LA.)*

was an English aristocrat, beauty, and intellectual who jousted with no less a figure than Alexander Pope, the 18th-century poet. In 1717 she wrote to a friend that "the smallpox, so fatal, and so general amongst us, is here entirely harmless, by the invention of ingrafting which is the name they give it." Lady Montagu described the procedure and quoted the French ambassador as saying that it is taken "by way of diversion." She went on to say in part that "I am patriot enough to take pains to bring this useful invention into fashion in England... (6)." Strong willed and intelligent, Lady Montagu was highly motivated with respect to smallpox. She had lost her brother to smallpox, and her own attack of smallpox had taken her beauty, leaving her with a pockmarked face. She had her own children inoculated, and on her return to England the royal family took note. The successful inoculation of the two daughters of the Prince of Wales in 1722 "began the firm establishment of inoculation as acceptable medical practice in England" (6). However, there was early resistance to the practice in France, for example, where it was officially accepted finally in 1769.

The risks and benefits of inoculation against smallpox and the attitude change in the population from resistance to acceptance were nicely documented by John B. Blake for colonial Boston (10). In the epidemic of 1721, theological and political considerations complicated the medical concerns. The death rate rose to 105 per 1,000, while that for the whole period from 1701 to 1774, including the epidemic, averaged 35 per 1,000. Inoculation was introduced in Boston in the 1721 outbreak by Zabdiel Boylston on the urging of Cotton Mather (6). Mather had learned of the practice in West Africa from his slave Onesimus. According to Blake, in the outbreak of 1729–1730 in Boston, of 2,600 persons with natural smallpox, 500 died (19%). In contrast, of 400 people inoculated, only 12 died (3%) (10).

The danger in variolation was the use of infectious "pus" taken directly from an individual with infected pustules or the use of ground, infected scabs. These materials were scratched into the skin of a healthy person in hopes of conferring resistance to disfiguring disease. Despite the marked reduction in mortality, there were several disadvantages attendant to inoculation. While less severe than natural smallpox, many experienced illness of various degrees of intensity, with some mortality. In addition, there was a costly preparation period of mercury and antimony administration. Even Edward Jenner experienced an arduous preparation period. Hence, the poor could not afford protection until 1764, when inoculation was first provided by the government (10). Most distressingly, inoculated persons were a source of virulent infection to their contacts. As a result, inoculation hospitals were necessary for supervision of administration. Blake notes that the experience of the American Revolutionary War resulted in greater acceptance of the practice. By tabulating deaths and cases from natural versus inoculated smallpox, Blake demonstrated a progressive acceptance of smallpox inoculation.

During the Revolutionary War smallpox played a devastating role, leading General George Washington to order the inoculation of the regular troops. In 1775–1776 the American attack on Quebec, to prevent its use as a British base, was thwarted by smallpox. As Hopkins put it, "But for the epidemic, the Continental troops would have captured the city, and hence, control of Canada" (see legend to Plate 31 in reference 6). In 1777 General Washington decided that the Continental Army needed to be inoculated to halt the spread of smallpox. "Should the disease rage with its usual virulence, we should have more to dread from it than the sword of the enemy…" (11). Various commentators have noted the crucial importance of Washington's decision in the outcome of the war (12). Thus, adaptive immunity by variolation for protection against disease played a significant role in the history of American independence.

Hard on the heels of American independence came the description by Edward Jenner (Fig. 2) of vaccination against smallpox. At a stroke, it significantly reduced or eliminated the risks attendant to inoculation (variolation): illness and mortality and transmission of virulent disease to unfortunate and unsuspecting contacts. Jenner's discovery was also revered in France. General Napoleon Bonaparte ordered the vaccination of his army soon after the turn of the 19th century. In fact, Jenner appealed to Napoleon for the pardon of an Englishman held in France. Considering the request, Napoleon was quoted as saying to the Empress Josephine, "What that man asks is not to be refused" (13).

Inoculation with smallpox, or variolation, was to provide an essential link in the later demonstration of the protective action of cowpox against smallpox infection. Edward Jenner was ideally prepared to be the vehicle of that demonstration. He had had first-hand experience with smallpox inoculation as a child of 8 years of age. He had undergone extensive preparation by purging and bleeding before inoculation (13). As an apprentice in surgery and pharmacy, he had learned of the folk knowledge that exposure to cowpox prevented infection with smallpox. As a medical practitioner, he apparently was an inoculator. Interestingly in light of the later use of cowpox to prevent smallpox, he inoculated his young son with swinepox in 1789. A few years later, he inoculated his son with smallpox, which failed to take, implying protection against smallpox (14) conferred by the earlier treatment.

Jenner's powers of observation as a naturalist were strong; he was elected to the Royal Society based on observations on the nesting behavior of the cuckoo. Those talents are on display in his self-published pamphlet on cowpox in 1798 (15). Jenner noted that "… but what renders the Cow-pox virus so extremely singular, is that the person who has been affected is for ever after secure from the infection of the Smallpox; neither exposure to the various effluvia, nor insertion of the matter into the skin, producing the distemper." Jenner's Case I was Joseph Merret, who had tended horses with sore heels, an infectious disease called "the grease." Merret developed sores on his hands and swelling and stiffness in his axillae. It was Jenner's belief that the grease

FIGURE 2 *Edward Jenner. Jenner was an English physician with an intense interest in natural science. He demonstrated the truth in the folk belief that previous infection with cowpox prevented smallpox. The description of the inoculation of a boy, James Phipps, with material from a sore on the hand of a dairymaid, Sara Nelms, has achieved iconic status. Jenner published his results of vaccination in 1798. Vaccination eliminated the scourge of smallpox through the WHO Global Eradication Programme by 1979. (Courtesy of the National Library of Medicine.)*

was transmitted by farm workers to cows which then transmitted cowpox to milkers, thereby protecting them from smallpox. Twenty-five years later, Merret was inoculated with "variolous matter," but it did not take. Jenner noted that since the population was thin and any case of smallpox recorded, it was certain that Merret had not had smallpox in the intervening years. Apparently the relationship between the grease

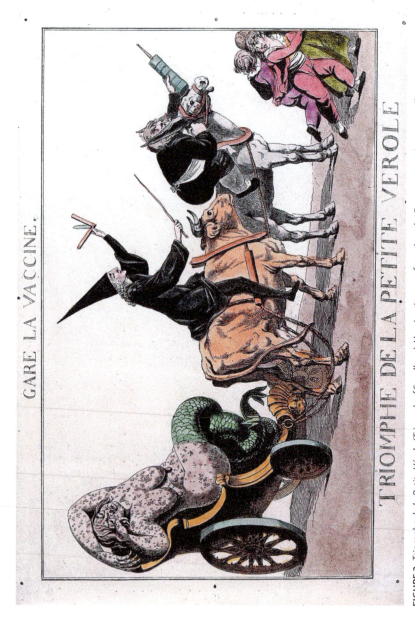

GARE LA VACCINE.

TRIOMPHE DE LA PETITE VEROLE

FIGURE 3 *Triomphe de la Petite Vérole (Triumph of Smallpox). Vaccination was feared on the European continent as well as in England. This French caricature satirized that fear. It shows a woman with smallpox turning into a mermaid, a physician riding a cow, and an apothecary with a giant syringe pursuing frightened children. (Courtesy of the Wellcome Library, London, United Kingdom.)*

and cowpox was discredited (16). However, the link between cowpox and immunity to smallpox was sustained. Then came what history has recorded as the decisive event.

Jenner attempted artificial infection by cowpox of a boy, James Phipps, with matter taken from a sore on the hand of a dairymaid, Sarah Nelms, who had been infected by her master's cows. Lesions of the incisions "were much the same as when produced in a similar manner by variolous matter," and a week after inoculation he had a brief episode of axillary discomfort and systemic symptoms. The crucial test came on 1 July following when he was inoculated with variolous matter: "No disease followed." Several months later he was again challenged with variolous matter, "but no sensible effect was produced on the constitution" (17). Phipps was the same age, 8 years, as Jenner had been when Jenner experienced the arduous preparation for variolation (13).

It is a paradox that this first experimental observation of what was the most beneficial medical public health intervention ever devised was first rejected by the Royal Society in 1797. Hopkins has aptly put it, "Because his evidence was so slim and his conclusion so audacious, the paper was quietly returned . . ." (6). Despite resistance among some (Fig. 3), the value of vaccination was very promptly recognized and vaccination was taken up by others. In just a decade after James Phipps was vaccinated, Thomas Jefferson was able to write to Edward Jenner in 1806, "Further generations will know by history only that the loathsome smallpox has existed" (6). It is an extraordinary achievement that came to pass. In 1966 the WHO undertook the Global Eradication Programme to rid the world of smallpox, which was accomplished and certified in 1979, closing the magnificent chapter in smallpox history. Further accounts of this momentous accomplishment are detailed in "the big red book," *Smallpox and Its Eradication* (18).

START OF THE SCIENCE OF IMMUNOLOGY: PHAGOCYTOSIS AND HUMORAL IMMUNITY

Theories of humoral immunity, the basis of serological testing, were forged against Elie Metchnikoff's (Fig. 4) theory of phagocytosis in host defense. Metchnikoff's studies of phagocytosis evolved from his interest in digestion by invertebrates in the latter 1870s (19). In her biography of Metchnikoff, Olga, his second wife, gives a charming quote describing the inception of her husband's phagocyte theory: "One day when the whole family had gone to a circus to see some extraordinary performing apes, I remained alone with my microscope, observing the life in mobile cells of a transparent star-fish larva, when a new thought suddenly flashed across my brain. It struck me that similar cells might serve in the defense of the organism against intruders" (20). Metchnikoff devised a simple experiment in which he "introduced them

FIGURE 4 *Elie Metchnikoff. Metchnikoff's studies of phagocytosis initiated the science of immunology, specifically cellular immunity. With Paul Ehrlich, who developed the theoretical basis for the action of antibodies or humoral immunity, Metchnikoff received the Nobel Prize in 1908.*

(rose thorns) at once under the skin of some beautiful star-fish larvae as transparent as water." The following morning he confirmed that the thorns were surrounded by mobile cells. "That experiment formed the basis of the phagocyte theory, to the development of which I devoted the next twenty-five years of my life."

On showing his experiments to Rudolf Virchow, the father of modern pathology, Metchnikoff was advised to proceed with caution. As opposed to Metchnikoff's view of inflammation as "a curative reaction," contemporary medicine viewed leukocytes as supporting the growth of microbes (20). Metchnikoff left Messina and on subsequent travels through Vienna, the word "phagocytes" was suggested by zoologists as a Greek translation of "devouring cells" (20).

In 1884, Metchnikoff published a seminal work on phagocytosis in *Virchow's Archive* (21). Regretfully, however, as his wife recorded, ". . . the memoir passed unnoticed; the full significance of it had not been grasped" (20). Over a century later, A. M. Silverstein, in his history of immunology, amplified that observation: "One may conclude that the cellular theory of immunity advanced by Elie Metchnikoff in 1884 did not constitute just one further acceptable step in a well-established tradition; rather it represented a significant component of a conceptual revolution with which contemporary science had not yet fully learned to cope" (19).

As described in the 1884 work, Metchnikoff investigated a fungal disease of *Daphnia*, a transparent water flea (22). *Daphnia* ingested asci, sac-like structures in which spores are formed. Upon release into the digestive tract, spores traversed the intestinal wall to the body cavity, where they were attacked by blood cells, ultimately disintegrating into granules. Host giant cells were seen to have formed from the fusion of ameboid cells; however, the host did not always win out.

Using Metchnikoff's other studies of anthrax bacilli inoculated subcutaneously into frogs and rabbits as a take-off point, George Nuttall confirmed that phagocytosis was observed but also noted extracellular destruction of anthrax (reference 23, translated in reference 22). With blood and other body fluids *in vitro* from several different species of animals, either immunized or not, Nuttall demonstrated degeneration of free anthrax bacilli and other bacteria. Hence, phagocytosis was not the entire explanation for destruction of bacteria. "These investigations have shown that independently of leucocytes, blood and other tissue fluids may produce morphological degeneration of bacilli" (22). The studies of Metchnikoff on phagocytosis and Nuttall on humoral mechanisms demonstrated the two arms of host defense. However, they did not demonstrate immunization, specific host responses after particular challenges. That was most clearly achieved in studies of immunity to tetanus and diphtheria toxins by Emil Behring and Shibasaburo Kitasato.

In his masterful *The History of Bacteriology*, published in 1938, William Bulloch succinctly described the stepwise series of discoveries leading up to the landmark work of Behring and Kitasato (24): "F. Loeffler (1884) discovered the diphtheria bacillus and Kitasato (1889) proved that the tetanus bacillus is the cause of lockjaw. Roux and Yersin (1889) proved that the diphtheria bacillus operates by virtue of a poison (toxin) which it elaborates, and K. Faber (1889) showed that the tetanus bacillus acts in a similar manner." The stage was set for Behring and Kitasato to describe the mechanism of immunity to diphtheria and tetanus (25). Observing that blood from immune animals neutralized diphtheria toxin, they used this concept to devise experiments to demonstrate that treated animals would be insensitive to tetanus. These experiments are described in Behring and Kitasato's paper translated in Brock's collection of landmark papers in microbiology (17). They present experiments in support of four conclusions: (i) blood of a rabbit immune to tetanus neutralizes or destroys tetanus toxin, (ii) that property is found in cell-free serum, (iii) the property is stable in other animals when used for therapy, and (iv) the property does not occur in animals not immune to tetanus. The importance of the paper can be found in Brock's comment, "The science of serology can be said to have begun with this paper." Behring followed this paper with a similar report focusing on diphtheria (26). He won the Nobel Prize in 1901.

While Silverstein characterized the 1890 Behring and Kitasato paper as "[t]he most telling blow to the cellular theory of immunity . . ." (19), the interplay of phagocytic and humoral mechanisms was studied by A. E. Wright and S. R. Douglas. They presented experiments allowing them to state, "We have here conclusive proof that the body fluids modify the bacteria in a way which renders them ready prey to the phagocytes . . . We may speak of this as an opsonic effect." They termed the elements in the blood exerting this effect as "opsonins" (27).

The theoretical basis for the activity of antibodies was developed by Paul Ehrlich in his side chain theory (28). With the stimulation of an antigenic challenge, host cells produced side chain receptors that specifically combined with the antigen. With sufficient stimulation, side chain receptors could be released into the blood as circulating antibodies (29). Perhaps more to the point in the current context, Ehrlich established principles of quantitation (24). Paul Ehrlich and Elie Metchnikoff shared the Nobel Prize in 1908, an early recognition of the importance of both humoral and cellular immunity (3).

The above-described studies, which were of great importance to establishing mechanisms of host defense, were heavily dependent on microscopic observations of the infecting organism. There was little direct application to submicroscopic viruses until the studies of Jules Bordet (Fig. 5), another Nobel Prize winner (1919); he was recognized for his work on complement fixation, which became one of the key mechanisms with which to document antiviral immunity. During the period in which this work was done, Bordet worked in the laboratory of Elie Metchnikoff at the Pasteur Institute in Paris. There were two preliminary steps which Bordet addressed in an 1895 paper in *Annales de l'Institut Pasteur* (translated and condensed in reference 17). Building on the work of R. F. J. Pfeiffer, he confirmed that the granulation followed by lysis of *Vibrio cholerae* bacteria exposed to serum of immunized rabbits is strain specific (Pfeiffer phenomenon). Testing several strains of vibrios *in vitro*, he showed that the greatest destruction was of bacterial strains against which the rabbits had been vaccinated. Such immunological specificity is a cornerstone of serological diagnosis.

Bordet demonstrated that there were separate immunizing and bacteriolytic components. While the bactericidal property is destroyed by heating, he built on the work of C. Fraenkel and G. Sobernheim, who showed that the immunizing component is resistant to heating. In a system in which counts of *Vibrio cholerae*

FIGURE 5 *Jules Bordet. With his brother-in-law, Octave Gengou, Bordet demonstrated the fixation of complement by reacting with bacteria and immune serum. The complement was then no longer available to participate in a hemolysis reaction. The assay, known as complement fixation, became a mainstay of diagnostic virology by demonstrating the development of antibodies in serum after infection. Bordet received the Nobel Prize in 1919. (Courtesy of the National Library of Medicine.)*

were the markers, he demonstrated that neither heated immune serum from goats nor fresh serum from nonimmune guinea pigs alone would inhibit growth. However, the combination of heated immune goat serum and unheated guinea pig serum abolished bacterial growth. This phenomenon occurred whether or not cells were present in the guinea pig serum. From a diagnostic perspective, two features deserve emphasis: (i) the source of complement, then known as alexine, need not be the species tested for immunity, and (ii) complement, but not the immune function of serum, is destroyed by heating to 60°C. Therefore, the components can be added separately. The dissection of the bacteriolytic system into two components was a remarkable accomplishment on its own merits. The experimental challenge had been made further complicated by at least two other factors. Some nonimmunized animals possessed bacteriolytic capacity in their serum, as shown in Nuttall's work. In addition, not all bacteria were equally susceptible to immune-mediated bacteriolysis. Hence, the clarity of Bordet's experimental design is all the more remarkable.

Next came the crucial paper in the scientific foundation of the complement fixation reaction. With his brother-in-law Octave Gengou, Bordet demonstrated the deletion or fixation of complement, as currently understood, by a combination of immune serum and the target bacilli (reference 30, translated in reference 22). With this combination, the complement was unavailable to facilitate hemolysis as a marker system. As Bordet and Gengou pointed out, the work was dependent on the previous demonstration of two concepts. First, red cells and microbes could each delete alexine (complement). Second, the same alexine could participate in either hemolysis or bacteriolysis. With appropriate controls, in a two-step experimental setup, they first mixed plague antiserum and a suspension of plague bacilli with alexine-containing serum from animals. In the second step, into the first mixture they introduced heated guinea pig serum immunized against rabbit's blood together with rabbit's blood. Hemolysis occurred in all tubes except those containing the bacilli, the specific antiserum, and alexine (Fig. 6). Hence, the antiserum conferred the capacity to fix alexine (complement), making it unavailable to participate in hemolysis of sensitized red cells. At a stroke, Bordet and Gengou had demonstrated a marker system that did not require microscopic examination for the presence of the pathogen. Complement fixation to detect viral antibodies became the cornerstone of viral diagnosis for many years.

There was an interesting parallel of scientific reports in 1901, each of importance to the beginnings of clinical virology. In that year, Bordet and Gengou demonstrated the experimental basis of the complement fixation reaction, while W. Reed and J. Carroll demonstrated that the submicroscopic pathogen for yellow fever passed through a filter which blocked bacteria (31).

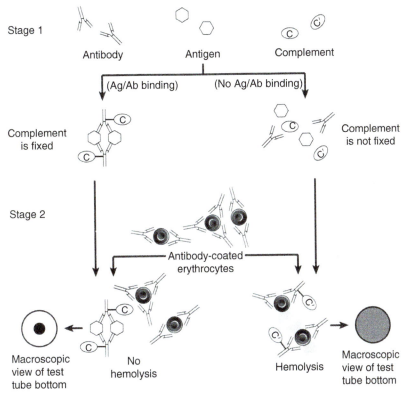

FIGURE 6 *Complement fixation. In stage 1, complement, antigen, and antibodies are mixed together. If antibody is present for the antigen, complement will be bound (fixed). In stage 2, if the complement has been fixed in the first stage, it will be unavailable to combine with antibody-coated erythrocytes. Therefore, a bull's-eye pellet of cells will appear in the bottom of the tube, as shown on the bottom left. However, if complement has not combined with antigen and antibody in the first stage, it will be available to lyse antibody-coated red cells. In that case, no bull's-eye pellet will be seen at the bottom of the tube, as shown on the bottom right (53). (From* Diagnostic Virology, *courtesy of the author, Diane S. Leland, Indiana University School of Medicine.)*

ANTIVIRAL NEUTRALIZATION AND PROTECTION

In years to come, neutralization tests, with complement fixation assays, would become serological mainstays of diagnostic virology. Neutralization of vaccinia virus was perhaps the earliest demonstration of principle. In the context of reviewing advances in bacteriological research in 1892, George Sternberg, a pioneer American bacteriologist, laid particular stress on antitoxin and its therapeutic implications (32). He also reported an experiment on vaccinia virus neutralization that he had conducted with William Griffiths, an expert in the production of vaccine in calves. Vaccinia is the animal poxvirus used for smallpox vaccination.

The intent was a preliminary step in antiserum production, but its usefulness was as a prototype for a diagnostic neutralization test. Vaccinia virus-containing lymph was incubated with serum from a recovered calf; separately, a lesion crust from a child was incubated with immune serum. Each mixture was inoculated into the skin of a nonimmunized calf. No lesions developed. In another controlled experiment, nonimmune serum was mixed with vaccinia virus-containing lymph and was compared with a mixture of vaccinia virus and immune serum. Skin lesions appeared at the site of inoculation of the mixture of virus and nonimmune serum but not at the site of inoculation of virus plus immune serum. While preliminary in scope, the prototype of neutralization was established. The marker system here was the skin of the calf, and in years to come it would be broadly implemented in experimental animals, in embryonated eggs, and in tissue culture. At the end of the published discussion, Sternberg commented, "I believe that there is something in the blood of the immune calf that neutralizes the vaccine virus." The year of Sternberg's paper, published in 1892, is usually marked as the start of the science of virology. Coincidentally, in a paper published in the same year in St. Petersburg, Dmitri Ivanowski described the passage of tobacco mosaic disease infectivity through a bacterial filter (33). These landmark events signify that virology and viral serology had their birth in the same year, albeit on different continents. Protection by neutralization was soon demonstrated in other virus-host systems. In April 1910 in Paris, A. Netter and C. Levaditi demonstrated protection of monkeys against experimental polio with serum from human subjects who had experienced the illness from 6 weeks to 3 years earlier (34). The monkeys revealed no signs of illness. In comparison, a monkey that received virus mixed with serum from a healthy person became paralyzed and died. The investigators noted that these studies demonstrated the identity of the human disease and the experimental disease in monkeys. In a following paper in May 1910, they also demonstrated antiviral activity from a subject who had not been paralyzed but had had symptoms of abortive illness (35). Thus, they identified the virus causing the abortive form of polio as well as the paralytic form. This supported the epidemiological findings of Ivar Wickman on the role of the abortive form in disseminating the illness (36). Also in May 1910, Simon Flexner and Paul A. Lewis in New York City demonstrated that serum from monkeys that had recovered from experimental poliovirus infection, when mixed with active virus, prevented paralysis in intracerebrally inoculated monkeys (37).

Similar findings were later reported for yellow fever. In their 1928 report from Nigeria on the experimental transmission of yellow fever to *Macacus rhesus* monkeys, Adrian Stokes et al. demonstrated that convalescent-phase human serum protected monkeys against experimental infection (38). Max Theiler greatly facilitated further laboratory studies of yellow fever infection by demonstrating the susceptibility of white mice to intracerebral inoculation of the virus (39). Virus passed in mice was neutralized by convalescent-phase serum from humans or monkeys. Parenthetically, in a later study of the protection test in mice, Theiler distinguished the terms "protection" as *in vivo* survival after inoculation of virus mixed with immune serum, whereas "neutralization" occurred *in vitro* (40) (Fig. 7). This seemed a useful distinction at the time, but it was frequently ignored

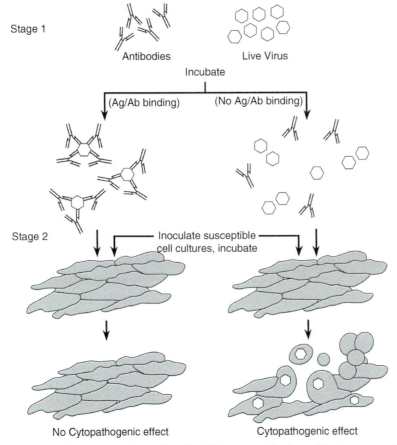

FIGURE 7 *Neutralization test in tissue culture. In the first stage, antibodies and live virus are mixed together. In the second stage, the mixture is added to susceptible cells in tissue culture. If antibodies specific for the virus are present, no cytopathic effects will occur; otherwise, the cells will be attacked and reveal cytopathic effects, as shown on the right (53). (From* Diagnostic Virology, *courtesy of the author, Diane S. Leland, Indiana University School of Medicine.)*

in the literature. For example, in tabulating various serological reactions, Joseph Edwin Smadel listed "neutralization" for *in vivo* tests for many viruses (41). In the present context, Theiler's assay was termed the mouse protection test, and it was an important step in laboratory diagnosis. The application of serology to the epidemiology of yellow fever was emphasized by W. A. Sawyer and W. Lloyd. They wrote of mapping "large areas with respect to endemicity, epidemicity, or absence of yellow fever . . ." (42). They pointed out the difficulty of adapting monkeys to such studies because of the great number of animals required. Instead, they modified Theiler's use of white mice and found the technique useful for epidemiological studies.

Another major step in the laboratory measurement of antiviral immunity was the development of the hemagglutination inhibition assay. George Hirst of the Rockefeller Foundation first observed and reported hemagglutination of red cells by allantoic fluid of influenza virus-infected embryonated chicken eggs. He reported, "When the allantoic fluid from chick embryos previously infected with strains of influenza A virus was being removed, it was noted that the red cells of the infected chick, coming from ruptured vessels, agglutinated in the allantoic fluid" (43). The observation was also made by L. McClelland and R. Hare in the same year (44). M. Burnet had observed the phenomenon but rued that he had not followed up. "Then came a discovery which I should have made but did not" (45). Inhibition of agglutination by serum from a recovered individual would be a demonstration of immunity that could be measured.

In the first publication, Hirst described hemagglutination inhibition as an efficient and immune-specific method to determine antibody titers. Compared to the more cumbersome mouse neutralization tests, hemagglutination results were "of the same order of magnitude." In a subsequent report, Hirst wrote, ". . . the mouse test is very complicated and involves the interplay of many forces over a period of 10 days, while the *in vitro* test is relatively simple. Because of the complexity of the mouse test, it seems probable that the agglutination inhibition test gives a more accurate picture of the *in vitro* combining ratios of virus and antibody" (46).

An assay related to hemagglutination, hemadsorption (Fig. 8), was later developed by Alexis Shelokov and his colleagues at the U.S. National Institutes of Health (47). Several viruses, particularly influenza virus, were isolated and identified by "selective attachment of erythrocytes onto the monolayer surface of tissue culture cells." In the absence of cytopathic effect, infection was recognized by the specific attachment of red cells to the infected, intact cell monolayer. One of the methods reported by the investigators to verify the type of viral isolates was hemadsorption-inhibition.

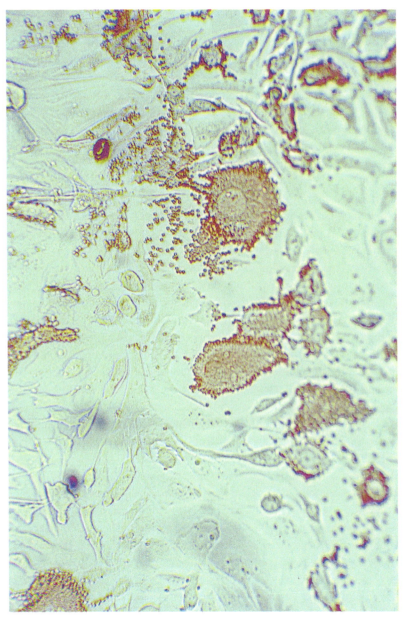

FIGURE 8 *Hemadsorption. Tissue culture cells infected with certain viruses produce receptors with an affinity for red blood cells which attach to the surface and are visible microscopically. In this illustration, influenza B virus has infected rhesus monkey kidney cells, facilitating the specific attachment of red blood cells which outline only the virus-infected cells. (Collection of Marilyn J. August.)*

STANDARDIZATION OF REAGENTS FOR THE FIRST DIAGNOSTIC LABORATORIES

By the 1940s, the technical advances made it possible to establish viral diagnostic laboratories. Evidence is marked by the appearance of first editions of two books that would run through several editions: Rivers's *Viral and Rickettsial Infections of Man* (48) and Francis's *Diagnostic Procedures for Virus and Rickettsial Diseases*, published by The American Public Health Association (49). Because the techniques used were often the same for both types of infectious agents, early labs were set up for viral and rickettsial diseases. The development of diagnostic capacities occurred none too soon, because the world was to embark on the massive conflict of World War II. The U.S. Army established the first general diagnostic virology and rickettsiology lab. It was set up at Walter Reed Army Medical Center in Washington, DC, in January 1941 under Colonel Harry Plotz. Plotz's investigative background was in rickettsial and viral diseases.

In 1940, commenting on serological techniques, C. E. van Rooyen and A. J. Rhodes wrote, "The principal difficulty, in fact, involves preparation of suitable antigens" (50). When the U.S. Army diagnostic lab was established, there was an urgent need to develop diagnostic reagents and standardize procedures. The Army did this as Smadel described in 1948 concerning the hemagglutination inhibition assay for the diagnosis of influenza: ". . . the United States Army adopted a single procedure, and, in addition, supplied its laboratories with standard antigens and antisera" (41). Smadel, who had worked at the Rockefeller Institute developing complement fixation assays (51) and who directed the viral and rickettsial diagnostic lab in the European Theater of Operations in World War II, returned to assume leadership of the diagnostic lab at Walter Reed in 1946 (Smadel Archives, Walter Reed Army Institute of Research, organized by Andrew Rogalski). Possessed of a keen analytic temperament and a personality at once forceful and supportive (D. Carleton Gajdusek and M. B. A. Oldstone, personal communications), Smadel was to remain a major figure in virology for several years.

The foundation was in place for serology in diagnostic virology laboratories with Sternberg's neutralization of vaccinia virus on the skin of a calf in 1892, by Bordet and Gengou's description of complement fixation in a hemolytic assay in 1901, and, to a lesser extent, by Hirst's 1941 study of the inhibition of hemagglutination of influenza virus. By 1956, in a consideration of diagnostic procedures in laboratory diagnosis, Edwin H. Lennette was to write, "By far the greatest proportion of examinations conducted in a diagnostic laboratory consists of serologic tests. These comprise the complement fixation, agglutination, hemagglutination, hemagglutination-inhibition, and *in vitro* neutralization techniques. The complement fixation method finds the greatest application, with hemagglutination and neutralization tests next, and the agglutination tests the least commonly used" (52).

REFERENCES

1. **Harrison TR, et al.** 1962. *Principles of Internal Medicine.* McGraw-Hill Book Co, Inc, New York, NY.
2. **Paul JR, White C.** 1973. *Serologic Epidemiology.* Academic Press, New York, NY.
3. **American Medical Association.** 2009. The Nobel Prize for 1908: Metchnikoff-Ehrlich. *JAMA* **301:**335 http://dx.doi.org/10.1001/jama.2008.965. Editorial.
4. **Rivers TM.** 1937. Viruses and Koch's postulates. *J Bacteriol* **33:**1–12 http://dx.doi.org/10.1128/jb.33.1.1-12.1937.
5. **Rhazes ABMIZ.** 1848. *A Treatise on the Small-Pox and Measles.* (Greenhill WA, translator.) Sydenham Society, London, United Kingdom.
6. **Hopkins DR.** 2002. *The Greatest Killer: Smallpox in History.* University of Chicago Press, Chicago, IL. http://dx.doi.org/10.7208/chicago/9780226189529.001.0001.
7. **Clendening L.** 1960. *Source Book of Medical History.* Dover Publishing, Inc, New York, NY.
8. **McNeill WH.** 1976. *Plagues and Peoples.* Anchor Press, Garden City, NY.
9. **Duffy J.** 1951. Smallpox and the Indians in the American colonies. *Bull Hist Med* **25:**324–341.
10. **Blake JB.** 1953. Smallpox inoculation in Colonial Boston. *J Hist Med Allied Sci* **8:**284–300 http://dx.doi.org/10.1093/jhmas/VIII.July.284.
11. **Stark RB.** 1977. Immunization saves Washington's army. *Surg Gynecol Obstet* **144:**425–431.
12. **Thursfield H.** 1940. Smallpox in the American War of Independence. *Ann Med Hist* **2:**312–318.
13. **Bailey I.** 2011. Edward Jenner, benefactor to mankind, p 21–25. *In* Plotkin SA (ed), *History of Vaccine Development.* Springer, New York, NY. http://dx.doi.org/10.1007/978-1-4419-1339-5_4.
14. **Spencer WG.** 1927. Eighteen letters written by Edward Jenner to Alexander Marcet between the years 1803-1814, presented to the Library of the Royal Society of Medicine by Dr. William Pasteur. *Proc R Soc Med* **20:**587–590 http://dx.doi.org/10.1177/003591572702000552.
15. **Jenner E.** 1798. *An Inquiry into the Causes and Effects of the Variolae Vaccinae.* Sampson Low, London, United Kingdom.
16. **Riedel S.** 2005. Edward Jenner and the history of smallpox and vaccination. *Proc Bayl Univ Med Cent* **18:**21–25 http://dx.doi.org/10.1080/08998280.2005.11928028.
17. **Brock TD (ed).** 1975. *Milestones in Microbiology.* American Society for Microbiology, Washington, DC.
18. **Fenner F, Henderson DA, Arita I, Jezek Z, Ladnyl ID.** 1988. *Smallpox and Its Eradication.* World Health Organization, Geneva, Switzerland.
19. **Silverstein AM.** 1989. *A History of Immunology.* Academic Press, New York, NY.
20. **Metchnikoff O.** 1921. *Life of Elie Metchnikoff 1845–1916.* Houghton Mifflin, Boston, MA.
21. **Metchnikoff E.** 1884. Ueber eine Sprosspilzkrankheit der Daphnien. Beitrag zur Lehre uber den Kampf der Phagocyten gegen Krankheitserreger. *Arch Pathol Anat Physiol Klin Med* **96:**177–195 http://dx.doi.org/10.1007/BF02361555.
22. **Lechevalier HA, Solotorovsky M.** 1974. *Three Centuries of Microbiology.* Dover Publishing, Inc, New York, NY.
23. **Nuttall G.** 1888. Experimente über die bacterienfeindlichen Einflusse des thierischen Korpers. *Z Hyg* **4:**353–394 http://dx.doi.org/10.1007/BF02188097.
24. **Bulloch W.** 1938. *The History of Bacteriology.* Oxford University Press, New York, NY. http://dx.doi.org/10.1097/00000441-193812000-00025.
25. **Behring E, Kitasato S.** 1890. Uber das Zustandekommen der Diphtherie-Immunitat und der Tetanus-Immunitat tei Thieren. *Dtsch Med Wochenschr* **16:**1113–1114 http://dx.doi.org/10.1055/s-0029-1207589.
26. **Behring E.** 1890. Untersuchungen uber das Zustandekommen der Diphtherie-Immunitat bei Thieren. *Dtsch Med Wochenschr* **16:**1145–1148 http://dx.doi.org/10.1055/s-0029-1207609.
27. **Wright AE, Douglas SR.** 1903. An experimental investigation of the role of the blood fluids in connection with phagocytosis. *J R Soc Med* **72:**357–370.
28. **Nuttall GHF.** 1924. Biographical notes bearing on Koch, Ehrlich, Behring and Loeffler, with their portraits and letters from three of them. *Parasitology* **16:**214–238 http://dx.doi.org/10.1017/S0031182000020023.

29. **Ehrlich P.** 1899. Croonian Lecture: on immunity with special references to cell life. *Proc R Soc Lond* **66:**424–448.

30. **Bordet J, Gengou O.** 1901. Sur l'existence de substances sensibilisatrices dans la plupart des sérums antimicrobiens. *Ann Inst Pasteur (Paris)* **15:**289–302.

31. **Reed W, Carroll J.** 1902. The etiology of yellow fever: a supplemental note. *Am. Med* **3:**301–305.

32. **Sternberg GM.** 1892. Practical results of bacteriological researches. *Trans Assoc Am Physicians* **7:**68–86.

33. **Ivanowski D.** 1942. Concerning the mosaic disease of the tobacco plant, 1892, p 25–30. *In* Johnson J (trans. ed.), *Phytopathological Classics, No. 7*. American Phytopathological Society, St Paul, MN.

34. **Netter A, Levaditi C.** 1910. Action microbiocide exercée par le sérum des malades atteints de paralysie infantile sur le virus de la poliomyélite aigüe. *C R Soc Biol* **68:**617–619.

35. **Netter A, Levaditi C.** 1910. Action microbiocide exercée sur le virus de la poliomyélite aigüe par le sérum des sujets antérieurement atteints de paralysie infantile. Sa constatation dans le sérum d'un sujet qui a présénte une forme abortive. *C R Soc Biol* **68:**855–857.

36. **Wickman I.** 1913. *Acute Poliomyelitis (Heine-Medin's Disease)*. (Maloney WJAM, translator.) The Journal of Nervous and Mental Diseases Publishing Company, New York, NY.

37. **Flexner S, Lewis PA.** 1910. Experimental poliomyelitis in monkeys. *J Exp Med* **12:**227–255 http://dx.doi.org/10.1084/jem.12.2.227.

38. **Stokes A, Bauer JH, Hudson NP.** 1928. Experimental transmission of yellow fever to laboratory animals. *Am J Trop Med* **8:**103–164 http://dx.doi.org/10.4269/ajtmh.1928.s1-8.103.

39. **Theiler M.** 1930. Susceptibility of white mice to the virus of yellow fever. *Science* **71:**367 http://dx.doi.org/10.1126/science.71.1840.367.a.

40. **Theiler M.** 1930. A yellow fever protection test in mice by intracerebral injection. *Ann Trop Med Parasitol* **27:**57–77 http://dx.doi.org/10.1080/00034983.1933.11684739.

41. **Smadel JE.** 1948. Serologic reactions in viral and rickettsial infections, p 67–96. *In* Rivers TM (ed), *Viral and Rickettsial Infections of Man*. J B Lippincott & Co, Philadelphia, PA.

42. **Sawyer WA, Lloyd W.** 1931. The use of mice in tests of immunity against yellow fever. *J Exp Med* **54:**533–555 http://dx.doi.org/10.1084/jem.54.4.533.

43. **Hirst GK.** 1941. The agglutination of red cells by allantoic fluid of chick embryos infected with influenza virus. *Science* **94:**22–23 http://dx.doi.org/10.1126/science.94.2427.22.

44. **McClelland L, Hare R.** 1941. The adsorption of influenza virus by red cells and a new method for measuring antibodies to influenza virus. *Can J Public Health* **32:**530–538.

45. **Burnet M.** 1969. *Changing Patterns. An Atypical Autobiography*. American Elsevier Co, New York, NY.

46. **Hirst GK.** 1942. The quantitative determination of influenza virus and antibodies by means of red cell agglutination. *J Exp Med* **75:**49–64 http://dx.doi.org/10.1084/jem.75.1.49.

47. **Shelokov A, Vogel JE, Chi L.** 1958. Hemadsorption (adsorption-hemagglutination) test for viral agents in tissue culture with special reference to influenza. *Proc Soc Exp Biol Med* **97:**802–809 http://dx.doi.org/10.3181/00379727-97-23884.

48. **Rivers TM (ed).** 1948. *Viral and Rickettsial Infections of Man*. J B Lippincott & Co, Philadelphia, PA.

49. **Francis T Jr.** 1956. *Diagnostic Procedures for Virus and Rickettsial Diseases*. American Public Health Association, New York, NY.

50. **van Rooyen CE, Rhodes AJ.** 1940. *Virus Diseases of Man*. Oxford University Press, London, United Kingdom.

51. **Smadel JE, Baird RD, Wall MJ.** 1939. Complement-fixation in infections with the virus of lymphocytic choriomeningitis. *Proc Soc Exp Biol Med* **40:**71–73 http://dx.doi.org/10.3181/00379727-40-10309P.

52. **Lennette EH.** 1956. General principles underlying laboratory diagnosis of virus and rickettsial infections, p 1–51. *In* Francis T Jr (ed), *Diagnostic Procedures for Virus and Rickettsial Diseases*, 2nd ed. American Public Health Association, New York, NY.

53. **Leland DS.** 1996. *Clinical Virology*. W B Saunders Co, Philadelphia, PA.

4 What Can Be Seen

FROM VIRAL INCLUSION BODIES TO ELECTRON MICROSCOPY

Seeing comes before words. The child looks and recognizes before it can speak.

Berger, Ways of Seeing (1)

INTRODUCTION

Generally too small to be seen under the light microscope, viruses often leave footprints, inclusion bodies, found when examined in properly fixed and stained preparations (2). Some inclusions were originally interpreted as protozoa, but in certain cases, such as Negri bodies in rabies, they became the hallmark for diagnosis. Inclusions were originally described in the first decade of the 20th century, only a few years after the first descriptions of tobacco mosaic disease, foot-and-mouth disease, and yellow fever as being caused by novel types of infectious agents, filterable viruses. Decades later, toward the middle of the 20th century, substitution of electron beams for rays of light in the electron microscope allowed much greater resolution and with it the actual visualization of individual viruses. Application of antisera to specimens for electron microscopic examination allowed recognition by aggregation of viral particles. This method, immunoelectron microscopy, facilitated discovery of viral agents not previously identified and became a potent viral diagnostic tool.

The capacity to recognize viral inclusions was dependent on crucial technical advances in cytopathology in the second half of the 19th century, both in the

To Catch a Virus, Second Edition. Authored by John Booss and Marie Louise Landry.
© 2023 American Society for Microbiology. DOI: 10.1128/9781683673828.ch04

perfection of the light microscope and in the preparation, cutting, and staining of tissue sections, as well as in photomicroscopy. Moreover, the use and interpretation of microscopic changes rested on a paradigm shift from macroscopic observations to cell-based pathology in the mid-19th century. Hence, we begin our consideration with an overview of Rudolf Virchow's promulgation of cell theory, that the cell must be seen as the fundamental seat of disease pathology.

From the pathologist's perspective, inclusion bodies were among the most important stigmata with which to make a diagnosis of viral disease. Of longstanding importance, the Negri body of rabies, described in 1903 by Adelchi Negri as a protozoan (3), holds a preeminent place in the history of diagnostic pathology. Likewise, Giuseppe Guarnieri (4) and E. E. Tyzzer (5, 6) described distinct inclusions associated with smallpox and with varicella, respectively, which were to become important in distinguishing the lesions of smallpox (variola) from those of chicken pox (varicella). Meanwhile, work proceeded on other herpes infections, including herpes febrilis, herpes genitalis, and herpes zoster (7).

Another member of the herpes family of viruses, cytomegalovirus, was first recognized through its characteristic cytopathology of large cell bodies and intranuclear inclusions, seen for some time as parasitic in origin and an incidental finding in salivary glands (8, 9). However, a series of pathological observations led to the definition of generalized cytomegalic inclusion disease, a major cause of fetal damage.

It remained for the electron microscope to give definition to the physical form of viruses, with the first electron micrographs published in the 1940s of bacteriophage, tobacco mosaic virus, and mammalian viruses. Refinements in techniques of electron microscopy allowed advances in the understanding of viral disease pathogenesis, identification of new viral pathogens, and contributions to the emerging field of rapid viral diagnostics.

RUDOLF VIRCHOW AND CELLULAR PATHOLOGY

The establishment of cellular pathology by Rudolf Ludwig Virchow (Fig. 1) was as important as the establishment of germ theory by Louis Pasteur and Robert Koch to understand the pathogenesis of disease, particularly viral diseases. While the concept of cell theory built slowly in the first half of the 19th century, it was Virchow who capped it with the idea that all cells come from other cells. Furthermore, before Virchow, pathology had been oriented toward organs and tissues, without a satisfactory unifying theory to bring diseases and tissues together. Virchow was to change that.

Cell theory developed from the work of the botanist M. J. Schleiden, who in 1838 asserted that plants consist of cells (10). T. Schwann extended the concept to animals in 1839. In 1831, the botanist Robert Brown had recognized the consistent presence of the nucleus in cells (11).

FIGURE 1 *Rudolf Virchow caricature. Virchow's* Cellular Pathology *revolutionized pathology, asserting that the cell is the fundamental focus of disease. Virchow helped facilitate the modern era of medicine yet was resistant to germ theory. (Courtesy of the National Library of Medicine.)*

What remained unclear, however, was the origin of cells. There was speculation that cells arose from inanimate substances. As Virchow put it, "At the present
time, neither fibers, nor globules, nor elementary granules can be looked upon
as histological starting points" (12, 13). These alternative views arose at least in
part from faulty observations. Virchow wrote, "This view was in part attributable
to optical illusions in microscopical observations." E. H. Ackerknecht noted the
possibility that a "... better cell theory was provided through the introduction of
much improved microscopes around 1830" (11).

Continued improvement in microscopes and histopathological techniques
continued through the 19th century, facilitating many of the key observations on
viral cytopathology. However, based on extensive microscopic observations, not
theory, Virchow asserted that all cells came from previous cells, "*omnis cellula e cellula*" (12, 13). Thus, Virchow was not the originator of the cell theory but put on
the capstone. It was very closely analogous to the later establishment of germ theory that all microbes arose from other microbes, not from inert putrefying matter.

The other major contribution of Virchow in the present context is the systematizing
of pathology based on cellular changes. As Fielding Garrison put it, "The rise of modern
medicine is inseparably connected with the name of Rudolf Virchow (1821–1902), the
founder of cellular pathology" (14). The seminal book on which that assessment was
based was Virchow's 1858 publication of 20 lectures, *Cellular Pathology as Based upon
Physiological and Pathological Histology* (13). The circumstances under which those
lectures arose are of some interest (10). Virchow had previously been removed from
a post in Berlin because of his remarks in a scientific report on an outbreak of typhus
that were highly critical of the government. However, after brilliant work at Würzburg,
he was invited back to the University of Berlin. On his return, he set out the latest principles of pathology in a series of lectures for physicians and arranged to have them
transcribed and published. As S. B. Nuland wrote, "What Virchow accomplished in
Cellular Pathology was nothing less than to enunciate the principles upon which medical research would be based for the next hundred years and more" (10). That proved to
be especially true for diseases caused by viruses, dependent on host cells for replication.

ADVANCES IN LIGHT MICROSCOPY

It is no accident that the first inclusions of importance for viral diseases, Negri bodies
in rabies, were described in 1903 (3). Visualizing these inclusions was dependent
on histological techniques which were perfected in the last third of the 19th century (15). There were two major developments: improvements in the microscope
and in tissue preparation. First, consider Antony van Leeuwenhoek (1632–1723),
who ground his own lenses for simple microscopes according to a method he kept
secret and who appears to have had visual acuity bordering on the miraculous. Also of
importance were other early microscopists, including Robert Hooke (1635–1703),

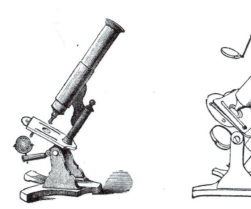

JAMES W. QUEEN & CO.,

Opticians,

NO. 924 CHESNUT STREET, PHILADELPHIA, PA.

MANUFACTURERS AND IMPORTERS OF FINE

Achromatic Microscopes,

With the most recent Improvements, Optically & Mechanically.

Microscopic objects for the study of Anatomy and Natural History, Achromatic object lenses, Eye pieces, Stanhope and Coddington lenses, Pocket magnifying lenses, Cabinets for Microscopic Objects, together with all articles used in the study of Microscopy.

The Most Complete priced and illustrated Catalogue of Microscopes and Optical, Mathematical and Philosophical Instruments ever offered, sent by mail, Free of Charge, to any part of the United States and Canada.

A Microscope magnifying 40 diameters, or 1,600 times can be furnished at $ 2.00
    "   "   100   "   10,000 " "   "   10.00
    "   "   150   "   12,500 " "   "   20.00
    "   "   600   "   360,000 " "   "   40.00
    "   "   800   "   640,000 " "   "   80.00

☞ Superior Microscopes from $100 to $300 each, varying in price according to their Optical and Mechanical perfection.

FIGURE 2 *Advertisement for achromatic microscopes, 1859. Optical aberrations, including chromatic aberration, impaired image resolution into the 19th century. This advertisement, from* The Microscopist's Companion: a Popular Manual of Practical Microscopy, *offers achromatic microscopes for sale in the second half of the 19th century. (Courtesy of the National Library of Medicine.)*

who used compound microscopes to increase magnification (Fig. 2). Hooke, incidentally, is credited with the first use of the word "cell" in the microscopic sense (11).

Image resolution, however, was hampered with the use of lenses in series in compound microscopes, and image clarity was reduced by optical aberrations (16). These included spherical aberration, which happens when "rays leaving the surface of the lens at different distances from the axis are brought to focus at different points; and chromatic aberration, which brings light of different colors to different foci and spreads the whole spectrum round each point in the image." General use of the microscope for scientific investigation was delayed pending correction of these aberrations. The problem of chromatic aberration was solved at the end of the 18th century.

It remained for Joseph Jackson Lister and Thomas Hodgkin to develop the concepts to overcome spherical aberration. Both men were also known for other historical connections and medical contributions. Joseph Jackson Lister was the father of Joseph Lister, the originator of antiseptic surgery based on principles developed by Pasteur. Thomas Hodgkin is best known for his description in 1832 of a cancer of the lymphatic system, which became known as Hodgkin's disease. Yet their collaboration in working out the problems of microscopic image distortion made possible early advances in histology. The advances reported were of sufficient importance that Hughes would observe, "With this paper animal histology may be said properly to begin" (16). Increases in the resolving power of the light microscope continued into the 1890s (17). Ernst Abbe's condenser and advances in oculars and objectives further contributed to improved resolution (15).

ADVANCES IN TISSUE PREPARATION

The solution of aberrations of the compound microscope and the increased power of resolution brought about the need for improved methods of tissue preparation. Sections cut by hand without fixation or embedded and unstained were of uneven and inconsistent quality and provided little contrast of cellular elements. Overcoming these problems was succinctly described by Esmond Ray Long (15). The development of the microtome allowed the tissue to be held firmly and regulated the precision of cutting sections of desired thickness. Firmness of cut, first achieved by freezing, was later attained by embedding tissue in a number of substances such as gum arabic, paraffin, collodion, and celloidin. Fixation solutions to prevent structural changes evolved from chromic acid to formaldehyde. As noted by Bracegirdle, fixation allowed tissues to be exchanged between laboratories (17).

A series of developments allowed the visualization of microscopic elements in tissues. "The first important staining procedure" was attributed by Long to Joseph von Gerlach, who in 1847 discovered that alkaline carmine stained nuclei (15). F. Böhmer is credited with employing hematoxylin as a nuclear stain in 1865, and it is still in wide use. Paul Ehrlich developed aniline dyes for histological use. Contributing to studies of the central nervous system, Camillo Golgi of Pavia (14) introduced stains containing metallic salts which assisted in cellular differentiation, particularly the use of the silver nitrate stain for nervous tissue.

Finally, the use of photomicroscopy to eliminate subjectivity by the artist-experimenter is credited to Alfred François Donné (17). His book of 1844 used photomicrographs for illustrations in an atlas. New histological staining methods and photomicroscopy would allow the characterization of intranuclear viral inclusions according to the acidophilic staining of the nuclear mass, the presence of a clear halo between it and the nuclear membrane, and margination of basophilic chromatin on the nuclear membrane. In 1934, E. V. Cowdry would comment, "No special technic is required for the identification of nuclear inclusions" (18). That observation rested on the secure foundation of histological technique established in the later 19th century, when such staining techniques were implemented.

RABIES—NEGRI BODIES

Death from rabies, and it is virtually uniformly lethal, is one of the most horrifying deaths known to humankind. It brings with it intense waves of fear and anxiety reflecting the anatomic sites in the brain in which the virus settles. Rather than massive waves of epidemic infection, it is the specter of the individual's horrible fate which gives rabies its place in human history (Fig. 3). Hence, the development of postexposure prophylaxis by Pasteur (see chapter 2) was greeted with universal acclaim. Yet the treatment was not utterly benign. It entailed a series of painful injections which occasionally produced complications in the central nervous system. Therefore, there was a vital need to establish a rapid diagnosis of rabies in the biting animal. The diagnosis of rabies in the dog which bit Joseph Meister in 1885 was based on the findings of wood fragments in the dog's stomach at necropsy. Meister, of course, was the first publicly acknowledged recipient of Pasteur's rabies vaccine (19). While of some value, observations of the animal's pica and agitated behavior were nonspecific means of diagnosis. Other conditions could be mimicked. As put by H. N. Johnson and T. F. Sellers, "Clinically, rabies in dogs must be differentiated from canine distemper; from fright; from disease due to vitamin deficiency, sometimes called running fits; and from pseudorabies due to the virus of infectious bulbar paralysis, also called Aujeszky's disease" (20). Hence, a specific diagnosis of rabies was of vital importance.

A definitive and rapidly performed diagnostic test emerged from the work of Adelchi Negri, who was working in Camillo Golgi's laboratory at the University of Pavia (21). On graduating from medical school, Negri was encouraged by Golgi to study the pathogenesis of rabies in the central nervous systems of infected animals. It was of course an appropriate project in the laboratory of Golgi, a pioneer of

FIGURE 3 *Poster of a rabid dog. Perhaps the oldest recognized and most feared transmissible disease, rabies was often incorrectly diagnosed until the description of the Negri body in 1903. (Courtesy of the Historical Library, Yale University School of Medicine/The Wellcome Collection.)*

neurohistology. As recounted by K. Kristensson et al., Negri started his study with infected rabbits but moved on to dogs (21). He had observed stained bodies in neurons of spinal ganglia of rabbits and in the hippocampus of dogs. While specific to rabies, the neuronal bodies were originally interpreted as parasitic in nature. This was not surprising in light of his mentor's work on malaria. Negri illustrated his paper with camera lucida drawings of the neuronal inclusions in tissue which had been fixed and stained (3).

Over the years, the procedure to identify Negri bodies was modified with respect to staining and fixation and the nature of the preparation studied. In what became a standard method, T. F. Sellers of the Georgia State Health Board Laboratory in the United States reported that less than 1 minute was required to prepare samples (22). The procedure was applicable to unfixed smears or impressions of brain tissue and was performed with a rapid dip into a prepared stain of basic fuchsin, methylene blue, and methyl alcohol. The Negri bodies were described as "bright cherry red and [to] stand out in strong relief."

The search for Negri bodies in the brains of the attacking animals remained the rapid diagnostic standard for decades. The 1923 edition of W. Osler and T. McCrae's textbook of medicine reported that "The recognition of the Negri bodies in smears of brain substance enables the diagnosis to be made promptly" (23). Two and a half decades later, the place of Negri bodies in the diagnosis of rabies in attacking animals was confirmed in the first edition of the American Public Health Association's *Diagnostic Procedures for Virus and Rickettsial Diseases*: "The definitive microscopic diagnosis of rabies is based on the finding of intracytoplasmic inclusion bodies, known as Negri bodies. " (20) (Fig. 4). Ultimately, the application of fluorescent antibody staining of suspect brain tissue as the diagnostic standard (see chapter 7) replaced the finding of Negri bodies. Sadly, the use of the test to find inclusions named for him significantly outlived Negri, who died at 37 of tuberculosis (21).

SMALLPOX—GUARNIERI BODIES AND ELEMENTARY BODIES

In rabies, the crucial diagnostic task is to determine whether the attacking animal is rabid. The finding of Negri bodies achieved that end, supporting the need for post-exposure prophylaxis of the bitten individual. The diagnostic task was different in the early stages of a pox outbreak. The challenge was to determine whether the skin lesions were those of an early stage or light case of smallpox or of some other pock-forming illness, especially chicken pox.

The importance of the distinction between chicken pox and smallpox had long-term consequences for the individual and also signaled a shift in the basis of medical understanding. Writing of the chicken pox, W. Heberden said, ". . . yet

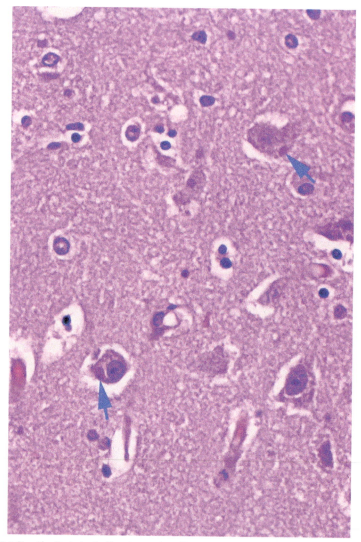

FIGURE 4 *Negri bodies in brain. The observation of intracytoplasmic inclusion bodies in the brains of animals was reported by Adelchi Negri in 1903. It became the diagnostic method of choice for decades. Arrows indicate Negri bodies in a brain tissue section. (Courtesy of J. H. Kim, Yale University School of Medicine.)*

it is of importance on account of the small-pox, with which it may otherwise be confounded, and so deceive the persons who have had it into a false security, which may prevent them either from keeping out of the way of small-pox, or from being inoculated" (24). Heberden's remarks, read to the College of Physicians in London in August 1767, were published in the first volume of *Medical Transactions* the following year. In the preface to that volume, the College noted that ". . . mere abstract reasonings have tended very little to the promoting of natural knowledge. By laying these aside, and attending carefully to what nature hath either by chance or upon experiment offered to our observation, a greater progress hath been made in this part of Philosophy, since the beginning of the last Century, than had been till that time from the days of Aristotle" (25). Thus, the preface sounded the transition from the Galen system of medicine, based on abstract reasoning, to an observation-based system. Those observations would be both clinical and histological.

Description of cytoplasmic inclusion bodies in smallpox and vaccinia lesions by Giuseppe Guarnieri in 1892 (4) actually preceded by a bit over a decade the description of the Negri body in rabies in 1903 (3). Like the Negri body, the Guarnieri body was originally depicted as protozoan in nature. The Guarnieri body, an eosinophilic, cytoplasmic inclusion body, was used to distinguish clinically mild smallpox from chicken pox. The observations on inclusion bodies were used in the 1904 work of W. T. Councilman et al. on smallpox. In autopsied cases, Guarnieri bodies were reported to occur in skin lesions in the first 10 days of the rash (26). Guarnieri also described scarification of the rabbit cornea, with instillation of test material, as a means of making the diagnosis. Small areas of cellular proliferation visible microscopically were found on staining to contain cells with cytoplasmic inclusions. The rabbit cornea test was modified to allow macroscopic examination of the lesions and later became known as the Paul test (27). As described by J. A. Toomey and J. A. Gammel, G. Paul found that if the rabbit eye is enucleated 48 hours after inoculation and placed in sublimate alcohol, the lesions would become visible as "intensely white elevations on a dull milky white background, with occasionally a distinctly visible small dark crater in the center of the elevation." In their own studies, Toomey and Gammel found the test to be positive in only 45 of 80 cases of smallpox tested. However, the major detriment of the Paul test to achieve diagnosis was the time to reading of the test, 3 days. In a 1948 review of diagnostic methods for smallpox concentrating on electron microscopy, C. E. van Rooyen and G. D. Scott noted that "the diagnosis is usually obvious when the result arrives" (28). Hence, neither the finding of Guarnieri bodies or cytoplasmic inclusions nor the Paul test was as effective or as rapid in establishing the diagnosis of smallpox in humans as was the Negri body in establishing the diagnosis of rabies in rabid animals.

The microscopic observation of what have been called elementary bodies or Paschen bodies has a long history, richly recounted by Lise Wilkinson (29) and by Mervyn Henry Gordon (30). Gordon's insightful work contrasts his early educational experience, when he was deemed "backward" and was nearly "sent to the colonies." However, Gordon's father felt his potential had not been tapped and sent him to a stimulating environment at Keble College, Oxford, where he thrived when exposed to a scientific curriculum not available in his previous school. He earned bachelor and medical degrees from Oxford, opened research opportunities leading to his study of viral diseases, that spawned his writings and observations on vaccinia and variola (31).

Yet it was Stanislaus von Prowazek who illustrated Giemsa-stained minute bodies in vaccinia material, calling them elementary bodies. Gordon argued for the priority of John Buist in describing the elementary bodies in Buist's 1887 book, *Vaccinia and Variola*, and reproduced plates from Buist's work. Buist identified them as spores of micrococci. Enrique Paschen, "the great apostle ... of elementary bodies," who used a Loeffler flagellum stain, maintained that they were the infectious particles in vaccinia and variola (30). That conclusion turned out to be correct. K. McIntosh has commented that the observation of these "spores" was the first rapid laboratory diagnostic test for a virus, noting that it occurred before the demonstration of viruses as filterable, infectious particles (K. McIntosh, Clinical Virology Symposium lecture, 2009). A. W. Downie, recounting electron microscopic work, wrote in 1965 that "Subsequent work has confirmed Paschen's view" (32). R. H. Green, T. F. Anderson, and Joseph Edwin Smadel described the elementary body of vaccinia as having "a high degree of regularity" and noted that the "particles are almost rectangular in shape" (33).

The recognition of elementary, or Paschen, bodies required an experienced virologist and expert histopathological preparation (34). There were several technical sources of error (35). For example, "masses of elementary bodies" were considered a requirement for diagnosis, and the appropriate size and color were rigidly prescribed. In addition, at least six smears were made to ensure adequate material for examination and verification. All efforts were made to avoid "disastrous conclusions." As of 1965, the findings of elementary bodies enabled a rapid presumptive diagnosis of variola but one which required confirmation by viral isolation and identification. Downie counseled that a negative finding should "be ignored if the clinical picture favors a diagnosis of smallpox" (32). Yet the identification of inclusions in stained smears remained in the laboratory diagnostic armamentarium of the WHO smallpox eradication effort (36) and was "open to misinterpretation by those not familiar with the technique. " Hence, the test "was not widely used."

VARICELLA—INTRANUCLEAR INCLUSIONS AND MULTINUCLEATED CELLS

The microscope provided another means of solving the clinical conundrum of varicella versus variola in ambiguous cases. It came about because of fortuitous circumstances early in the career of E. E. Tyzzer. As a young assistant in pathology at Harvard, he had been sent by W. T. Councilman, professor of pathological anatomy, to study variola in monkeys in the Philippines. Councilman had published the clinical and histopathological studies of an outbreak of smallpox in Boston in 1901 to 1903 (26), and Tyzzer had played an important role in those studies, undertaking independent work on vaccinia in experimental animals. Tyzzer's study of vaccinia (37) was published in the same issue of the *Journal of Medical Research* as the study of variola by Councilman et al.

Tyzzer concluded that the Guarnieri body, which he called the vaccine body, was "an organism and the etiological agent" of vaccinia. Using Guarnieri's term, cytoryctes, he thought them to be intracellular parasites. In retrospect, it is of interest that Councilman et al. had commented in their paper concerning variola that "There is little doubt that the discovery and the general confirmation of the protozoal origin of malaria at this time had a great influence in directing attention to the possible presence of similar bodies in other diseases" (26).

While in the Philippines, Tyzzer had the opportunity to study an outbreak of varicella among adult prisoners at the Bilibid Prison in Manila. In his review of the literature for his publication, Tyzzer was careful to credit P. G. Unna as having provided the only previous histopathological descriptions of varicella (38). Unna had described two types of cellular degeneration and had observed "peculiar giant cells" with multiple nuclei. As part of his studies, Tyzzer made limited attempts to transmit the disease to experimental monkeys and the corneas of rabbits. These transmission experiments were negative, in contrast to what would have been expected with variola.

Tyzzer reported clinical observations and histopathological findings in lesions from 11 of more than 300 cases of varicella (5, 6). He made detailed clinical observations and longitudinal histopathological studies of biopsied skin lesions from 12 hours to 7 days after their appearance (5, 6). The first changes noted were of "a swelling of both the cytoplasm and the nuclei" as well as the first descriptions of intranuclear inclusions in varicella skin lesions. He found that the chromatin was pulled away from these masses, leaving a clear space, giving what would later be characterized as a halo around the intranuclear inclusions. Tyzzer confirmed Unna's description of "ballooning degeneration" of large, multinucleated cells. Because he had found isolated lesions of the blood vessels, Tyzzer concluded that the infectious agent was disseminated by the blood. In contrast to the inclusions in vaccinia, he commented that in varicella "the nuclear inclusions are so structureless and irregular that it is difficult to consider them as living organisms."

Tyzzer set criteria for "diagnosis of cases in which there was a question of smallpox or varicella." He noted that the presence of large multinucleated cells in clear early vesicles was "consistent with varicella and against smallpox" and that preparation of lesions taken by incision could be performed at the bedside. In contrast, in using histological sections, "the simple presence of definite cytoplasmic or nuclear inclusions within the epithelial cells is not sufficient to establish the diagnosis of smallpox, for both nuclear and cytoplasmic inclusions are also found in varicella" (Fig. 5). Finally, while the method was slower, he noted that inoculation of rabbit cornea with vesicle fluid would produce a keratitis after 1 to 2 days in the case of smallpox but not so with varicella. As described above, the method of scarification of the rabbit cornea and swabbing with a test sample had been used by Guarnieri and

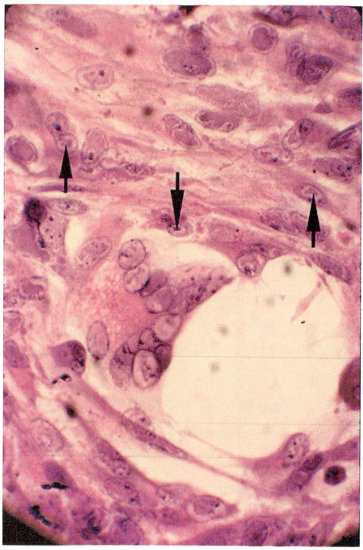

FIGURE 5 *Varicella-zoster virus inclusions in a monolayer of human diploid fibroblasts. Arrows indicate intranuclear inclusions in a field of multinucleated cells. Hematoxylin and eosin stain. (Collection of Marilyn J. August.)*

later modified as the Paul test and was well established to diagnose smallpox. The use of bedside cytopathological preparations for rapid diagnosis of dermatological conditions was reintroduced into the literature decades later by Tzanck (39). What has been known as the Tzanck smear has maintained its clinical usefulness to the present day, particularly, but not exclusively, for the diagnosis of infection by the herpes family of viruses (40, 41) (Fig. 6). There is another notable piece to the Tyzzer scientific

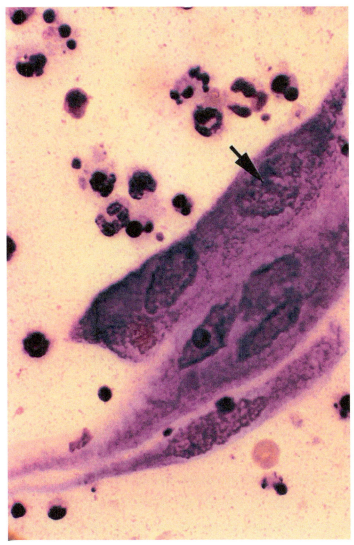

FIGURE 6 *Tzanck smear. Tzanck smears, performed at the bedside, offer a rapid diagnostic test of various skin lesions. For example, the presence of intranuclear inclusions suggests one of the herpes viruses, as in the example shown. (Collection of Marilyn J. August.)*

biography. He had come under the tutelage of Councilman at Harvard, and 3 decades later Thomas Weller worked in his department (42). Weller shared the Nobel Prize with John Enders and Frederick Robbins for their cultivation of poliovirus in tissue culture. Weller's lab was also able to isolate varicella virus in culture (43) and to serially pass it in culture (44). There is a satisfying continuity of important virological contributions by Councilman, Tyzzer, and Weller.

CYTOMEGALIC INCLUSION DISEASE OF THE NEWBORN

Virus-induced cellular enlargement and intranuclear inclusions are associated with a disease of the newborn. As in rabies and in smallpox, the inclusions were originally thought to be protozoan in nature. Unlike the Negri body, which was pathognomonic for the presence of rabies infection in the attacking animal, or the Guarnieri body, which identified smallpox in the afflicted person, cytomegalic inclusion-bearing cells in newborn infants were first found in syphilitic infants. Such cells were originally described in 1904 (8, 9), and findings accumulated so that by 1950 the microscopic findings could be found to define widespread organ disease of the newborn (45, 46). In the case of this disease, too, Weller was one of the investigators who isolated the virus and whose later studies defined the disease and its sequelae based on viral isolation.

In October 1904, A. Jesionek and B. Kiolemenoglou of Munich reported the finding of structures interpreted as protozoa in a syphilitic baby (8). The Jesionek and Kiolemenoglou publication prompted a report by P. Ribbert of Göttingen in December 1904, which also demonstrated masses which were interpreted as protozoa in a syphilitic baby and in the parotid glands of children (9). Three accompanying figures in that article showed several cells with the appearance, according to a present-day interpretation, of intranuclear inclusions with halos. In each study, the authors consulted with colleagues who concluded that the structures were protozoa of one variety or another. Figure 7 demonstrates what today would be interpreted as large, intranuclear inclusions with a surrounding halo and marginated chromatin, often described as owl's eye inclusions.

Until the early 1920s, most subsequent reports concurred with that likely etiology (47). However, E. W. Goodpasture and F. B. Talbot reported a case presenting numerous figures to challenge that interpretation (48). They favored instead the explanation of an "abnormal cytomorphosis" of tissue cells. Similar changes were noted in the salivary glands of guinea pigs. Likewise, Tyzzer described "a somewhat similar structural variation in the intranuclear body . . . in cutaneous lesions in varicella," but they favored the cause as an effect of chronic inflammation on cells in infancy, offering the descriptive name "cytomegalia."

It remained for W. C. VonGlahn and A. M. Pappenheimer to argue effectively in another case report for a viral etiology (49) to account for these unusual cells.

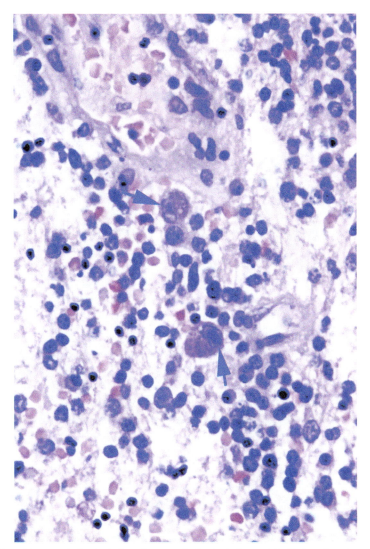

FIGURE 7 *Cytomegalovirus inclusions. Cytomegalovirus inclusions, originally thought to be parasitic in nature, identified cytomegalic inclusion disease of infants. This image is from the brain of a cytomegalovirus-infected fetus. (Courtesy of J. H. Kim, Yale University School of Medicine.)*

They dismissed out of hand the possibility of protozoan parasites. They also noted that "it seems safe to infer that the inclusions are not produced by a banal, nonspecific nuclear degeneration." While being cautious, they favored the possibility "that the inclusions in this case are caused by a virus identical with or closely related to the herpetic group." They cited the observations of B. Lipschutz of similar structures produced by herpes viruses (7) and of similar transmissibility to the rabbit cornea (50). The agent causing herpes febrilis had also been shown to be a filterable virus by that time (51). However, before cytomegalovirus was isolated, more work remained to define the incidence, the clinical syndrome, and the usefulness of urine cytopathology in its diagnosis.

In a postmortem series, S. Farber and S. B. Wolbach demonstrated intranuclear and cytoplasmic inclusion bodies in the submaxillary glands in 22 of 183 (12%) infants in their study reported in 1932 (52). In two cases, inclusions were found in viscera. In 1950, J. P. Wyatt distinguished an incidental localized infection of the salivary glands from "generalized cytomegalic inclusion disease" (46). Prominent renal, hepatic, pulmonary, and bleeding tendencies were found, and the lethal nature of the generalized form was emphasized. They concluded that the microscopic findings were pathognomonic of disease of specific viral etiology. The failure to isolate a virus from human materials in animals was attributed to species specificity and the inability of the virus to replicate in nonhuman cells. Presciently, they suggested urine cytology as a means of making the diagnosis in life based on the autopsy findings of large inclusion-bearing, desquamated cells in renal tubules of afflicted infants. Wyatt et al. were attempting to move recognition of the disease into the clinical arena, declaring that it was not simply a "pathologist's" disease.

In the same year, 1950, and from the same city, St. Louis, M. G. Smith and F. Vellios reported another 20 cases of cytomegalovirus infection, detailing clinical and autopsy findings from 3 of those cases (45). In addition, their literature review of 69 cases revealed frequent liver damage and blood dyscrasias in the study group. They noted that multiple-organ involvement could cause variations in clinical presentation. They, too, concluded that the appearance of the inclusion-bearing cells was pathognomonic and that there was "reasonable evidence" that the salivary gland virus was the cause. That comment was prescient, since Smith was to report the isolation of human cytomegalovirus from salivary gland 6 years later (53).

Two years following the suggestion of Wyatt et al. in 1950 to examine urinary cytology for diagnosis of generalized cytomegalic inclusion disease during life, G. H. Fetterman did so (54). A smear was made of urine, "murky brownish fluid," and several inclusion-bearing cells were found. The afflicted infant died shortly thereafter, and the diagnosis of generalized disease was confirmed. As predicted by Wyatt et al., intranuclear inclusion-bearing cells were demonstrated both in urine sediment and in a kidney section in a tubule. The following year, R. D. Mercer et al. presented the findings in another case (55). They concurred "... that the diagnosis of the syndrome can be made clinically in the neonatal period by study of the urine sediment."

By the 1950s, a devastating clinical syndrome of newborns was defined by histological criteria of cytomegaly with intranuclear and cytoplasmic inclusions. Soon thereafter, three laboratories would isolate cytomegalovirus in cell culture

(53, 56, 57). Weller would go on with his colleagues to further document the clinical spectrum of cytomegalic inclusion disease in virologically confirmed cases (58). In addition to the previously recognized hepatosplenomegaly with jaundice and bleeding, microcephaly with cerebral palsy and mental retardation was recognized in survivors. It has been estimated that 27,000 cases of cytomegalovirus infection occur in pregnant women annually in the United States. Given the potentially devastating effects on the fetus, the development of a cytomegalovirus vaccine has great urgency (59).

THE BEGINNINGS OF ELECTRON MICROSCOPY AND VIROLOGICAL STUDIES

The electron microscope transformed the science of virology, provided a powerful approach for rapid viral diagnosis, and became an exquisitely sensitive tool with which to define new viral agents, especially when other methods had failed. Until the 1940s, however, the forms, structures, and assembly of viruses were unknown. Electron microscopy made the forms visible, provided the basis on which to distinguish one class of viruses from another, and helped to define the processes of infection and replication. The development of the electron microscope was extraordinarily rapid. The crucial theoretical insights on which the use of electrons to form images was based were articulated in the 1920s, the first instrument was constructed in the early 1930s, and its application to the study of viruses was first achieved in the late 1930s.

A few individuals pioneered the development of electron microscopy, including its application to biological studies. Ernst Ruska, working first as a student and then as a colleague of Max Knoll with fellow student Bodo von Borries (Fig. 8), received the Nobel Prize in 1986 for development of the electron microscope. The story, as told by Ruska himself in his Nobel Prize lecture, reveals a certain amount of serendipity (60). Ruska did 2 years of study of electrotechnical engineering in Munich. However, when his father, a historian of sciences, accepted a position in Berlin, Ruska transferred his studies to the Technische Hochschule in Berlin. He worked on high-voltage techniques and joined a research group headed by Max Knoll. The goal of the group was to develop high-speed cathode ray oscilloscopes for research on lightning and other electrical surges in the electrical power lines (61). Based on theory articulated by Hans Busch in 1927, Ruska's task concerned ". . . the electron beams emerging divergently from the cathode [that] had to be concentrated in a small writing spot on the fluorescent screen of the cathode-ray oscillograph" (60). The mechanism to concentrate electron beams was a short magnetic coil which served like a convex lens for light, and the "focal length of this 'magnetic electron lens' [could] be changed continuously by means of the coil current." Knoll and Ruska enhanced the focusing by adding iron encapsulation of

FIGURE 8 *Bodo von Borries (left) and Ernst Ruska in the early 1930s on vacation on the island of Ruegen. (Courtesy of the Ernst Ruska Archive, Berlin, Germany.)*

the magnetic field. D. Gabor had previously shortened the magnetic field by iron encapsulation (61), and Knoll and Ruska enhanced that effect by increasing the iron encapsulation (Fig. 9).

Having first shown that sharp images could be produced by electrons focused through a short coil, Ruska demonstrated that additional magnification could be achieved with a second coil. That was demonstrated with an apparatus in 1931 which was termed the first electron microscope (60). The capacity of electrons to function in a manner similar to light waves was based on Louis de Broglie's (1925) concept of moving electrons functioning as a wave (60). The capacity to

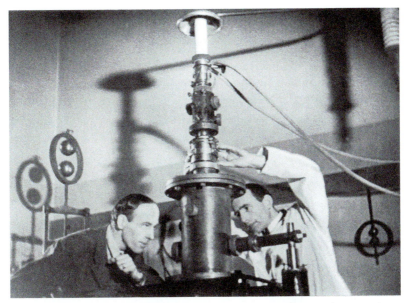

FIGURE 9 *Max Knoll (left) and Ernst Ruska with an early transmission electron microscope, the 1933 Uebermikroskop. The photo was taken in the early 1940s. (Courtesy of the Ernst Ruska Archive, Berlin, Germany.)*

discriminate two points as separate by a light microscope (resolution) was shown by Ernst Abbe in 1878 to be dependent on wavelength (61). It turned out that the electronic waves were significantly shorter than light waves; hence, as Ruska put it, "There was no reason to abandon the aim of electron microscopy surpassing the resolution of light microscopy" (60).

While proof of concept of the superior magnification had been achieved, numerous problems remained before the successful application of electron microscopy to biological specimens. Among the most important to be confronted was the intense heat generated in the specimen by the focused electron beam. Other crucial issues concerned the nature and enhancement of contrast in the images and specimen prepared of sufficient thinness and free of distortion. In solving these problems, many other investigators played crucial roles.

Ladislaus Marton in Brussels developed his own electron microscope by December 1932 (62). He estimated "that biological objects would offer the richest fields for this new tool." Elsewhere in reporting his studies he noted that "This high resolving power cannot be applied in biological research, however, without developing a new histological technique to prevent the destruction of organic cells by the intense electronic bombardment" (63). The effect of the energy on the sample could "'burn it to a cinder'" (62). Using a leaf of *Drosera intermedia* for study, he found the best results by impregnation with osmium (63). At about the same time, he determined that contrast in an object studied by electron microscopy was due to "differences in the scattering properties of the object" rather than by absorption (62). In 1941, Marton wrote a clear exposition for microbiologists not only on the construction of the electron microscope, comparing it with the light microscope, but also on techniques for sample preparation (64). Helmut Ruska (Fig. 10), the younger brother of Ernst Ruska, played a

FIGURE 10 *Helmut Ruska, circa 1930. As the younger brother of Ernst Ruska and trained in medicine, Helmut Ruska pioneered the electron microscopy of viruses. Together with Ernst and his brother-in-law, Bodo von Borries, they published the first electron micrographs of viruses in 1939. This was the first visualization of the particulate nature of viruses. (Courtesy of Ernst Ruska Archive, Berlin, Germany.)*

pioneering role in the application of electron microscopy to the study of viruses, and he was extremely prolific in publishing his exciting new observations. Trained in medicine, Helmut Ruska encouraged his brother, Ernst, to study disease processes (65). With Ernst and brother-in-law Bodo von Borries, in 1939 Helmut published the first electron micrographs of viruses, including tobacco mosaic virus (66). Additional work in collaboration with others was published that same year (67). A study of bacteriophage was also published by Helmut Ruska in 1941 (68). Having studied several viruses by 1943, Helmut was able to offer an overview of virus types organized by morphology (69). However, advanced studies of viral morphology or virus architecture were not seen until the development of negative staining by S. Brenner (70) and R. W. Horne one and a half decades later (71). Morphology and symmetry would serve as key components for viral classification, crucial in viral diagnostic work (72). These breakthroughs are highlighted in more detail in chapter 7. In the early 1940s, too, the foundations were laid for what would become immunoelectron microscopy. Tobacco mosaic virus was used in an early immunological study of antigen-antibody reactions. Starting with the precipitin reaction, visualization of a precipitate with the unaided eye, and the implied lattice formation of an antigen-antibody precipitate, T. F. Anderson and W. M. Stanley studied the reaction of tobacco mosaic virus with its antiserum (73). While directed at the study of an immunological reaction, lattice formation in the precipitin reaction, these electron microscopic observations were precursors to immunoelectron microscopy, so important later in diagnostic virology.

An investigation into a significant virus of humans at this time was a study by R. H. Green et al. in 1942 of the structure of vaccinia virus (33). This study was a precursor to van Rooyen and Scott's evaluation in 1948 of the electron microscope for the diagnosis of smallpox (28) (Fig. 11). The latter investigators noted five previously established means of diagnosis: isolation in rabbits, isolation on chorioallantoic membranes, immunological flocculation reaction, complement fixation, and microscopic examination of skin scrapings for Paschen bodies; they added one more.

For electron microscopic diagnosis of smallpox, both laboratory and field samples were studied by van Rooyen and Scott (28). The story of field sample collection is remarkable. Samples were prepared by inoculating electron microscopic mounts at the bedside in "Arab huts" in North Africa, "where owing to the intense heat, it was not possible to employ even the most elementary refinement in technique." Even under these adverse field conditions, "clumps and masses of elementary bodies" could be recognized when samples were returned to the laboratory for electron microscopic examination. These reports described early claims of

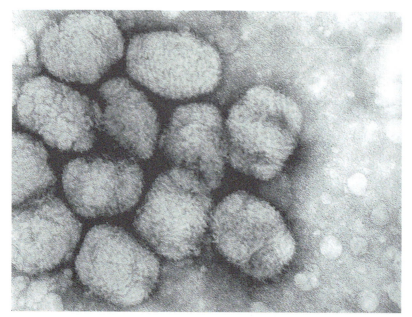

FIGURE 11 *Electron micrograph of poxvirus. Vaccinia and smallpox were early subjects of electron microscopic study, and the brick shape was readily identified. This electron micrograph utilized more advanced techniques than were originally available, such as negative staining, allowing visualization of structural detail. (Courtesy of CDC-PHIL [#2292]/Fred Murphy.)*

electron microscopy for "the rapid and early diagnosis of smallpox." In the coming decades, these rudimentary electron microscopic developments would be complemented by further progress in other rapid diagnostic techniques.

One must note with awe that the development of the electron microscope—from the theoretical concept of an electromagnetic lens to the first application for diagnosis of viral diseases—occurred over a span of slightly more than 30 years. The application of electron microscopy to the understanding of viruses was revolutionary. Although the sizes of viruses had been previously determined by filtration and ultracentrifugation, no one knew what a virus looked like. The electron microscope achieved that in very short order, revealing the shapes, repeating capsid structures, and nucleoprotein cores. It became a very powerful tool for research and diagnostic virology (Fig. 12).

FIGURE 12 *"1887–1987: A Century of Science for Health." Illustration demonstrating the armamentarium of research tools, including light and electron microscopes, experimental animals, and illustrative items. Each of the tools has been applied to the study and diagnosis of viruses and their infections. (Courtesy of the National Library of Medicine.)*

REFERENCES

1. **Berger J.** 1977. *Ways of Seeing.* Penguin Books, New York, NY.
2. **Craighead JE.** 2000. *Pathology and Pathogenesis of Human Viral Disease.* Academic Press, San Diego, CA.
3. **Negri A.** 1903. Beitrag zum Stadium der Aetiologie der Tollwuth. *Z Hyg Infektionskr* **43:**507–528.
4. **Guarnieri G.** 1892. Ricerche sulla patogenesi ed etiologia dell'infezione vaccinica e variolosa. *Arch Sci Med (Torino)* **16:**403–424.
5. **Tyzzer EE.** 1906. The histology of the skin lesions in varicella. *J Med Res* **14:**361–392, 7.
6. **Tyzzer EE.** 1906. The histology of the skin lesions in varicella. *Philipp J Sci* **1:**349–372.
7. **Lipschutz B.** 1921. Untersuchungen uber die Atiologie der Krankheiten der Herpesgruppe (Herpes Zoster, Herpes Genitalis, Herpes Febrilis). *Arch Dermatol Syph* **136:**428–482 http://dx.doi.org/10.1007/BF01843151.
8. **Jesionek A, Kiolemenoglou B.** 1904. Ueber Einen Befund von protozoënartigen Gebilden in den Organen eines Hereditär-luetischen Fötus. *Munch Med Wochenschr* **51:**1905–1907.
9. **Ribbert P.** 1904. Ueber Protozoenartige Zellen in der Niere eines Syphilitischen Neugeborenen und in der Parotis von Kindern. *Zentralbl Allg Pathol* **15:**945–948.
10. **Nuland SB.** 1995. *Doctors: a Biography of Medicine,* 2nd ed. Vintage Books, New York, NY.
11. **Ackerknecht EH.** 1953. *Rudolf Virchow: Doctor, Statesman, Anthropologist.* Arno Press, New York, NY.
12. **Clendening L.** 1960. p 628. *In Source Book of Medical History.* Dover, New York, NY.
13. **Virchow R.** 1978. *Cellular Pathology as Based upon Physiological and Pathological Histology,* special edition. The Classics of Medicine Library, Birmingham, AL.
14. **Garrison FH.** 1963. *An Introduction to the History of Medicine,* 4th ed. W B Saunders, Philadelphia, PA.
15. **Long ER.** 1981. *A History of Pathology.* Dover, New York, NY.
16. **Hughes A.** 1959. *A History of Cytology.* Abelard-Schuman, New York, NY.
17. **Bracegirdle B.** 1977. The history of histology: a brief survey of sources. *Hist Sci* **15:**77–101 http://dx.doi.org/10.1177/007327537701500201.
18. **Cowdry EV.** 1934. The problem of intranuclear inclusions in virus diseases. *Arch Pathol (Chic)* **18:**527–542.
19. **Geison GL.** 1995. *The Private Science of Louis Pasteur.* Princeton University Press, Princeton, NJ.
20. **Johnson HN, Sellers TF.** 1948. Rabies, p 219–242. *In* Francis TF Jr (ed), *Diagnostic Procedures for Virus and Rickettsial Diseases.* American Public Health Association, New York, NY.
21. **Kristensson K, Dastur DK, Manghani DK, Tsiang H, Bentivoglio M.** 1996. Rabies: interactions between neurons and viruses. A review of the history of Negri inclusion bodies. *Neuropathol Appl Neurobiol* **22:**179–187 http://dx.doi.org/10.1111/j.1365-2990.1996.tb00893.x.
22. **Young CC, Sellers TF.** 1927. A new method for staining Negri bodies of rabies. *Am J Public Health (N Y)* **17:**1080–1081 http://dx.doi.org/10.2105/AJPH.17.10.1080.
23. **Osler W, McCrae T.** 1923. *The Principles and Practice of Medicine,* 9th ed. D Appleton & Co, New York, NY.
24. **Heberden W.** 1768. On the chicken-pox, p 427–436. *In Medical Transactions.* College of Physicians, London, United Kingdom.
25. **College of Physicians.** 1768. Preface, p iii–ix. *In Medical Transactions.* College of Physicians, London, United Kingdom.
26. **Councilman WT, Magrath GB, Brinckerhoff WR.** 1904. The pathological anatomy and histology of variola. *J Med Res* **11:**12–135.
27. **Toomey JA, Gammel JA.** 1927. The Paul test in the diagnosis of smallpox. *J Infect Dis* **41:**29–31 http://dx.doi.org/10.1093/infdis/41.1.29.
28. **Van Rooyen CE, Scott GD.** 1948. Smallpox diagnosis with special reference to electron microscopy. *Can J Public Health* **39:**467–477.
29. **Wilkinson L.** 1979. The development of the virus concept as reflected in corpora of studies on individual pathogens. 5. Smallpox and the evolution of ideas on acute (viral) infections. *Med Hist* **23:**1–28 http://dx.doi.org/10.1017/S0025727300050997.

30. **Gordon M.** 1937. Virus bodies: John Buist and the elementary bodies of vaccinia. *Edinburgh Med J* **44:**65–71.

31. **Garrod LP.** 1954. Mervyn Henry Gordon. 1872–1953. *Obituary Notices Fellows R Soc* **9:**153–163 http://dx.doi.org/10.1098/rsbm.1954.0011.

32. **Downie AW.** 1965. Poxvirus group, p 932–964. *In* Horsfall FL, Tamm I (ed), *Viral and Rickettsial Infections of Man,* 4th ed. J B Lippincott Co, Philadelphia, PA.

33. **Green RH, Anderson TF, Smadel JE.** 1942. Morphological structure of the virus of vaccinia. *J Exp Med* **75:**651–656 http://dx.doi.org/10.1084/jem.75.6.651.

34. **Parker RF.** 1948. Variola and vaccinia, p 83–92. *In* Francis TF Jr (ed), *Diagnostic Procedures for Virus and Rickettsial Diseases.* American Public Health Association, New York, NY.

35. **van Rooyen CE, Illingworth RS.** 1944. A laboratory test for diagnosis of smallpox. *BMJ* **2:**526–529 http://dx.doi.org/10.1136/bmj.2.4372.526.

36. **Fenner F, Henderson DA, Irita I, Jezek Z, Ladnyi ID.** 1988. *Smallpox and Its Eradication.* World Health Organization, Geneva, Switzerland.

37. **Tyzzer EE.** 1904. The etiology and pathology of vaccinia. *J Med Res* **11:**180–185, 224–229.

38. **Unna PG.**1896. *The Histopathology of the Diseases of the Skin,* p 635–639. William F Clay, New York, NY.

39. **Tzanck A.** 1948. Le cytodiagnostic immédiat en dermatologie. *Ann Dermatol* **8:**205–218.

40. **Kelly B, Shimoni T.** 2009. Reintroducing the Tzanck smear. *Am J Clin Dermatol* **10:**141–152 http://dx.doi.org/10.2165/00128071-200910030-00001.

41. **Yang RA, Nodine S, Anderson JB, Laughter MR, Zangara T, Dinkel R, Dunnick CA.** 2022. Tzanck smear in dermatologic practice. *J Derm Nurs Assoc* **14:**28–32.

42. **Weller TH.** 1978. *Ernest Edward Tyzzer, 1875–1965: a Biographical Memoir.* National Academy of Sciences, Washington, DC.

43. **Weller TH, Stoddard MB.** 1952. Intranuclear inclusion bodies in cultures of human tissue inoculated with varicella vesicle fluid. *J Immunol* **68:**311–319.

44. **Weller TH.** 1953. Serial propagation *in vitro* of agents producing inclusion bodies derived from varicella and herpes zoster. *Proc Soc Exp Biol Med* **83:**340–346 http://dx.doi.org/10.3181/003797 27-83-20354.

45. **Smith MG, Vellios F.** 1950. Inclusion disease or generalized salivary gland virus infection. *AMA Arch Pathol* **50:**862–884.

46. **Wyatt JP, Saxton J, Lee RS, Pinkerton H.** 1950. Generalized cytomegalic inclusion disease. *J Pediatr* **36:**271–294, illust http://dx.doi.org/10.1016/S0022-3476(50)80097-5.

47. **Jackson L.** 1922. An ameba-like organism in the kidneys of a child. *J Infect Dis* **30:**636–642 http://dx.doi.org/10.1093/infdis/30.6.636.

48. **Goodpasture EW, Talbot FB.** 1921. Concerning the nature of "protozoan-like" cells in certain lesions of infancy. *Am J Dis Child* **21:**415–425.

49. **Vonglahn WC, Pappenheimer AM.** 1925. Intranuclear inclusions in visceral disease. *Am J Pathol* **1:**445–446.

50. **Lowenstein A.** 1919. Aetiologische Untersuchungen über den fieberhaften Herpes. *Munch Med Wochenschr* **66:**769–770.

51. **Luger A, Landa E.** 1921. Zur Atiologie des Herpes Febrilis. *Z Gesamte Exp Med* **24:**289–321 http://dx.doi.org/10.1007/BF02739198.

52. **Farber S, Wolbach SB.** 1932. Intranuclear and cytoplasmic inclusions ("protozoan-like bodies") in the salivary glands and other organs of infants. *Am J Pathol* **8:**123–139.

53. **Smith MG.** 1956. Propagation in tissue cultures of a cytopathogenic virus from human salivary gland virus (SGV) disease. *Proc Soc Exp Biol Med* **92:**424–430 http://dx.doi.org/10.3181/003797 27-92-22498.

54. **Fetterman GH.** 1952. A new laboratory aid in the clinical diagnosis of inclusion disease of infancy. *Am J Clin Pathol* **22:**424–425 http://dx.doi.org/10.1093/ajcp/22.5.424.

55. **Mercer RD, Luse S, Guyton DH.** 1953. Clinical diagnosis of generalized cytomegalic inclusion disease. *Pediatrics* **11:**502–514 http://dx.doi.org/10.1542/peds.11.5.502.
56. **Rowe WP, Hartley JW, Waterman S, Turner HC, Huebner RJ.** 1956. Cytopathogenic agent resembling human salivary gland virus recovered from tissue cultures of human adenoids. *Proc Soc Exp Biol Med* **92:**418–424 http://dx.doi.org/10.3181/00379727-92-22497.
57. **Craig JM, Macauley JC, Weller TH, Wirth P.** 1957. Isolation of intranuclear inclusion producing agents from infants with illnesses resembling cytomegalic inclusion disease. *Proc Soc Exp Biol Med* **94:**4–12 http://dx.doi.org/10.3181/00379727-94-22841.
58. **Weller TH, Hanshaw JB, Scott DME.** 1962. Virologic and clinical observations on cytomegalic inclusion disease. *N Engl J Med* **266:**1233–1244 http://dx.doi.org/10.1056/NEJM196206142662401.
59. **Dekker CL, Arvin AM.** 2009. One step closer to a CMV vaccine. *N Engl J Med* **360:**1250–1252 http://dx.doi.org/10.1056/NEJMe0900230.
60. **Ruska E.** 1993. The development of the electron microscope and of electron microscopy. Nobel Lecture, 1986, p 355–380. *In* Frangsmyr T, Ekspang G (ed), *Nobel Lectures, Physics 1981–1990.* World Scientific Publishing, Singapore.
61. **Freundlich MM.** 1963. Origin of the electron microscope. *Science* **142:**185–188 http://dx.doi.org/10.1126/science.142.3589.185.
62. **Marton L.** 1968. *Early History of the Electron Microscope.* San Francisco Press, San Francisco, CA.
63. **Marton L.** 1934. Electron microscopy of biological objects. *Nature* **133:**911 http://dx.doi.org/10.1038/133911b0.
64. **Marton L.** 1941. The electron microscope: a new tool for bacteriological research. *J Bacteriol* **41:**397–413 http://dx.doi.org/10.1128/jb.41.3.397-413.1941.
65. **Kruger DH, Schneck P, Gelderblom HR.** 2000. Helmut Ruska and the visualisation of viruses. *Lancet* **355:**1713–1717 http://dx.doi.org/10.1016/S0140-6736(00)02250-9.
66. **Ruska H, von Borries B, Ruska E.** 1939. Die Bedeutung der Übermikroskopie für die Virusforschung. *Arch Gesamte Virusforsch* **1:**155–169 http://dx.doi.org/10.1007/BF01243399.
67. **Kausche VGA, Pfankuch E, Ruska H.** 1939. Die Sichtbarmachung von Pflanzlichem Virus im Übermikroskop. *Naturwissenchaften* **27:**292–299 http://dx.doi.org/10.1007/BF01493353.
68. **Ruska H.** 1941. Über ein neues bei der Bakteriophagen Lyse auftretendes Formelement. *Naturwissenschaften* **29:**367–368 http://dx.doi.org/10.1007/BF01479367.
69. **Ruska H.** 1943. Versuch zu einer Ordnung der Virusarte. *Arch Virol* **2:**480–498 http://dx.doi.org/10.1007/BF01244584.
70. **Brenner S, Wolpert S.** 2001. *My Life in Science.* Biomed Central Ltd, London, United Kingdom.
71. **Brenner S, Horne RW.** 1959. A negative staining method for high resolution electron microscopy of viruses. *Biochim Biophys Acta* **34:**103–110 http://dx.doi.org/10.1016/0006-3002(59)90237-9.
72. **Almeida JD.** 1983. Uses and abuses of diagnostic electron microscopy. *Curr Top Microbiol Immunol* **104:**147–158 http://dx.doi.org/10.1007/978-3-642-68949-9_9.
73. **Anderson TF, Stanley WM.** 1941. A study by means of the electron microscope of the reaction between tobacco mosaic virus and its antiserum. *J Biol Chem* **139:**339–344 http://dx.doi.org/10.1016/S0021-9258(19)51390-4.

5 The Turning Point
CYTOPATHIC EFFECT IN TISSUE CULTURE

. . . they provide criteria by which the presence of the virus can be recognized in vitro and hence may afford a basis of technics for isolating virus from patients or animals. . . .

Robbins, Enders, and Weller, 1950 (1)

INTRODUCTION

In 1949, John Enders, Thomas Weller, and Frederick Robbins published a report of their soon-to-be-recognized Nobel Prize-winning discovery that poliomyelitis virus could be cultivated in cell cultures of nonneural origin (2). As Weller would write years later, "This was an electrifying result. It appeared to show that the Lansing strain of poliovirus had successfully grown in non-nervous human tissue culture—a result that had not been previously reported" (3). The report contained preliminary observations of the virus effect *in vitro*; however, the virus was confirmed by its paralytic consequences after inoculation into mice and monkeys.

Another paper that would transform diagnostic virology was a subsequent report by Robbins et al. in 1950 in which the term cytopathogenic effect (CPE) was coined (1). Later shortened to cytopathic effect, this phrase was used to describe degenerative changes in tissue cultures that were specifically associated with viral infection and visible on microscopic examination. Before that report of CPE, virus isolation and identification constituted a complex, highly specialized, and costly operation.

To Catch a Virus, Second Edition. Authored by John Booss and Marie Louise Landry.
© 2023 American Society for Microbiology. DOI: 10.1128/9781683673828.ch05

Looking back from less than a decade later in 1958, A. J. Rhodes and C. E. van Rooyen would write, "It is small wonder, then, that virus diagnostic tests were restricted to special research groups, and were not much used by physicians in their practice." They continued, "Since about 1952 or 1953, the situation in regard to laboratory diagnosis has changed in a remarkable fashion, for the introduction of the newer tissue culture technics has simplified and accelerated the methods of isolation of viruses" (4). From the 1950s onward, diagnostic virology was propelled as a specific laboratory science, no longer simply an appendage to research labs.

THE BEGINNINGS OF TISSUE CULTURE

Between W. Reed and J. Carroll's transmission of yellow fever to human volunteers by filtered blood in 1902 (5) and the transmission of polio to experimental monkeys (6), a scientific advance in experimental embryology was reported in 1907 which would transform diagnostic virology decades later. Ross Granville Harrison (Fig. 1) was an experimental embryologist whose studies addressed a controversy in the development of the nervous system on the origin of peripheral nerves. Harrison's work was the first to firmly establish that nerve fibers in the peripheral nervous system grow out from central neurons (7). Harrison dissected out pieces of embryonic tissue from frogs and put them onto a coverslip with fresh lymph which clotted, thereby holding the tissue in place (8). The preparation was turned over and sealed onto a hollow slide. The preparations could be observed by microscopy under high magnification for about a week. Harrison reported, ". . . fibers were observed which left the mass of nerve tissue and extended out into the surrounding lymph clot." He observed enlarged ends of the fibers with the "mass of protoplasm undergoing amoeboid movements." Because he was unable to make satisfactory permanent specimens of intact nerves, he resorted to another technique, placing a cylindrical clot into an excavated section of the embryo and observing that axons had grown into the clot. While the latter technique of experimental manipulation of the embryo was essential to consolidating his proof of the outgrowth of neurons, the former technique of growing animal tissue outside of the body was to transform several fields in biology, including diagnostic virology in the years to come.

Michael Abercrombie wrote of Harrison, "He simultaneously unleashed a new technique of tremendous power, and solved a hotly contested problem" (7). Successful implementation of the new technique required solving at least two issues not achieved by others at the time. First, his original explants in salt solution died until frog lymph was added and clotted, giving a substrate for cell movement. Second, the problem of infection, which would plague cell culturists until the advent of antibiotics, was solved by washing the embryos with sterile distilled water. The groundbreaking report was published in 1907, the same year that

FIGURE 1 *Ross Granville Harrison. In 1907, Harrison reported the outgrowth of nerve fibers from embryonic tissue* in vitro. *While Harrison did no further work with the technique, tissue culture was to transform many areas of biology. In years to come, tissue culture became the standard technique for virus isolation and identification. (Courtesy of the National Library of Medicine.)*

Harrison moved from Johns Hopkins to Yale to become a professor of comparative anatomy. He assisted others, such as Montrose Burrows from the Rockefeller Institute, to advance tissue culture techniques, but Harrison himself did not pursue further developments on his own.

It seems remarkable in retrospect that Harrison was not awarded the Nobel Prize for this advance. He died in 1959, over half a century after his momentous publication, after a full scientific life in experimental embryology and in advancing the study of zoology at Yale and nationally. Another memoirist addressed the controversy of the absence of the award (9). "In 1917, a majority of the Nobel Committee recommended that the prize should be given to him." However, it happened that the Institute was not to award a prize that year, a war year, a sad commentary on the vicissitudes of scientific honors. Nor would the Prize be awarded in 1933, when a special investigation decided against his work because of ". . . the rather limited value of the method and the age of the discovery. . ." (9). Thus, Harrison was passed over for his discoveries of neural outgrowth *in vitro*, which had very broad implications in many areas of research such as cancer, physiology, and preventive medicine, as well as in virology.

While Harrison did not further develop his technique of tissue culture, it was taken up and furthered by an investigator at the Rockefeller Institute, Alexis Carrel, who did win the Nobel Prize, albeit not for tissue culture. Carrel, born in a suburb of Lyon, France, in 1873, was a highly imaginative and dexterous surgeon and biologist (http://www.nobelprize.org/nobel_prizes/medicine/laureates/1912/). He was awarded the Nobel Prize in 1912 ". . . for his work on suturing of vessels and transplantation of organs" (10). There was no mention of his work in tissue culture, which he had recently begun.

In the spring of 1910, Montrose Burrows, an associate of Carrel at the Rockefeller Institute, joined Harrison at Yale to learn ". . . the method of growing tissues outside the body" (11). During that time, Burrows was able to substitute blood plasma for lymph, which Harrison recognized was not the optimal growth medium. Burrows took the technique one step further when he adapted the technique to the chicken embryo. It is interesting to note, particularly from the present-day perspective of attaching multiple authors to publications, that Harrison was not a coauthor. Rather, Burrows expressed his "great obligation" and his appreciation for Harrison's "ready personal assistance."

On Burrows' return to the Rockefeller Institute, Carrel and Burrows reported the successful cultivation of adult mammalian tissues and organs *in vitro* (12). Presciently, in their conclusion they observed, ". . . it may render possible the cultivation of certain microorganisms in conjunction with living tissue cells. " Tissue culture was to be used for viral cultivation, particularly vaccinia virus, the smallpox vaccine virus, well before its diagnostic use. In fact, Carrel and Thomas Rivers reported growth of vaccinia virus *in vitro* in animal corneas in 1927 (13).

Carrel continued making contributions to tissue culture techniques in subsequent years. According to a later commentator, he ". . . was very much the public face of tissue culture in the U.S.A." (14). In 1911 with Burrows, perhaps in response to a challenge that they had not succeeded in growing tissues outside the body, Carrel published details of their techniques, with several photographic plates illustrating tissue growth *in vitro* (15). Carrel turned his attention to prolonging tissue culture indefinitely. He examined various experimental features and was able to report that connective tissue fragments had survived for at least 85 days. He also reported that isolated cells could proliferate in culture and subsequently undergo several passages *in vitro* (16). By 1923, he was able to report leaving the cells undisturbed in special containers which allowed medium changes (17). Carrel retired from the Rockefeller Institute in 1938 and returned to France during World War II to set up an Institute for the Study of Human Problems (http://www.nobelprize.org /nobel_prizes/medicine/laureates/1912/).

EARLY APPLICATIONS OF TISSUE CULTURE TO VIRAL GROWTH

The application of Harrison's technique to the propagation of viruses was taken up by Edna Steinhardt and her associates starting in 1912 at the Research Laboratory of the New York Board of Health. Their first studies to demonstrate Negri bodies, the marker of rabies virus infection, produced ambiguous results. Cultivation of brains inoculated with rabies virus or incubation of brain tissue from animals dying with strains of rabies failed to reveal the development of structures different from controls (18).

Following the inconclusive work with rabies *in vitro*, Steinhardt and her coinvestigators were successful in demonstrating replication of vaccinia virus *in vitro* by "Harrison's method." They put pieces of corneas from rabbits or guinea pigs in an emulsion of virus, placed the corneal tissue on cover glasses with blood plasma, inverted in hollow slides, and incubated them (19). On fixation and staining, while the corneal sheets showed active spreading, no inclusion bodies were found. They also used an *in vivo* assay in which virus rubbed onto shaved skin of rabbits produced a vaccinial eruption. The number of vesicles produced by various dilutions of samples allowed an approximate estimate of the amount of virus present. They demonstrated an increase in the amount of virus after incubations of 7 to 18 days. In a subsequent study, Steinhardt and R. A. Lambert, using the assay system of vesicles on shaved rabbit skin, again demonstrated replication *in vitro* of vaccinia virus in Harrison-type cultures of corneas taken from uninfected animals (20). No replication was found when the cornea and plasma were taken from an immune animal.

Thus, from 1912 to 1914, Steinhardt and her colleagues initiated studies of viral replication in cell culture. Using an *in vivo* assay system of vesicles on shaved rabbit skin, they found replication of vaccinia virus *in vitro*, demonstrated immunity *in vitro*, and found tissue specificity for viral replication. Despite an impressive body of work, their discoveries appeared to garner little attention and acceptance.

For example, in S. Benison's masterful oral history memoir of Tom Rivers (Fig. 2), the dean of early studies of virology in the United States, Rivers does not note the work of Steinhardt et al. in response to a specific question about "growing vaccine virus in tissue culture" (21). Rather, it was left for Benison to report in a footnote that the first work on vaccinia virus cultivation in culture was undertaken by Steinhardt and her colleagues. Such a significant achievement was, in a sense, treated almost as an afterthought.

FIGURE 2 *Thomas Rivers. Rivers established laboratory studies of virological diseases at the Rockefeller Institute in New York City. He edited the first comprehensive textbook of clinical virology and rickettsiology in the United States. (Courtesy of Rockefeller Archive Center.)*

In a review by John Enders in 1948 describing virus growth in tissue culture, while acknowledging that Steinhardt et al. had "revealed the possibilities of this new technic for the cultivation of viruses," Enders cast some doubt on the reliability of their findings, maintaining that they did not "... obtain unequivocal evidence that multiplication of the agent occurred . . ." *in vitro* (22). Instead, Enders credited Frederic Parker and Robert N. Nye with demonstrating viral multiplication *in vitro* (23). Materials harvested from successive *in vitro* cultures were inoculated intradermally into rabbits so that a high multiplication rate was found, demonstrated by the appearance of vesicular eruptions on the rabbit skin.

In a brief report in 1928, Rivers and his coinvestigators reported observations *in vitro* (24) which can be viewed as anticipating the CPE described by the Enders laboratory two decades later (1). They studied the effect of infecting rabbit corneas *in vitro* by scarification with a virus emulsion and immersion in virus for 1 hour with either vaccinia or herpes simplex virus. The infected cornea pieces were incubated in centrifuge tubes between clots of plasma and tissue extract. After 24 to 48 hours, the clots with cornea bits were fixed and stained. In vaccinia virus-infected cultures, "abundant typical Guarnieri bodies appeared in the cytoplasm," whereas "numerous acidophilic inclusions occurred in the nuclei of cells injured by herpetic virus." Hence, in a modification of tissue culture, tissue layered between clots of plasma and tissue extract, Rivers et al. found "definite evidence" of cell growth. The work represented an advance on the findings of Steinhardt et al., not specifically referenced in their report, with inclusion bodies found *in vitro*. Importantly, it added to the work of Steinhardt et al. in foreshadowing the Enders lab's observation of CPE in that transfer to an animal for assay was not required.

Another forerunner of the *in vitro* work of the Enders lab on poliovirus was that of C. Levaditi (25). Levaditi was born in Romania in 1874, and his long career was spent mainly at the Pasteur Institute in Paris (26). He was said to have "spanned the eras of Pasteur and modern times" (27). John Paul, a polio researcher who compared him to Tom Rivers in the United States, noted that Levaditi "was almost the first, and for a long time the only person, in France to study viruses systematically." Levaditi attempted to culture poliovirus in spinal ganglion cells rather than in non-neural cells (2). His many contributions to the study of polio included immunity, intestinal infection, and infection by the oral route, as well as demonstration of survival of the virus *in vitro*. Among numerous other microbiological subjects that Levaditi studied was the elusive cause of encephalitis lethargica. He died at 79 in 1953 on the eve of the demonstration of the conquest of polio by vaccination (27).

The study of vaccinia virus in culture remained an important focus for virologists in the first four decades of the 20th century. In 1928, Hugh Bethune Maitland and Mary Cowan Maitland at the University of Manchester claimed to have grown vaccinia virus without tissue culture (28). However, the growth medium contained minced chicken kidney cells which they claimed had autolysed. Whereas significant multiplication of the virus was observed, no tissue growth was found, and autolysis was noted to be extensive by the third day. The finding of vaccinia virus growth in the absence of cells was disputed by Rivers, who confirmed virus growth but also found surviving cells (21). Rivers acknowledged the Maitlands as superb investigators and credited them with the development of a medium for virus growth that came to be called Maitland's medium. Hugh and Mary Cowan Maitland were graduates of the University of Toronto and married in 1926. They worked together at the University of Manchester in England, where Hugh held the Chair of Bacteriology. The Maitlands passed away in 1972 (29). Much of the interest in growing vaccinia virus in culture rested on the hope that it could replace lymph vaccine against smallpox raised in calves. However, in a comprehensive review of cultivation of viruses *in vitro* in 1939, it was concluded that ". . . tissue culture vaccine may today be best used as an adjunct to lymph vaccine" (30). This prediction was, in fact, erroneous.

Murray Sanders, who produced that comprehensive review, made several points. He noted that there were two types of *in vitro* tissue cultivation: hanging-drop slides and flask cultures. Another type, roller cultures, in which culture tubes were mounted on a rotating drum, would be applied to the culture of vaccinia virus a year later by the Enders lab (31). Sanders also noted that quantitation of viruses could be achieved by animal inoculation, counting of lesions on cho-rioallantoic membranes of developing eggs, or cutaneous testing. The crucial step of quantitation *in vitro* would come a decade later. Underlying the success of *in vitro* tissue culture was the maintenance of sterility to exclude bacterial and fungal infection, which was paramount. Even in the post-antimicrobial era, the first thing a virology trainee was taught was sterile technique. The advent of sul-fonamides and penicillin promised to diminish but not eliminate the risk of con-tamination (32, 33).

Great strides in tissue culture growth of viruses made in the latter 1940s and in the1950s were of signal importance for diagnostic virology. The key obser-vations would emerge from the John Enders lab in Boston in the context of the battle against polio. As Nobelist Tom Weller put it, "The major stimulus for the development of the poliovirus vaccine came not from a scientist but from a promi-nent political figure: Franklin Delano Roosevelt" (3). The most notable person to endure paralytic polio, Franklin Delano Roosevelt (FDR) was stricken as an adult in August 1921. FDR's remarkable resilience allowed him to run for the Presidency and to be elected to the first of four terms in 1932 (34). He guided the country out of the Great Depression and through most of World War II, tragically dying of a cerebral hemorrhage just before declaration of victory over Germany.

FDR . . . ". . . BY FRIDAY EVENING HE LOST THE ABILITY TO WALK OR MOVE HIS LEGS . . ."

At 39 years of age, FDR was remarkably fit and vigorous. Photographs from the time just before his polio show him to be slender and athletic looking. However, in the summer of 1921 at the family's vacation retreat at Campobello Island, he appeared tired yet he launched into vigorous activity (35). In his own words in a letter from 1924 first published in 1946, he wrote, "First symptoms of illness appeared in August 1921 when I was thoroughly tired from overwork" (36). Roosevelt noted a chill one evening in August and continued, "The following morning the muscles of the right knee appeared weak and by afternoon I was unable to support my weight on my right leg. That evening the left knee began to weaken also and by the following morning I was unable to stand up. This was accompanied by a continuing temperature of about 102 and I felt thoroughly achy all over." At first, the condition was diagnosed by a local physician as a bad cold; the noted elderly surgeon William Keen, vacationing nearby, ventured a diagnosis of a blood clot in the lower spinal cord. It was finally diagnosed as polio by the Harvard polio specialist R. W. Lovett (35). There ensued marked nursing needs: to be catheterized, fed, and bathed. In mid-September, he was transported to New York's Presbyterian Hospital, where he remained until the end of October.

From the perspective of diagnostic virology, it is important to note that the diagnosis was based entirely on the clinical picture, and no laboratory virological studies were performed at that time. The diagnosis of polio has been challenged by A. S. Goldman et al., who concluded by analysis that an inflammatory neuropathy, Guillain-Barré syndrome, was most likely (37). The point must remain moot, however, in the absence of diagnostic, virological laboratory studies. FDR, his doctors, and the public believed that he had polio, and it was the diagnosis of polio in FDR which propelled subsequent events (Fig. 3).

For Roosevelt, the illness steeled his resolve, and he went through years of rehabilitation. He purchased an old resort at Warm Springs, GA, which he converted into a therapeutic facility for persons who had experienced polio (35). Under the guidance of his advisor Louis Howe and with the assistance of FDR's wife, Eleanor, his political career resumed. He was first elected Governor of New York State and then to four terms as President of the United States.

Prior to his first nomination for the Presidency, Roosevelt suffered rumors that circulated saying that polio made him unfit for that office. Earle Looker, a journalist, arranged for a medical evaluation by specialists to assess FDR's physical condition, and he was pronounced fit (38, 39). Years later, after FDR's passing, Eleanor Roosevelt wrote, "Perhaps the experience, above all others, which shaped my husband's character along more definite lines and gave him a strength and depth that he did not have as a young man was the long struggle with infantile paralysis" (40).

FIGURE 3 *FDR at Hill Top Cottage on his family's estate in Hyde Park, NY. FDR, who experienced polio as an adult, is shown with his dog, Fala, and Ruthie Bie. The National Foundation for Infantile Paralysis, which FDR founded with his partner, Basil O'Connor, underwrote the costs of developing the Salk polio vaccine. Less well known is that the Foundation supported important investigations in basic science. (Courtesy of the Franklin D. Roosevelt Presidential Library and Museum, Hyde Park, NY.)*

Roosevelt's illness fueled the fight against polio. Even as President during World War II, he wrote passionately of "formidable enemies at home and abroad." He wrote further, "Not until we have removed the shadow of the Crippler from the future of every child can we furl the flags of battle and still the trumpets of attack. The fight against infantile paralysis is a fight to the finish, and the terms are unconditional surrender" (41). Although FDR did not live to see that battle won, the polio vaccine achieved that goal. The organization that would organize the fight and provide resources to "furl the flags of battle" was the National Foundation for Infantile Paralysis. It was led by FDR's indefatigable law partner, Basil O'Connor, and a research coordinator, D. Harry Weaver. The development of the National Foundation under O'Connor has been well chronicled (27, 42) (Fig. 4).

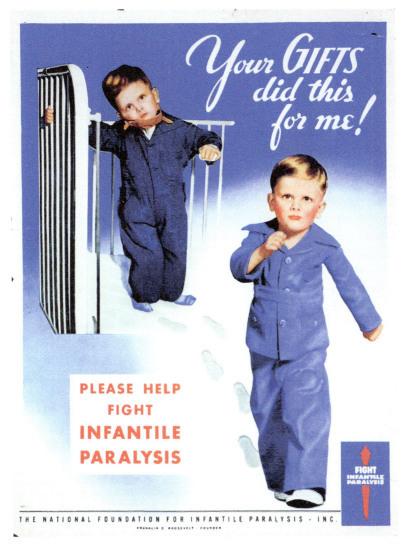

FIGURE 4 *Polio poster. This fund-raising poster shows a vigorous and fully intact, young boy striding toward the viewer. He is shown on a background featuring the same boy exhibiting polio, with a weak right leg, unable to hold his head upright, and with his left hand in a brace; footsteps lead from illness to health. (Photo provided courtesy of March of Dimes.)*

Tissue culture would be adapted to grow poliovirus, allowing the production of the vaccine. Tissue culture would also facilitate viral diagnosis, not requiring assay in animals or in embryonated eggs. That ability to grow virus in tissue culture and to simultaneously measure its amount was the pivotal event, the turning point, after which the study of diagnostic virology accelerated immeasurably (43).

THE GROWTH OF POLIOVIRUS *IN VITRO*: "IT WAS ALMOST AN AFTERTHOUGHT"

John Enders, the leader of the team which demonstrated that poliovirus would grow in nonneural cells *in vitro* and which showed that cytopathology in tissue culture could be used to titrate virus, had a long professional gestation. Enders was born into a financially comfortable family, and his interests in preparatory school tended toward the humanities (44). His undergraduate career at Yale was interrupted by World War I, in which he served as a flight instructor. Following his return to Yale and graduation with a B.A. degree, he "drifted" for a few years, seeking a career. Enders ultimately found his calling. After abandoning real estate and the teaching of English, he took a master's degree in philology at Harvard.

By a marvelous quirk of fate at Harvard, Enders shared common living quarters at "Mrs. Patches' house" and became friendly with Hugh Ward, an instructor in the Department of Bacteriology and Immunology. Ward, as recounted by Enders, was a Rhodes Scholar, battalion surgeon, and outstanding oarsman (45). Enders also went on to praise Ward's sterling qualities of mind as a bacteriologist and teacher and credits Ward for turning him to bacteriology. "It was largely because of Hugh Ward's example and encouragement that I gave up graduate study of English literature and philology and entered the field of bacteriology."

The department was chaired by Hans Zinsser, a charismatic figure in science with an exceptionally broad cultural range. In a memoir of Zinsser, Enders opened with, "Various were the attributes of Hans Zinsser's luminous mind and spirit" (46). It is no surprise that Enders, a man of broad intellectual interests, should be attracted by Zinsser. Enders was to say of Zinsser's lab, "Under such influences, the laboratory became much more than a place just to work and teach; it became a way of life."

Enders took his doctorate with Zinsser in 1930 at age 33 and stayed on in the department at Harvard. His initial research until 1937 concerned bacteria. He then shifted to studies in virology and rickettsiology. In another coincidental encounter during that period, Harry Plotz, an American who had been for most of his career at the Pasteur Institute in Paris, came to the lab as a visiting fellow to work on typhus. The goal of that work by Zinsser, Plotz, and Enders was to create a vaccine, for which they grew the typhus bacillus in tissue culture (47). Enders would ultimately focus on viruses, and Plotz would become the initial director of the first laboratory devoted exclusively to the diagnosis of viral and rickettsial diseases.

The Nobel Prize-winning team led by Enders had two junior members, Tom Weller and Fred Robbins. Their initial career paths appeared to be more focused than was that of Enders's initial ventures. Each sprang from scientific stock. Weller's father was an academic pathologist; Weller stated in his autobiography, ". . . during my pre-collegiate years, my father was the major influence in molding my life" (3). Weller first developed an interest in birds and then in parasites at the University of Michigan. He took a master's degree in parasitology before entering medical school at Harvard in 1936. He, too, came under the influence of Hans Zinsser. By yet another fortuitous personal interaction, Weller roomed with Fred Robbins in his fourth year of medical school. In that year, Weller did a tutorial on cell cultures in Enders's lab, originally to learn to grow parasites. Together with Alto Feller and Enders, Weller applied the roller tube culture method, a novel concept at the time, to the first step in growing vaccinia virus. In addition to sustaining and likely proliferating virus, they demonstrated intracytoplasmic inclusions (31).

Weller's further work in Enders's lab growing viruses in tissue culture was interrupted by World War II, when he spent time at an Army laboratory in Puerto Rico (3). One principle demonstrated in this body of work was the prolonged maintenance of cultures, which could be crucial to allow virus replication. Although Weller had chosen pediatrics as a clinical specialty in his medical training, he preferred to pursue his interests investigating infectious diseases. He accepted the Chairmanship of the Department of Tropical Public Health at Harvard in 1954. He was to go on to make highly important contributions with isolations of cytomegalovirus, varicella-zoster virus, and rubella virus. Quite appropriately, Weller titled his autobiography *Growing Pathogens in Tissue Cultures: Fifty Years in Academic Tropical Medicine, Pediatrics, and Virology* (3).

Fred Robbins's parents were both botanists; his father was at one time the Director of the New York Botanical Garden and later became the Chairman of Botany at the University of Missouri (48). Fred Robbins obtained A.B. and M.S. degrees in premedical studies from the University of Missouri and completed his medical studies at Harvard in 1940. Robbins's training, too, was interrupted by military service in World War II. Like Joseph Edwin Smadel, who is discussed further below, he had wanted to work in Tom Rivers's special laboratory unit in the Navy. However, Robbins's request was denied by the Army. Robbins served in North Africa and Naples in an Army laboratory section, making significant contributions to the study of Q fever as a cause of pneumonia in American troops in the Mediterranean area. With colleagues he published a series of papers characterizing the clinical features, epidemiology, laboratory outbreak, and isolation of the etiologic agent. The work on Q fever was undertaken with Smadel at the Viral and Rickettsial Diseases

Laboratory at the Walter Reed Army Medical Center (49). That laboratory was originally directed by Harry Plotz and after World War II by Smadel.

Following his return to civilian life, Robbins worked in Enders's lab under a fellowship funded by the National Foundation for Infantile Paralysis. His first assignment was to assist Enders in reviewing viral tissue culture techniques. He reviewed work demonstrating changes in virus-infected tissue cultures. That review sensitized him to later observe changes in poliovirus-infected tissue culture in his own work (48).

After his work in Enders's lab, Robbins moved to Cleveland in 1952. He was appointed Professor of Pediatrics at the Western Reserve School of Medicine and at the Cleveland Metropolitan Hospital in Pediatrics and Infectious Diseases. He went on to become Dean of the School of Medicine at Western Reserve and in 1980 became President of the Institute of Medicine (50) (Fig. 5).

Enders and his two young colleagues, Weller and Robbins, were termed "the three princes of Serendip" by author Greer Williams (51). In the writings of John Enders, one sees that he was deeply intellectual and reflective. He thrived in an environment in which he and his colleagues could work, reflect, and write without the distractions of a heavy teaching load, committee work, and departmental politics.

Enders's loyalty and lack of personal ambition in the award of the Nobel Prize were later recounted by Robbins: "He made it clear from the outset that his two junior colleagues were full participants and that any recognition that might result from the work should be shared, and so it was" (48). Some years following Hans Zinsser's death in 1940, an opportunity was presented to establish a research laboratory in infectious diseases at the Children's Hospital in Boston. Enders moved there in January 1947 and invited Tom Weller to help run the laboratory. The National Foundation had awarded Harvard a 5-year grant to study viruses and directed some of these funds to Enders's lab (3, 51). When Fred Robbins joined the lab at the start of 1948, the team, setting, and funding were in place to explore the use of tissue culture to study viruses.

The origins of Enders's work with viruses reached back to the late 1930s, when he collaborated with W. M. Hammon on a study of a viral disease of cats, malignant panleukopenia, and with H. F. Shaffer on the use of the chorioallantoic membrane to study infectivity and inactivation by antibody of herpes simplex virus (44). Commenting on the work with A. E. Feller and Enders on prolonged survival of vaccinia virus in roller tube cultures (31), Weller credited a Harvard Medical School classmate, Larry Kingsland, as having become interested in roller cell culture methods, which had been introduced by George Otto Gey at Johns Hopkins Medical School. Kingsland and Weller also set up a tissue culture laboratory at the Children's Hospital in Boston when they were interns in 1940–1941 (27). During that time, Weller and Kingsland studied cultures of varicella-zoster virus and

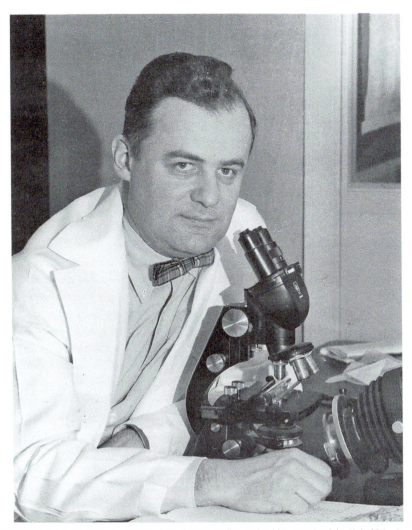

FIGURE 5 *Frederick Robbins. John Enders, Thomas Weller, and Robbins received the Nobel Prize in 1954 for the discovery of the capacity to grow poliovirus in tissue culture of nonneural origin. With Enders and Weller, Robbins demonstrated that CPE in tissue culture could be used to detect the growth of virus for isolation from clinical specimens, to measure the amount of virus present, and to determine the presence of antibodies. These crucial discoveries led to the rapid and widespread development of diagnostic virology laboratories. (Image 00716, property of Case Western Reserve University Archives.)*

attempted isolation from an autopsied case of polio. The cultures from the latter case deteriorated, and they assumed it was due to contamination. Years later in his autobiography, Weller speculated, perhaps ruefully, whether they had observed "... the cell destroying activity of the polioviruses for the first time" (3).

During the war years, Enders, who served as a member of the Commission on Measles and Mumps of the Army Epidemiological Board (27), worked on humoral and cutaneous immunity to mumps virus (52). Among Weller's first projects on joining Enders's lab in 1947 was the attempt to grow egg-adapted mumps virus in tissue culture. With Enders, Weller modified what had been the standard technique since the Maitlands' report of 1928 (28) by replacing only the nutrient fluid and leaving the cell tissue culture in place (53). Using detection of hemagglutination in the tissue culture fluid as the marker system, Weller and Enders showed that the egg-adapted mumps virus rapidly replicated in suspended chick embryo cell cultures. They found a similar result with the PR8 strain of influenza A virus. They commented that the assay of hemagglutinin in tissue culture fluid provided a "swift and convenient" method of tracking viral multiplication which was "far less complex than in chick embryos or in the mouse." Thus, they were on track to develop tissue culture systems as fundamental techniques in diagnostic virology to serve for both isolation and assay without the need to titrate in eggs or animals.

There followed one of the great examples of what the author Greer Williams had termed serendipity and which occurred, according to Weller, as "... almost an afterthought ..." (51). Following the successful growth of mumps virus in culture, Weller turned to one of his principal interests, varicella-zoster virus. He obtained aborted human embryonic tissue from which 12 cultures were made. Weller inoculated four cultures with throat washings from a case of chicken pox, retained four cultures as uninoculated controls, and had four cultures remaining. Those four cultures were inoculated with a mouse-passed Lansing strain of poliovirus which had been stored in the freezer because of Weller's "... interest in viruses paralyzing mice. " On day seven, no evidence of varicella replication was found by complement fixation, the endpoint of the experiment. Fluids from the "afterthought" poliovirus cultures were inoculated intracerebrally into five mice. Over the course of 17 days, four of the five mice became sick or died. It was, as Weller put it, "an electrifying result" (3).

Much more work needed to be done before the result could be accepted, as it ultimately was, demonstrating the successful growth of poliovirus *in vitro* in non-neural human tissue culture. Weller replicated the findings. Importantly, he demonstrated that the results were not a consequence of contamination with a murine pathogen, Theiler's virus.

The first report by Enders et al. in *Science* in January 1949 described the growth of the Lansing strain of poliovirus in culture derived from human embryonic tissues devoid of intact neurons (2). They also observed a loss of staining properties of infected cultures, a step toward the description of cytopathogenicity but an observation on which they exercised caution. A subsequent publication reported that each of the three antigenic types of poliovirus would grow in suspended cell cultures (54) derived from several human embryonic, mature, and tumor tissues. At the same time, Robbins et al. reported that the three strains would grow in roller tubes containing several different types of human tissues (55). The finding that each of the three strains would grow in tissue culture was important because of the potential for vaccine production. Hence, these findings opened the door to a poliovirus vaccine, a project taken up in short order by the National Foundation for Infantile Paralysis but one in which the Enders lab did not want to participate (3, 51).

The crucial contribution to diagnostic virology by the Enders group, however, was the recognition of cytopathogenic effect, shortened to cytopathic effect (CPE) (3), as a visual aid in the detection and assay of virus infections in tissue culture. While the use of tissue culture to grow virus reached back decades, it was burdened with the need to assay culture fluids for virus growth by inoculating animal or egg systems. Steinhardt and her colleagues grew vaccinia virus in tissue culture as early as 1913 but had to confirm culture infectivity by inoculating the shaved skin of rabbits to generate vesicles (19). There were early reports of cell changes *in vitro* in virus-infected cultures, but these reports had little impact on assay methods. Most notably, in 1932, C. Hallauer reported morphological changes in cultures infected with fowlpox virus (56). Was this in fact CPE, and had its importance failed to be appreciated and exploited by subsequent tissue culture virologists?

C. H. Huang reported an approach to *in vitro* assay of viral infection in which western equine encephalomyelitis virus was applied to minced chicken embryo skeletal muscles (57, 58). The endpoint of the assay system was the failure of outgrowth of infected tissue compared to control, uninfected tissues or the outgrowth of tissue of infected preparations neutralized by antibody. Another approach described by Weller and Enders in 1948 was the demonstration of hemagglutinin in suspended chick embryo cell cultures for *in vitro* assay of mumps and influenza viruses (53).

The observation of cell degeneration after poliovirus infection, reported cautiously in the 1949 *Science* report, was to be termed "cytopathogenic effect" by Enders. Further study was needed to validate the observation (59). The results of those studies were reported by Robbins, Enders, and Weller in 1950 (1). They demonstrated cell degeneration in suspended cell cultures and in roller tube cultures infected with poliovirus. Examination of cells from suspended cultures required fixation and staining, which were somewhat labor-intensive.

Using plasma hanging-drop cultures derived from flask cultures, they demonstrated a failure of cell migration in highly infected samples. The inhibition could be titrated by virus dilution. Thus, there was no inhibition of migration in more diluted samples. A consequence of virus-induced cell destruction was reduced tissue metabolism reflected by diminished acid production and signaled by colorimetric determination using phenol red indicator in the culture medium. With cell death, the medium changed from healthy pink-orange to failing pink-purple as the pH increased as a result of cell death. Inhibition of the CPE by type-specific antisera was also documented. The authors described roller tube cultures as "a more convenient and rapid procedure for the demonstration of cell injury" and found that poliovirus rapidly destroyed growing cells.

Another advance in tissue culture technique for diagnostic virology was the use of penicillin and streptomycin in the nutrient medium. Not only did antibiotics reduce contamination during preparation of cell cultures but also they were critical for testing clinical specimens such as feces and throat washings for virus content, preventing bacterial overgrowth which would destroy the cell cultures. Thus, in 1951 Robbins et al. reported the successful isolation and typing of poliomyelitis virus primarily from fecal specimens (60). In his "Reminiscences of a Virologist," Robbins recounted, ". . . how excited I was when I was able to isolate a virus directly from a patient, identify it as poliovirus, type it, and demonstrate an antibody response in the patient's blood, all in tissue culture" (48). Thus began the tissue culture era of diagnostic virology (Fig. 6).

The findings of Enders, Weller, and Robbins, in addition to providing the means to create the polio vaccine and to merit recognition with the 1954 Nobel Prize in Physiology or Medicine (61), opened the floodgates for the isolation of viral agents in the 1950s and beyond. Their work transformed diagnostic virology from a largely serologically based enterprise to one in which virus isolation became feasible for laboratories attached to public health departments and academic hospitals.

The first diagnostic laboratory was devoted primarily to the diagnosis of viral and rickettsial infections, in contrast to research labs which did occasional diagnostic work. The laboratory was set up at the Walter Reed Army Medical School in January 1941 in anticipation of World War II, which had started in Europe. The lab was established well before the advent of tissue culture systems for the isolation and assay of viruses. Hence, many of the operational concepts and structures of diagnostic virology laboratories were established by the time the technology of tissue culture would swell the need for such diagnostic labs.

FIGURE 6 *CPE. The upper panel shows a monolayer of uninfected, human diploid fibroblasts. The lower panel shows a focus of viral infection, described as CPE, marked by enlarged, rounded cells seen in the center of the photograph surrounded by normal, uninfected cells. CPE resulted from cytomegalovirus replication in the monolayer. (Courtesy of Diane S. Leland and Indiana Pathology Images.)*

THE FIRST VIRAL DIAGNOSTIC LABORATORY

In 1939, as the war in Europe was gathering, Harry Plotz, an expatriate American scientist, returned to New York City. He had worked for 18 years in the laboratory originally established by Elie Metchnikoff at the Pasteur Institute in Paris (62), where Metchnikoff was a pioneer in studies of cell-mediated immunity.

Plotz's story is an interesting one. Having made a sustained series of contributions to the understanding of typhus, a rickettsial disease, including studies in Europe during World War I, he worked at the Pasteur Institute on a variety of infectious disease problems. When America entered World War I, he served as an advisor to William Gorgas, a hero of yellow fever eradication. Seventeen months after Plotz's return to American shores, he was appointed to direct the first laboratory devoted principally to the diagnosis of viral and rickettsial diseases at the Walter Reed Army Medical School in Washington, DC, in January 1941 (63); (T. O. Berge, Laboratory specialties, p. 1–26. *In* History of the U.S. Army Medical Service Corps, unpublished manuscript). His legacy is significant.

Harry Plotz was born in 1890. He obtained his M.D. degree in 1913 from the Columbia College of Physicians and Surgeons, where his interest in infectious diseases was sparked (64). He went to the Mount Sinai Hospital in New York City for house staff training and came under the influence of Nathan Brill, an investigator of typhus fever and for whom recurrent epidemic typhus was named. In 1914, Plotz published a report of the isolation of an organism thought to be the cause of typhus fever (65). As a member of an expedition to Serbia in 1915 sponsored by the Red Cross and funded by a Mount Sinai trustee, Plotz investigated an outbreak of typhus with colleague George Baehr. They were captured by the Bulgarians but managed to carry on their research (Typhus Research at the Mount Sinai Hospital, 1913–1917, unpublished manuscript; personal communication, B. Niss, Mount Sinai Hospital archivist). Plotz was graduated from the Mount Sinai house staff in 1917 but soon returned to Europe as a lieutenant colonel in the U.S. Army to continue his study of typhus fever (66). A further contribution to the field was development of a delousing apparatus for the Army to protect troops against the vector of typhus (64).

Shortly after returning to the United States in 1920, Plotz married Ella Sachs, the daughter of banker Samuel Sachs, the head of Goldman Sachs. Years later, a column in a Washington, DC, newspaper noted that his young wife had passed away in pregnancy (67). Further, ". . . the youthful husband, feeling that life had lost its meaning, buried himself in science." Plotz returned to Europe and took up an unpaid full-time position at the Pasteur Institute. His publication record for the next 18 years reveals that he worked on a variety of microorganisms, with a major focus on fowl plague virus. Toward the end of the Pasteur Institute years, he worked on measles virus, demonstrating inapparent infection in monkeys (68) and multiplication *in vitro* in chicken embryo cells (69). In 1938, he contributed a major review of cell culture methods to Levaditi and Lepine's textbook *Les Ultra-virus des Maladies Humaines* (70). As noted previously, Plotz collaborated with Zinsser and Enders at Harvard as a guest worker on tissue culture production for a vaccine against typhus fever (47). With experience in clinical diseases and in the laboratory, he was exceptionally well prepared to lead the first diagnostic laboratory in the United States for viral and rickettsial diseases (Fig. 7).

After war had begun in Europe but before the entrance of the United States, the U.S. Army began preparations to deal with the threat that infectious diseases posed to troops in foreign as well as domestic locations. Based on the experiences of World War I, a Board for the Investigation and Control of Influenza was established in January 1941 (71). In that same month, the Army established the first diagnostic laboratory for viral and rickettsial diseases in two basement rooms of the Army Medical School (63; Berge, unpublished). Under the direction of Plotz, who was now a colonel, the lab was staffed by two enlisted technologists,

FIGURE 7 *Harry Plotz. Colonel Harry Plotz (left) is shown receiving a World War II medal from Briga-dier General George Callender. Plotz was the founding chief of the first virus and rickettsia lab set up primarily for diagnosis. It was established in January 1941 at the Walter Reed Army Medical Center. (From* Borden's Dream. *Courtesy of the Borden Institute, Fort Detrick, MD.)*

Kenneth Wertman and Reginald Reagan (72). Impressive in their own right, the two technologists were later commissioned in the Sanitary Corps; both went on to doctoral degrees and significant publication records in viral and rickettsial studies.

Shortly before his premature passing in January 1947, Plotz described the activities of the Army's diagnostic lab to a meeting in November 1946 of the American Public Health Association (APHA). The activities included serological testing and isolation of agents of disease, vaccine development, and the training of clinicians and laboratory technologists (73). Plotz noted the deficient teaching of viral and rickettsial diseases in schools of medicine and encouraged greater emphasis. He noted that some universities and public health services were planning diagnostic labs similar to those developed by the Army. His final comment in that paper served as a spur for what was to come in diagnostic virology: "These efforts should be encouraged in every way possible." Harry Plotz retired as head of the laboratory in 1945, turning over the leadership to Joseph Edwin Smadel. Plotz held the position of Consultant to the Secretary of War until January 1947, when he passed away from a heart attack suffered while working in his laboratory (64).

Smadel (Fig. 8) assumed leadership of the Walter Reed Laboratory on his return from the European Theater of Operations (ETO) and prior to the passing of Harry Plotz. He was one of the leaders of American clinical virology in its formative years. Said Tom Rivers, "Joe Smadel is one of the most knowledgeable all-round virologists in the country—far better than I was in my heyday" (21). Smadel was born into a medical family in 1907 in Vincennes, IN; his father was a general practitioner and his mother a former nurse (74). After attending the University of Pennsylvania, he received his M.D. from Washington University in St. Louis. Staying on at the Barnes Hospital of Washington University for training in pathology, he participated in studies which isolated and characterized the virus in an outbreak of St. Louis encephalitis. His first published papers concerned pathological changes in the central nervous system of mice and neutralization of the virus. He then moved to the Rockefeller Institute for Medical Research in New York City and ultimately joined the virology laboratory of Tom Rivers. The Rockefeller years, 1934 to 1942, were highly productive, particularly in the investigations of vaccinia virus and lymphocytic choriomeningitis virus.

At the start of World War II, prior to his departure to the ETO, Smadel went to the Walter Reed Army Diagnostic Laboratory, then under the direction of Harry Plotz. In the ETO, Smadel served in the 1st Medical General Laboratory, as Chief of the Virus Department (75). Smadel had originally wanted to serve in Tom Rivers's unit in the Navy, but he was rejected by the Navy. The ETO labs were located in buildings erected in England for the Red Cross-Harvard Field Hospital. As described by Smadel in the

FIGURE 8 *Joseph Edwin Smadel. One of America's premier medical virologists, Smadel directed the European Theater of Operations (ETO) virus and rickettsial diagnostic lab for the Army during World War II. On returning to the United States, he succeeded Harry Plotz as chief of the diagnostic lab at the Walter Reed Army Medical Center. (Courtesy of the Lasker Foundation.)*

annual report, the laboratory menu consisted of serological tests, including complement fixation for numerous infectious agents, cold agglutinin tests for primary atypical pneumonia, and agglutination inhibition tests for influenza A and B. An outbreak of influenza A in troops was investigated, and an outbreak of psittacosis in birds was documented (76).

Smadel later organized and commanded a laboratory in support of the troops in northern France following the landing at Normandy. Special duties during this time included investigation of typhus fever in Africa, Italy, and Egypt. Toward the end of the conflict, he served with the Intelligence Division, investigating research developments in German universities in infectious diseases, particularly typhus fever (75).

Smadel served in the ETO until VE Day, 8 May 1945, after which he returned to Walter Reed Army Medical Center as Chief of the Department of Viral and Rickettsial Diseases to replace Harry Plotz. He concluded his military service in 1946 but stayed on as chief in a civilian capacity. While at Walter Reed, he initiated studies of infectious diseases in various international locations, including work in Kuala Lumpur in 1948. There he demonstrated the effectiveness of chloramphenicol treatment for typhus and typhoid. He also played a major role in commissions of the Armed Forces Epidemiological Board on Immunization, Rickettsial Diseases, and Influenza (74). In 1956, he moved to the National Institutes of Health in Bethesda, MD, and was recognized with the Albert Lasker Clinical Research Award in 1962. Smadel played important advisory roles, including consulting on the development of the polio vaccine. Like Harry Plotz, he died young, at 56 years of age.

Smadel's legacy includes his role in mentoring numerous virologists. While at Walter Reed, he trained laboratory and clinical professionals. He helped teach a virology course in the graduate school of the University of Maryland with Joel Warren and Maurice Hilleman (F. M. Bozeman and B. L. Elisberg, personal communication). This was one of the first courses of its type anywhere. He was a dynamic individual, outspoken and with a strong motivating effect on those with whom he worked or counseled. D. Carleton Gajdusek, the Nobel laureate, credited Smadel for encouraging him to follow his interests and wrote that their relationship was one of protégé and protector (D. C. Gajdusek, personal communication). Gajdusek published his correspondence with Smadel, writing, "Above all else, the publication of these letters is a tribute to the late Dr. Joseph E. Smadel. " (77).

Smadel's writings and published presentations provide an instructive window into the development of clinical virology in the latter 1940s and early 1950s. His orientation in many papers is toward the clinician and the help that could be rendered by the laboratory in patient care. In a 1948 paper in *JAMA*, just before the epochal work of Enders et al. on growing poliovirus *in vitro*, he emphasized the use of serology, noting that isolation and identification were "still research procedures" (78). While the Walter Reed lab had the capacity to grow viruses, even so, a period of 10 days to 3 weeks was required for the isolation and identification of many agents, in most cases too long for clinical application. For the performance of serological assays, diagnostic antigens were needed and could be prepared at the Walter Reed lab and distributed worldwide to Army laboratories; however, reagents could not be provided to nonmilitary laboratories. Smadel expressed optimism that these diagnostic antigens would become "generally available within a few years" because of the interest of commercial manufacturers. Some tests were already being performed by state and city laboratories and by a number of laboratories associated with teaching institutions. Smadel cited a symposium, the results of which were to be published as the first edition of *Diagnostic Procedures for Virus and Rickettsial Diseases* by the APHA in 1948 (79). That venerable resource, which went through numerous editions over several decades, became a benchmark for diagnostic virology and rickettsiology techniques.

A mere 2 years later and still before the force of the tissue culture revolution was felt, Smadel was writing of ". . . the new entity, the 'Virus Diagnostic Laboratory', which was conceived and developed during the past decade" (80). Pointing out that research labs "occasionally condescended to do routine clinical diagnostic tests," the impetus to the establishment of the virus diagnostic laboratory was supplied by the Army. Smadel described the Walter Reed lab as the first virus diagnostic laboratory and noted the development of laboratories to serve civilian medicine. By this point, he described diagnostic antigens and control materials as available through commercial biological suppliers. Even so, he noted that "all the present methods provide the diagnosis too late in the course of the individual's disease for the information to be of value to the physician responsible for his care." Smadel noted the need for "an entirely new line of approach which will provide a specific diagnosis at the time the physician first visits the acutely ill patient." He described assays which had been developed to detect minute amounts of antigen in rickettsial diseases. As a clinically oriented virologist, Smadel had articulated the need for rapid viral diagnosis as early as 1950.

Another remarkable individual who worked at Walter Reed was Maurice Hilleman (Fig. 9). He was to become the world's foremost vaccinologist (81). It was said of Hilleman at his passing in 2005 that he had ". . . probably saved more lives than any other scientist in the 20th century . . ." (82). Born and raised on a farm in Montana, he obtained his Ph.D. in microbiology from the University of Chicago. He worked at the pharmaceutical manufacturer Squibb from 1941 to 1948, where he developed the vaccine against Japanese B encephalitis, important to the war effort in the Pacific in World War II (82). Hilleman moved to the Walter Reed Army Medical Center in 1948 and stayed until 1957, moving to the pharmaceutical firm Merck, where he spent the remainder of his remarkably fruitful career. At Walter Reed, he identified "antigenic drift," or minor changes in the antigenic nature of influenza virus isolates requiring annual reformulation of the influenza vaccine. He also identified major changes, "antigenic shift," which are periodically associated with influenza pandemics when the virus no longer resembles influenza virus strains known to the susceptible population. He correctly predicted the 1957

FIGURE 9 *Maurice Hilleman. America's premier vaccinologist, Hilleman was at the Walter Reed Army Medical Center from 1948 until 1957. Among his accomplishments at Walter Reed, Hilleman described shifts in the antigenic nature of influenza viruses resulting in worldwide pandemics, and he isolated adenovirus from military troops. At the time of his death, he had created 8 of the 14 vaccines recommended for use in the United States. (Courtesy of the National Library of Medicine.)*

Hong Kong pandemic, and his efforts resulted in the production of vaccine before the arrival of the virus in the United States. That record of scientific productivity at Walter Reed, which also included discovery of an adenovirus (83), would sustain any reputation. In the long view of history, his legacy was remarkable. At the time of his death, Hilleman had produced 8 of the 14 vaccines recommended in the United States (84).

THE FIRST DIAGNOSTIC VIROLOGY LABORATORY
IN THE U.S. CIVILIAN SECTOR

The first nonmilitary diagnostic virology laboratory was established with support from the Rockefeller Foundation in Berkeley, CA, as part of the California Department of Health. Edwin Herman Lennette (Fig. 10) was designated the Chief of the

FIGURE 10 *Edwin Herman Lennette. With an extensive background in the study of numerous viruses, Lennette was designated the chief of the first state public health laboratory in the United States for viral and rickettsial diagnosis. He held the position in California from 1947 until retiring in 1978. With Nathalie J. Schmidt, he had a major role in establishing the field of diagnostic virology; he has been called "the father of diagnostic virology." (Courtesy of Edwin Paul Lennette.)*

Viral and Rickettsial Disease Laboratory in 1947 and held that position until his retirement in 1978. His career spanned the formative years in diagnostic virology, and he has been called the "father of diagnostic virology." He was joined in the lab in 1954 by Nathalie J. Schmidt. Until her passing in 1986, they formed a remarkable team in developing and refining diagnostic techniques, training virologists in those techniques, and disseminating knowledge of diagnostic virology through publications, conference presentations, and coediting the core APHA diagnostic virology and rickettsiology procedures text starting in 1964 (85).

Born in 1908 in Pittsburgh, PA, Lennette obtained his undergraduate and doctoral degrees from the University of Chicago (86). Raised in limited financial circumstances, Lennette took 2 years out of his undergraduate education to earn income as an accountant. As an undergraduate, he had started to work in the lab of Paul Hudson and continued with his doctoral work on poliovirus and cellular immunity, long before it was fashionable. Lennette was awarded his Ph.D. in Hygiene and Bacteriology from the University of Chicago in 1935. Working simultaneously on his M.D. degree, which he received the following year, Lennette decided to remain in Chicago to do his internship, at the urging of Hudson. During much of this time, he was doing both clinical and lab work, a dual perspective which was to serve him well in years to come in diagnostic virology. He understood well the relationships between the characteristics of the patient's illness, the appropriate laboratory assays, and the importance of a properly collected and transported specimen, the crucial "triad" which he was to espouse throughout his career (87).

A bit of serendipity followed in which his future evolved as a result of an intervention by Tom Rivers, the doyen of virology at the Rockefeller Institute in New York City. According to Lennette, Joe Smadel had gone to the Rockefeller laboratories from Washington University, creating the need for a virologist in St. Louis. Rivers arranged for Lennette to fill that need, and Lennette and his wife departed for St. Louis after the completion of his internship (86). At Washington University, he was an instructor in Pathology and came to work with Margaret Smith. While that tenure was brief, 1938–1939, it was productive. He published several papers with Margaret Smith on St. Louis encephalitis virus and the first paper on herpes simplex virus isolation in a case of acute encephalitis (88). In 1939, he was invited to join Frank Horsfall's lab at the Rockefeller Institute as part of the International Health Division (IHD), where he worked on the complement fixation assay for influenza. That work contributed to the recognition of the antigenic diversity of influenza virus and to the designations of influenza A and B as nomenclature (89). Lennette's productivity was again demonstrated as he produced a dozen papers on influenza with Horsfall's group.

With the outbreak of World War II there was a reorganization of staff, and Lennette was sent by the IHD to its Yellow Fever Service in Rio de Janeiro, Brazil, from 1941 to 1944. As recounted by Lennette, by chance, he encountered Hilary Koprowski, who had just emigrated with his family from Europe to Brazil to escape the German armies (86). Koprowski, who was to go on to a long and highly productive career in virology, was at that time a recent graduate from medical school, a biochemist, and a trained pianist but not yet a virologist. Lennette tells of giving Koprowski a stack of "reprints on medical virology" which, although a biochemist, he mastered impressively. They worked together for 3 years, producing papers on diverse topics, including yellow fever virus, Venezuelan encephalitis virus, and the serological and antigenic relationships of several encephalitis-causing arboviruses. It was also during his work on Venezuelan encephalitis that Lennette, among others, became quite ill. As he described the experience, he said, "Yes, they wrote me off. They were ready to bury me. I was very, very ill. And then Hilary Koprowski came down with the disease. By that time it dawned on us that there was something strange and wrong. " It was one of the first instances of transmission of an arthropod-borne virus via aerosolization within the laboratory (86). They also examined the interference phenomenon of viruses *in vitro*. Hence, the Brazil experience was another productive period in which several clinically important viruses were studied and in which Lennette collaborated with another outstanding, albeit junior, virologist.

In 1944, Lennette was assigned to the Rockefeller Foundation Laboratories in Berkeley, CA, a warm-up to what would be his long relationship with the California State Department of Health. His assignments were to work on the problem of hepatitis associated with the yellow fever vaccine and to establish a viral diagnostic laboratory. According to Lennette, "In effect the Berkeley Laboratory was the first civilian viral diagnostic laboratory, and as such, together with Joe Smadel's laboratory, pioneered much of the field" (86). "Joe Smadel's laboratory" was the lab at Walter Reed which had been led originally by Harry Plotz and then by Joe Smadel. In the Berkeley laboratory, Monroe Eaton described the agent responsible for primary atypical pneumonia. There was controversy at first about the finding, but Eaton was ultimately vindicated. Apparently, Lennette's work on hepatitis associated with yellow fever vaccine came to naught, but many of the papers with Koprowski were published during this period, as was the description of a fatal case of St. Louis encephalitis virus in California.

In 1946, Lennette accepted the position of Chief, Medical-Veterinary Division, at Fort Detrick, MD, and moved, resigning his employment with the IHD of the Rockefeller Foundation, with which he had worked since 1939. He took the position in part for economic reasons; the "Rockefeller Foundation had never been noted for being overgenerous to its staff." Yet on reviewing Lennette's career to that point, it is clear that the association with the Rockefeller Foundation had served him well. He had the opportunity to work on viral disease problems of great importance in fine labs with excellent investigators and had been highly productive, establishing himself as a recognized authority. In any event, the time at Fort Detrick proved to be frustratingly administrative in nature, and he accepted the opportunity to return to Berkeley as the Chief of the Viral and Rickettsial Disease Laboratory of the California Department of Health in the latter half of 1947. He was broadly prepared to implement a diagnostic virology and rickettsiology lab in which he would be presented with a wide array of human infectious material.

In its historical development, the State Department of Public Health in California had early recognized the continued prevalence of viral diseases against which little progress had been made in comparison to significant gains made against other types of microbial diseases. In 1936, the Department accepted funds from the IHD of the Rockefeller Foundation to establish the laboratory as one of three locations in the United States as "listening posts" to monitor influenza (90). A newly constructed laboratory had been opened under the direction of Monroe Eaton in 1939. In 1942, the department sought to initiate some routine diagnostic virology services, and state funding of the effort began. However, at this time the diagnostic work was still heavily dependent on research development. Three types of serological assays were performed: complement fixation, neutralization, and cold agglutination, as well as animal inoculation for a survey of typhus. With the allocation of funds by the legislature in 1946 for the study of encephalitis, the virus laboratory had become established as a part of the Division of Laboratories. The recognition of Q fever in California in 1947 also prompted special appropriations from the legislature. Since Q fever was of particular interest to Lennette, it was another reason that persuaded him to leave Fort Detrick and return to Berkeley (86).

Thus, not only was Lennette highly qualified by experience and temperament to take over as the director of the diagnostic lab but also the Department of Public Health and the Legislature had for two decades recognized the importance of viral and rickettsial diseases. The Chief of the Division of Laboratories who helped recruit Lennette, Malcolm Merrill, had worked in virology at the Princeton Rockefeller Institute. There Merrill had known Wendell Stanley, Nobel Prize winner for crystallization of viruses, who was recruited to the University of California at Berkeley soon after Lennette's return. The governor of California at the time, Earl Warren, according to Lennette was "very public health minded, very helpful, and always very supportive of the Department of Public Health" (86). Although Lennette's salary was paid by the state, the Rockefeller Foundation continued its support of the laboratory effort for a few more years. The circumstances augured well for the development of diagnostic virology.

Soon after his return to Berkeley in 1947, Lennette organized the personnel who remained in the department after Monroe Eaton's departure for Harvard, and he set about planning new laboratories (86). The number of assays performed for viral and rickettsial agents jumped to over 73,000 in 1949 from just under 6,000 in 1946. The range of serological assays, too, was enhanced, but viral isolates in tissue culture were not yet reported (90). Recognizing the importance of tissue culture, Lennette sent a member of his staff, Anna Weiner, to the Enders lab to learn tissue culture techniques. She returned to the lab and taught others (86).

Lennette actively pursued studies of Q fever in Northern California, with at least 30 publications on the subject between 1948 and 1953. Many of these studies were performed in collaboration with Gordon Meiklejohn, who was to go on to be a leader in academic medicine at the University of Colorado. However, one of Lennette's greatest talents was to recognize ability and to hire a young and as-yet-unrecognized microbiologist who had been only recently trained. Nathalie J. Schmidt came to work with Lennette in 1954. In contrast to Lennette, who became director of the Berkeley laboratory a dozen years after the award of his degree and with experience in diverse settings, Schmidt joined Lennette in the Berkeley lab only 1 year after receiving her Ph.D. degree. She had been born in Flagstaff, AZ, in 1928 to a father who was a jeweler and a watchmaker and a mother who had worked for the Arizona State Department of Health (91). Schmidt was described as a lapidarist, a cutter of precious stones, and an accomplished leather worker. It does not seem much of a cognitive stretch to recognize the ready application of the necessary artistic and technical facilities to her skilled laboratory work. She did her undergraduate work at the University of Arizona, from which she graduated in 1950. She moved on to Northwestern University, where she was awarded a master's degree followed by a doctorate in 1953 in Bacteriology-Immunology. She worked under Harry Harding and was highly productive over a brief (3-year) period. Ten publications resulted from those years, most dealing with various aspects of complement fixation assays. Following her doctoral studies, she went on to work for a year as a bacteriologist before accepting Lennette's offer of a position in the Berkeley laboratory.

The encounter seemed fortuitous. As Lennette described it, he had chaired a scientific session in San Francisco in 1953 in which Schmidt presented data on the serology of psittacosis (86). He was impressed with her capacities and recognized that she could develop an assay for polio as part of a project funded by the National Foundation for Infantile Paralysis. After accepting the offered position, Schmidt spent several weeks at Johns Hopkins with Manfred Mayer, an expert on complement fixation. Lennette noted to Sally Smith Hughes in his oral biography that he needed an alter ego, someone to work with, and that Schmidt was his first full-time staff member. She did successfully develop a complement fixation assay for polio and went on developing and refining assays for over three decades in the Berkeley laboratory. She was to become an internationally recognized expert with 225 scientific publications. She passed away from breast carcinoma in 1986. Her major area of expertise was the enteroviruses, yet she also made significant contributions to the diagnosis and understanding of rubella, hepatitis, and the herpes family of viruses. According to Bagher Forghani, who was originally a postdoctoral fellow and then a research scientist in the lab, Schmidt rapidly adapted new techniques. Many of these were immunologically based such as immunofluorescence, radioimmunoassay, enzyme immunoassay, and monoclonal antibody applications which represented advances in diagnostic virology (91). Among Lennette and Schmidt's multiple contributions to the field of diagnostic virology were the training of scientists and technologists, contributions to the scientific literature, and key roles as coeditors of the APHA's *Diagnostic Procedures for Viral and Rickettsial Diseases* starting with the third edition in 1964 (85).

Other state public health laboratories also began to develop diagnostic virology capabilities. For example, Gilbert Dalldorf became Director of the Division of Laboratories and Research of the New York State Department of Health in 1945. Working with Grace Sickles, he made the first isolates of the coxsackieviruses in suckling mice in 1948 (92) and continued studies of these viruses, which sometimes mimicked the clinical effects of poliovirus but had their own clinical spectrum (93). Dalldorf contributed to the understanding of diagnostic laboratory aspects of virology with the concise and practical *Introduction to Virology*, published in 1955 (94).

In the next chapter, the early diagnostic labs in university hospitals (95), the great advances brought about by the National Institutes of Health, the contributions of the Centers for Disease Control and Prevention (96), and the proliferation of virus isolates in various cell culture systems are examined.

REFERENCES

1. **Robbins FC, Enders JF, Weller TH.** 1950. Cytopathogenic effect of poliomyelitis viruses *in vitro* on, human embryonic tissues. *Proc Soc Exp Biol Med* **75**:370–374 http://dx.doi.org/10.3181/003797 27-75-18202.
2. **Enders JF, Weller TH, Robbins FC.** 1949. Cultivation of the Lansing strain of poliomyelitis virus in cultures of various human embryonic tissues. *Science* **109**:85–87 http://dx.doi.org/10.1126/science.109.2822.85.
3. **Weller TH.** 2004. *Growing Pathogens in Tissue Cultures: Fifty Years in Academic Tropical Medicine, Pediatrics, and Virology.* Science History Publications, Canton, MA.
4. **Rhodes AJ, van Rooyen CE.** 1958. *Textbook of Virology for Students and Practitioners of Medicine*, 3rd ed. Williams & Wilkins, Baltimore, MD.
5. **Reed W, Carroll J.** 1902. The etiology of yellow fever: a supplemental note. *Am Med* **3**:301–305.

6. **Landsteiner K, Popper E.** 1908. Mikroskopische Präparate von einem menschlichen und zwei Affenruckenmarken. *Wien Klin Wochenschr* **21**:1830.
7. **Abercrombie M.** 1961. Ross Granville Harrison 1870–1959. *Biogr Mem Fellows R Soc* **7**:111–126.
8. **Harrison RG.** 1907. Observations on the living developing nerve fiber. *Proc Soc Exp Biol Med* **4**:140–143 http://dx.doi.org/10.3181/00379727-4-98.
9. **Nicholas JS.** 1961. p 132–162. *In Ross Granville Harrison 1870–1959. A Biographical Memoir*. National Academy of Sciences, Washington, DC.
10. **Ackerman J.** 1912. The Nobel Prize in Physiology or Medicine 1912: presentation speech. *In Nobel Lectures, Physiology or Medicine 1901–1921*. Elsevier, Amsterdam, The Netherlands.
11. **Burrows MT.** 1910. The cultivation of tissues of the chick-embryo outside the body. *JAMA* **60**:2057–2058 http://dx.doi.org/10.1001/jama.1910.04330240035009.
12. **Carrel A, Burrows MT.** 1910. Cultivation of adult tissues and organs outside of the body. *JAMA* **60**:1379–1381 http://dx.doi.org/10.1001/jama.1910.04330160047018.
13. **Carrel A, Rivers TM.** 1927. La fabrication du vaccin *in vitro*. *C R Soc Biol* **96**:848–850.
14. **Wilson D.** 2005. The early history of tissue culture in Britain: the interwar years. *Soc Hist Med* **18**:225–243 http://dx.doi.org/10.1093/sochis/hki028.
15. **Carrel A, Burrows MT.** 1911. Cultivation of tissues *in vitro* and its technique. *J Exp Med* **13**:387–396 http://dx.doi.org/10.1084/jem.13.3.387.
16. **Carrel A.** 1912. On the permanent life of tissues outside of the organism. *J Exp Med* **15**:516–528 http://dx.doi.org/10.1084/jem.15.5.516.
17. **Carrel A.** 1923. A method of the physiological study of tissues *in vitro*. *J Exp Med* **38**:407–418 http://dx.doi.org/10.1084/jem.38.4.407.
18. **Steinhardt E, Poor DW, Lambert RA.** 1912. The production *in vitro* in the normal brain of structures simulating certain forms of Negri bodies. *J Infect Dis* **11**:459–463 http://dx.doi.org/10.1093/infdis/11.3.459.
19. **Steinhardt E, Israeli C, Lambert RA.** 1913. Studies on the cultivation of the virus of vaccinia. *J Infect Dis* **13**:294–300 http://dx.doi.org/10.1093/infdis/13.2.294.
20. **Steinhardt E, Lambert RA.** 1914. Studies on the cultivation of the virus of vaccinia. II. *J Infect Dis* **14**:87–92 http://dx.doi.org/10.1093/infdis/14.1.87.
21. **Benison S.** 1967. *Tom Rivers: Reflections on a Life in Medicine and Science*. The MIT Press, Cambridge, MA.
22. **Enders JF.** 1948. Propagation of viruses and rickettsia in tissue culture, p 114–127. *In* Rivers TM (ed), *Viral and Rickettsial Infections of Man*. J B Lippincott, Philadelphia, PA.
23. **Parker F Jr, Nye RN.** 1925. Studies on filterable viruses. I. Cultivation of vaccine virus. *Am J Pathol* **1**:325–335.
24. **Rivers TM, Haagen E, Muckenfuss RS.** 1928. A method of studying virus infection and virus immunity in tissue cultures. *Proc Soc Exp Biol Med* **26**:494–496 http://dx.doi.org/10.3181/00379727-26-4361.
25. **Levaditi C.** 1913. Symbiose entre le virus de la poliomyélite et les cellules des ganglions spinaux, à l'état de vie prolongée *in vitro*. *C R Soc Biol* **74**:1179.
26. **Kalantzis G, Skiadas P, Lascaratos J.** 2006. Constantin Levaditi (1874–1953): a pioneer in immunology and virology. *J Med Biogr* **14**:178–182 http://dx.doi.org/10.1258/j.jmb.2006.05-30.
27. **Paul JR.** 1971. *A History of Poliomyelitis*. Yale University Press, New Haven, CT.
28. **Maitland HB, Maitland MC.** 1928. Cultivation of vaccinia virus without tissue culture. *Lancet* **212**:596–597 http://dx.doi.org/10.1016/S0140-6736(00)84169-0.
29. **Downie AW.** 1973. Hugh Bethune Maitland: 15 March 1895-13 January 1972. *J Med Microbiol* **6**:253–258.
30. **Sanders M.** 1939. Cultivation of the viruses: a critical review. *Arch Pathol (Chic)* **28**:541–586.
31. **Feller AE, Enders JF, Weller TH.** 1940. The prolonged coexistence of vaccinia virus in high titre and living cells in roller tube cultures of chick embryonic tissues. *J Exp Med* **72**:367–388 http://dx.doi.org/10.1084/jem.72.4.367.

32. **Gey GO, Gey MK, Inui F, Vedder H.** 1945. The effects of crude and purified penicillin on continuous cultures of normal and malignant cells. *Bull Johns Hopkins Hosp* **77**:116–131.

33. **Koprowski H, Lennette EH.** 1944. Propagation of yellow fever virus in tissue cultures containing sulfonamides. *Am J Hyg* **40**:1–13.

34. **Ward GC.** 1989. *A First-Class Temperament: the Emergence of Franklin Roosevelt.* Harper & Row, New York, NY.

35. **Smith JE.** 2007. *FDR/Jean Edward Smith.* Random House, New York, NY.

36. **Roosevelt FD.** 1946. "A history of the case" in Franklin D. Roosevelt's own words. *J S C Med Assoc* **42**:1–2.

37. **Goldman AS, Schmalstieg EJ, Freeman DH Jr, Goldman DA, Schmalstieg FC Jr.** 2003. What was the cause of Franklin Delano Roosevelt's paralytic illness? *J Med Biogr* **11**:232–240 http://dx.doi. org/10.1177/096777200301100412.

38. **Looker E.** 1931. Is Franklin D. Roosevelt physically fit to be president? *Liberty* **25**:6–10.

39. **Looker E.** 1932. *This Man Roosevelt.* Brewer, Warren & Putnam, New York, NY.

40. **Roosevelt E (ed).** 1950. *F.D.R.: His Personal Letters, 1905–1928.* Duell, Sloan and Pearce, New York, NY.

41. **Roosevelt, F. D.** 1 December 1944. FDR letter to Basil O'Connor. Franklin Delano Roosevelt Presidential Library, PPF 4885.

42. **Oshinsky DM.** 2005. *Polio: an American Story.* Oxford University Press, New York, NY.

43. **Mortimer P, Weller TH, Robbins FC.** 2009. Classic paper: How monolayer cell culture transformed diagnostic virology: a review of a classic paper and the developments that stemmed from it. (*Science,* New Series, Vol. 109, No. 2822 (Jan. 28, 1949), p 85–87). *Rev Med Virol* **19**:241–249.

44. **Weller TH, Robbins FC.** 1991. John Franklin Enders: February 10, 1897-September 8, 1985. *Biogr Mem Natl Acad Sci* **60**:47–65.

45. **Enders JF.** 1963. Personal recollections of Dr. Hugh Ward. *Aust J Exp Biol Med Sci* **41**:L381–L384 http://dx.doi.org/10.1038/icb.1963.58.

46. **Enders JF.** 1940. Hans Zinsser in the laboratory. *Harv Med Alumni Bull* **15**:13–15.

47. **Zinsser H, Plotz H, Enders JF.** 1940. Mass production of vaccine against typhus fever of the European type. *Science* **91**:51–52 http://dx.doi.org/10.1126/science.91.2350.51.

48. **Robbins FC.** 1997. Reminiscences of a virologist, p 121–134. *In* Daniel TM, Robbins FC (ed), *Polio.* University of Rochester, Rochester, NY.

49. **Robbins FC, Rustigian R, Snyder MJ, Smadel JE.** 1946. Q fever in the Mediterranean area; report of its occurrence in Allied troops; the etiological agent. *Am J Hyg* **44**:51–63.

50. **Marko A, Marko K, Jackson KT (ed.).** 2007. Robbins, Frederick Chapman. *In The Scribner Encyclopedia of American Lives,* vol. 7: 2003– 2005. Charles Scribner's Sons, Detroit, MI. (Reproduced in *Biography Resource Center,* Gale, Farmington Hills, MI, 2009.)

51. **Williams G.** 1960. *Virus Hunters.* Alfred A Knopf, New York, NY.

52. **Enders JF.** 1964. The Nobel Prize in Physiology or Medicine 1954. *In Nobel Lectures, Physiology or Medicine 1942–1962.* Elsevier, Amsterdam, The Netherlands.

53. **Weller TH, Enders JF.** 1948. Production of hemagglutinin by mumps and influenza A viruses in suspended cell tissue cultures. *Proc Soc Exp Biol Med* **69**:124–128 http://dx.doi.org/10.3181/003797 27-69-16638.

54. **Weller TH, Enders JF, Robbins FC, Stoddard MB.** 1952. Studies on the cultivation of poliomyelitis viruses in tissue culture. I. The propagation of poliomyelitis viruses in suspended cell cultures of various human tissues. *J Immunol* **69**:645–671.

55. **Robbins FC, Weller TH, Enders JF.** 1952. Studies on the cultivation of poliomyelitis viruses in tissue culture. II. The propagation of the poliomyelitis viruses in roller-tube cultures of various human tissues. *J Immunol* **69**:673–694.

56. **Hallauer C.** 1932. Über das verhalten von Huhnerpestvirus in der Gewebekultur. *J Hyg Infektionskr* **113**:61–74 http://dx.doi.org/10.1007/BF02177065.

57. **Huang CH, Sanders M.** 1942. Titration and neutralization of the western strain of equine

encephalomyelitis virus in tissue culture. *Proc Soc Exp Biol Med* **51:**396–398 http://dx.doi.org/10.3181 /00379727-51-13990.

58. **Huang CH.** 1943. Further studies on the titration and neutralization of the western strain of equine encephalomyelitis virus in tissue culture. *J Exp Med* **78:**111–126 http://dx.doi.org/10.1084/ jem.78.2.111.

59. **Enders JF.** 1952. General preface to studies on the cultivation of poliomyelitis viruses in tissue culture. *J Immunol* **69:**639–643.

60. **Robbins FC, Enders JF, Weller TH, Florentino GL.** 1951. Studies on the cultivation of poliomyelitis viruses in tissue culture. V. The direct isolation and serologic identification of virus strains in tissue culture from patients with nonparalytic and paralytic poliomyelitis. *Am J Hyg* **54:**286–293.

61. **Gard S.** 1954. The Nobel Prize in Physiology or Medicine 1954: Presentation Speech. *In Nobel Lectures, Physiology or Medicine 1942–1962.* Elsevier, Amsterdam, The Netherlands.

62. **Time.** 13 March 1939. Pasteur's Pride. *Time.* http://www.time.com/time/printout/ 0,8816,760922,00.html.

63. **Ginn RVN.** 1997. *The History of the U.S. Army Medical Service Corps.* Office of the Surgeon General and Center of Military History United States Army, Washington, DC. http://dx.doi.org/10.1037/ e680482007-001

64. **New York Times.** 7 January 1947. Obituary, Harry Plotz. *New York Times.*

65. **Plotz H.** 1914. The etiology of typhus fever (and of Brills disease). Preliminary communication. *JAMA* **62:**1556.

66. **Plotz H.** 1920. Scientific notes and news. *Science* **52:**246.

67. **Oldfield, E.** 10 October 1946. Did you happen to see—Dr. Harry Plotz? *Times Herald,* Washington, DC, p 2.

68. **Plotz H.** 1938. La Rougeole Inapparente Chez le Singe (Macacus rhesus). *C R Hebd Seances Mem Soc Biol* **11:**141–142.

69. **Plotz H.** 1938. Culture *in vitro* du virus de la rougeole. *Bull Acad Med* **119:**598–601.

70. **Plotz H.** 1938. Culture des Ultravirus, p 1111–1151. *In* Levaditi C, Lepine P (ed), *Les Ultravirus des Maladies Humaines.* Librairie Maloine, Paris, France.

71. **Woodward TE.** 1998. *Make Room for Sentiment: a Physician's Story.* University of Maryland Medical Alumni Association and the Historical Society of Carroll County, Baltimore, MD.

72. **Reagan RL.** 1980. *One Man's Research: the Autobiography of Reginald L. Reagan.* Dan River Press, Stafford, VA.

73. **Plotz H.** 1947. Wartime army medical laboratory activities. *Am J Public Health* **37:**836–839 http:// dx.doi.org/10.2105/AJPH.37.7.836.

74. **Woodward TE.** 1964. Joseph E. Smadel, 1907–1963. *Trans Assoc Am Physicians* **77:**29–32.

75. **Smadel Archives, Walter Reed Army Institute of Research.** A. Rogalski, archivist. Smadel Reading Room, Forest Glen, MD.

76. **Virus Division, Headquarters First Medical General Laboratory.** 31 December 1943. Annual report for period June, 1943 through December, 1943. A.P.O. 519. *In* Smadel Archives, Walter Reed Army Institute of Research. A Rogalski, archivist.

77. **Gajdusek DC (ed).** 1976. *Correspondence on the Discovery and Original Investigations on Kuru: Smadel-Gajdusek Correspondence, 1955–1958.* National Institute of Neurological and Communicative Disorders and Stroke, National Institutes of Health, Bethesda, MD.

78. **Smadel JE.** 1948. The practitioner and the virus diagnostic laboratory. *J Am Med Assoc* **136:**1079–1081 http://dx.doi.org/10.1001/jama.1948.02890340005002.

79. **Francis T Jr (ed).** 1948. *Diagnostic Procedures for Virus and Rickettsial Diseases.* American Public Health Association, New York, NY.

80. **Smadel JE.** 1950. Laboratory diagnosis of viral and rickettsial diseases, a reappraisal after ten years. *Bacteriol Rev* **14:**197–200.

81. **Offit PA.** 2007. *Vaccinated: One Man's Quest To Defeat the World's Deadliest Diseases.* Harper Collins, New York, NY.

82. **Kurth R.** 2005. Obituary: Maurice R. Hilleman (1919-2005). *Nature* **434:**1083 http://dx.doi
.org/10.1038/4341083a.

83. **Hilleman MR, Werner JH.** 1954. Recovery of new agent from patients with acute respiratory illness.
Proc Soc Exp Biol Med **85:**183–188 http://dx.doi.org/10.3181/00379727-85-20825.

84. **Altman LK.** 12 April 2005. Maurice Hilleman, master in creating vaccines, dies at 85. *New York Times.*

85. **Lennette EH, Schmidt NJ (ed).** 1964. *Diagnostic Procedures for Viral and Rickettsial Diseases,* 3rd ed.
American Public Health Association, Inc, New York, NY.

86. **Lennette EH.** 1988. *Pioneer of Diagnostic Virology with the California Department of Public Health.* Oral
history conducted in 1982, 1983, and 1986 by S. Hughes. Regional Oral History Office, The Bancroft
Library, University of California, Berkeley, CA.

87. **California State Department of Health Services.** 18 November 2005. Dedication of memorial bust,
Edwin H. Lennette, M.D., Ph.D., Chief of the Viral and Rickettsial Disease Laboratory 1947–1978.
California State Department of Health Services, Richmond, CA.

88. **Smith MG, Lennette EH, Reames HR.** 1941. Isolation of the virus of herpes simplex and the
demonstration of intranuclear inclusions in a case of acute encephalitis. *Am J Pathol* **17:**55–68.1.

89. **Horsfall FL Jr, Lennette EH, Rickard ER, Andrewes CH, Smith W, Stuart-Harris CH.** 1940. The
nomenclature of influenza. *Lancet* **236:**413–414 http://dx.doi.org/10.1016/S0140-6736(00)98518-0.

90. **Merrill MH, Lennette EH.** 1953. Virus diseases and the public health. *Calif Health* **10:**105–109.

91. **Lennette EH, Emmons RW, Allen B, Forghani B, Grant K, Dennis J, Fox V, Hagens S, Martins
MJ, Schmidt V.** 1995. Nathalie Joan Schmidt. *Biography: Scientific Contributions in Virology.* California
Public Health Foundation, Berkeley, CA.

92. **Dalldorf G, Sickles GM.** 1948. An unidentified, filterable agent isolated from the feces of children with
paralysis. *Science* **108:**61–62 http://dx.doi.org/10.1126/science.108.2794.61.

93. **Rapp F.** 1994. Gilbert Dalldorf 1900–1979. *Biogr Mem Natl Acad Sci* **65:**95–105.

94. **Dalldorf G.** 1955. *Introduction to Virology.* Charles C Thomas, Springfield, IL.

95. **Horstmann DM, Hsiung GD.** 1965. Principles of diagnostic virology, p 405–424. *In* Horsfall FL,
Tamm I (ed), *Viral and Rickettsial Infections of Man,* 4th ed. J B Lippincott, Philadelphia, PA.

96. **Etheridge EW.** 1992. *Sentinel for Health: a History of the Centers for Disease Control.* University of
California Press, Berkeley, CA. http://dx.doi.org/10.1525/9780520910416.

6 A Torrent of Viral Isolates

THE EARLY YEARS OF DIAGNOSTIC VIROLOGY

> . . . the virologist must be just as much an epidemiologist and
> clinician when studying the effect of prevalent nonfatal viruses
> in man as he is a well grounded experimentalist or pathologist
> when studying similar effects in mice.
>
> Robert J. Huebner, 1957 (1)

INTRODUCTION

The demonstration by the Enders laboratory in 1950–1951 that cytopathic effect
(CPE) in tissue culture could be used to detect, isolate, and identify a virus from a
clinical specimen was the turning point for diagnostic virology (2, 3). There followed
in the 1950s and 1960s an outpouring of studies from a variety of federal, state, and
university-based hospital laboratories using tissue culture systems to isolate human
viruses. Many tissue culture types were found to have the capacity to isolate different spectra of viruses. In addition, neonatal and infant mice, easily maintained in the
laboratory, were found to be susceptible to certain viruses such as arboviruses and
coxsackieviruses. These tools were utilized to study a torrent of human viral isolates
and their relationships to human disease. In certain cases, no clear disease associations were found, producing what Robert J. Huebner of the U.S. National Institutes
of Health (NIH) termed "the virologist's dilemma" (1).

In parallel, the methodology, procedures, and personnel necessary to establish
and operate clinical diagnostic viral and rickettsial laboratories developed from

To Catch a Virus, Second Edition. Authored by John Booss and Marie Louise Landry.
© 2023 American Society for Microbiology. DOI: 10.1128/9781683673828.ch06

the pioneering work of Harry Plotz and Joseph Edwin Smadel at the Walter Reed Army Laboratory and from the work of Edwin Herman Lennette and Nathalie J. Schmidt at the State of California Laboratory in Berkeley (4). Many individuals and laboratories made important contributions to clinical virology while advancing new methods and training a cadre of diagnostic virologists.

DIAGNOSTIC VIROLOGY IN UNIVERSITY HOSPITAL LABORATORIES

Public health virology laboratories such as those run by Lennette in California and Gilbert Dalldorf in New York State served epidemiological purposes, alerting clinicians and public health officials to the presence of infections in the community. These laboratories served also to introduce and refine techniques in diagnostic virology that would become essential to hospital laboratories charged with the care of individual patients at the time of their illness (4). Hospital bacteriology laboratories were well positioned to provide that function, yet in the early years of clinical virology, from the 1940s into the 1960s, technical expertise in diagnostic virology was still highly specialized, and there was little crossover of personnel between bacteriology and virology laboratories. However, efforts to bridge the gap between public health and patient care started in hospitals associated with university schools of medicine. An early example was the Division of Virology at the Children's Hospital of Philadelphia (CHOP) at the University of Pennsylvania under Werner and Gertrude Henle.

The Henles were best known for their work linking Epstein-Barr virus with infectious mononucleosis, Burkitt's lymphoma, and nasopharyngeal carcinoma. Yet they were also among the early investigators working on clinical problems such as influenza and mumps. Werner Henle, born in Dortmund, Germany, in 1910, sprang from a distinguished medical and scientific heritage. His father was a surgeon and his grandfather was Jakob Henle, the anatomist and, important to the present context, the originator of the Henle-Koch postulates on the etiology of infectious diseases. Werner Henle received his medical training and graduated from the University of Heidelberg in 1934 and did his internship in that city (5). He immigrated to the United States in 1936 during that dark period of Germany's history and became an instructor in the Department of Medical Microbiology at the University of Pennsylvania. Gertrude Henle, née Gertrude Szpingier, also received her medical training in Heidelberg, where she met Werner and graduated in 1936. A year after Werner, she came to the United States, where they were married; she, too, joined the Department of Medical Microbiology at the University of Pennsylvania (6).

Werner Henle's (Fig. 1) interest in research in virus diseases began in 1940, when he learned tissue culture of chicken cells (5). He established a virology laboratory at CHOP, where he and Gertrude Henle studied a wide range of basic and clinical virological questions. Several studies refining the complement fixation assay for influenza and mumps were published (7, 8). The latter 1940s were still an

FIGURE 1 *Werner and Gertrude Henle of CHOP. Werner Henle was the grandson of Jakob Henle, the anatomist. Werner and Gertrude Henle were trained in medicine in Germany and immigrated to the United States in 1936 and 1937, respectively. At CHOP, where Werner established a virology laboratory, they were affiliated with the University of Pennsylvania. They had long and productive careers in which they made contributions to basic and diagnostic virology. (Courtesy of the Fritz Henle estate.)*

era, as Frank Horsfall put it, in which it was "highly impracticable to attempt the direct recovery of such agents from more than a relatively small number of cases." *In vitro* serological techniques such as complement fixation and hemagglutination inhibition were preferred over neutralization tests, which were performed *in vivo* using infectious virus and intact animals (9).

The Henles' report on the diagnosis of mumps in an epidemic of the fall of 1946 to the spring of 1947 (10) described the inoculation of 8-day-old chicken embryos for viral isolation and clinical skin testing as well as three assays for serological diagnosis: hemagglutination inhibition, agglutination of mumps virus antigen-coated red cells, and complement fixation. In the complement fixation assay, two antigens were prepared, a viral antigen and a soluble antigen. Antibodies to the soluble and viral antigens developed with different kinetics, allowing an estimate to be made of the chronological development of antibodies after mumps virus infection (7). Such studies represented the state of the art just before the start of the cell culture era. In the case of mumps, antibodies were particularly important, as the Henles noted, to diagnose complications of mumps virus infection such as meningoencephalitis, orchitis, and pancreatitis in the absence of salivary gland involvement.

In 1953, Klaus Hummeler, who had received his medical degree from the University of Hamburg, joined the Henles as the Associate Director of the Virus Diagnostic Laboratory. Hummeler helped to develop immunoelectron microscopy as a diagnostic technique (11). As the cell culture era in diagnostic virology evolved in the 1950s, the Henles' lab at CHOP investigated the relative sensitivities of various cell types for virus isolation and characterization (12, 13). Later, in the 1960s, Stanley Plotkin at the Wistar Institute started to do diagnostic virology work for CHOP, an arrangement which was formalized in 1972 (S. Plotkin, personal communication). Harvey Friedman, a fellow in infectious diseases, spent 3 months in Lennette's lab at Berkeley learning diagnostic techniques and then together with Plotkin opened an official clinical diagnostic virology laboratory at CHOP in 1975. At the end of the 1980s, the laboratory started a conversion to an entirely molecular basis for diagnostic virology under Richard Hodinka.

At the Yale University School of Medicine, Gueh-Djen (Edith) Hsiung entered the field of virology as a postdoctoral associate in research under Joseph L. Melnick in 1953 (14). For more than five decades, she made fundamental contributions to the development of diagnostic virology. These contributions included the development and refinement of cell culture techniques for virus isolation, identification, and quantitation. She also trained postdoctoral fellows and medical house staff, many of whom went on to direct viral diagnostic laboratories all over the United States. Edith Hsiung was known for her publication of four editions of a diagnostic virology manual, implementation of an intensive laboratory diagnostic virology course for visiting scientists, and creation of a national virology reference laboratory for the Department of Veterans Affairs. To her trainees and colleagues, she was best known for her insistence on virus isolation as the "gold standard" in diagnostic virology.

Hsiung was born in Hupei, China, in 1918 and received her bachelor of science degree in biology from Ginling College in 1942 (14). She would have gone to medical school in China except that the Peking Union Medical College closed during World War II. Instead, she worked in the Epizootic Prevention Bureau on veterinary vaccines. A predictive story from those years, demonstrating her work ethic and resourcefulness, concerned the transport of stock rinderpest vaccine virus. In the absence of refrigeration to transport the vaccine, she inoculated a goat, fed it, and sat with it during a 27-day transport by truck. She came to the United States following World War II and pursued graduate studies at Michigan State University, receiving her doctorate in microbiology in 1951. To pay medical bills for surgery to fuse a congenitally dislocated hip, Hsiung worked for 2 years at the Wene Poultry Laboratory in New Jersey, where she developed the first vaccine for infectious bronchitis virus in chickens. Then, after being told at age 35 that she was too old to attend medical school, she accepted a research position at Yale University in the Department of Preventive Medicine at the Yale Poliomyelitis Study Unit to work with Joseph Melnick.

The years with Melnick, from 1954 to 1960, prepared Hsiung well for her later work in diagnostic virology. Their first publication in 1955 simplified the cell culture plaque technique for viral quantitation introduced 3 years earlier by Renato Dulbecco (15). Working with enteroviruses, they demonstrated viral plaque formation in stoppered bottles, thus replacing unsealed petri dishes and the need for humidified CO_2 incubators (16). Cultured cells grew well in the closed bottles and survived for long periods under an agar overlay, allowing the study of slowly growing viruses. This technique offered a means to identify mixed viral infections by revealing plaques with different morphologies after various periods of incubation.

In her ensuing work with enteroviruses, Hsiung pioneered the recognition of plaque morphology and comparative cell susceptibility as a means for virus diagnosis and identification. Hsiung and Melnick studied the factors influencing viral growth in cell culture, differences in various types of cells, and alteration of virus characteristics on passage in mice after cell culture adaptation (17). Cell culture was to become the central feature of Hsiung's diagnostic virology work after she had studied virus-cell interactions intensively during the half dozen years with Melnick. It was a crucial preparation.

Hsiung's career at Yale-New Haven Hospital began in 1960, when she was appointed by John R. Paul, on the recommendation of Robert H. Green, as Director of the Diagnostic Virology Laboratory, working with Dorothy Horstmann (18) (Fig. 2). Green was an infectious diseases specialist who had worked with Joe Smadel at the Rockefeller Institute, publishing one of the first electron micrographic studies of vaccinia virus. He was to continue to play a fundamental role in Hsiung's career. She joined Green briefly at the NYU-Manhattan VA Medical Center (VAMC), from 1965 to 1967, before returning with him to a Yale University School of Medicine appointment in New Haven and at the VAMC in West Haven, CT. It was at the VAMC in West Haven that much of Hsiung's important work in developing the field of diagnostic virology was accomplished, including setting up the National Virology Reference Center in the VA in 1984. Internationally, she assisted in the establishment of labs in Caracas, Venezuela; in Basel, Switzerland; and in Taiwan (14).

Hsiung contributed to the advancement of diagnostic virology with numerous publications concerning the functions of the diagnostic lab and individual reports on viral isolates. Her papers included instructive, state-of-the-art reviews of diagnostic laboratory organization, techniques, and findings (19). In a recollection

FIGURE 2 *G.-D. Hsiung in the Section of Epidemiology and Preventive Medicine at Yale University, 1959. Seen at the end of the front row on the right, she stands next to Robert Green, her long-time colleague. Next to him is John Paul, the chairman and polio scholar. Dorothy Horstmann, with whom Hsiung established the diagnostic virology laboratory at the Yale New Haven Hospital a year later, is in the front row, third from the left. (Yale University, Harvey Cushing/John Hay Whitney/Medical Library.)*

in 1980, she discussed the marked changes in clinical virology in the preceding 20 years (18). By charting the number of publications in *Index Medicus*, she demonstrated the fall of interest in polioviruses following the development of the vaccine, a stable level of interest in influenza viruses presumably due to their annual occurrence, the burgeoning of interest in the hepatitis viruses beginning in the 1970s, and the steady rise in attention to the herpesvirus group over the two decades.

Reports on viral isolates included both the association of viral agents with particular clinical syndromes and also the discovery of unsuspected agents in cell cultures. With W. H. Gaylord, Hsiung described the vacuolating virus of monkeys in 1961 (20). It turned out to be simian virus 40 (SV40) (21), a virus of significant interest. The finding by Hilleman's group that SV40 caused tumors when inoculated into newborn hamsters coupled with the finding of SV40 contamination of both the Salk and the Sabin polio vaccines caused great public health concern. That was ameliorated by follow-up studies, which found no increased tumor incidence 30 years after polio immunization (22).

Hsiung became an acknowledged expert on viral latency and persistence in primate tissues (23). These observations came at a time when there was great clinical interest in persistent viral infections as the cause of certain chronic diseases. She developed several animal models of human diseases, including transplacental transmission of cytomegalovirus (CMV) in the guinea pig as a model of congenital human CMV infection. As the field of antiviral treatment emerged, Hsiung used animal models to explore the usefulness of antiviral compounds.

Another area of contribution to the development of diagnostic virology was educating professionals in the requisite viral laboratory techniques. "It soon became apparent that knowledge of, and experience with, methods for the recognition and characterization of individual virus groups were confined primarily to research laboratories. Consequently, the idea of a postgraduate course in 'Diagnostic Virology' was conceived and the first session was offered in 1962 at the Yale Medical School" (18). The intensive 2-week workshops were offered annually or every other year, usually at her home institution but also at other locations such as the American Type Culture Collection and, in later years, in China and Taiwan. To facilitate teaching the course in 1964, she published *Diagnostic Virology* in collaboration with J. R. Henderson (24). It went through four editions as the field advanced and new techniques were adopted. The fourth and final edition was published in 1994 as *Hsiung's Diagnostic Virology as Illustrated by Light and Electron Microscopy* with C. K. Y. Fong and M. L. Landry (25). Published originally as a manual for her course, the book became widely used as a practical handbook in the laboratory. The text included a straightforward method of characterizing isolates based on physical and chemical properties. It not only served as a practical guide but also emphasized the basic characteristics of viruses. Hsiung accepted numerous fellows as postdoctoral trainees and successfully funded an NIH contract for the training of fellows in diagnostic virology (Fig. 3). In later years, with the receipt of numerous awards, she became an ambassador for diagnostic virology. The Pan American Society for Clinical Virology established an annual award in her honor.

THE BEGINNINGS OF DIAGNOSTIC VIROLOGY AT THE CDC

The Communicable Disease Center (CDC; now called the Centers for Disease Control and Prevention) was established in 1946, succeeding a unit of the U.S. Public Health Service based in Atlanta, which had been set up to fight malaria and was called Malaria Control in War Areas (26). The CDC's first director, Joseph W. Mountin, saw the CDC as serving the public health needs of the states through various centers of expertise. He viewed epidemiology as crucial, as an "early-warning system." Alexander Langmuir took charge of the Epidemiology Division in 1949 and by 1950 had established the Epidemic Intelligence Service for surveillance of communicable diseases. Seward Miller, who was put in charge of the Laboratory Division, saw its role ". . . to assist the CDC's Epidemiological Division in emergencies, do research on diagnostic techniques, act as a reference diagnostic center for difficult specimens, and offer training for the states' laboratory technicians." Miller hired Morris Schaeffer, an experienced virologist and infectious diseases specialist, to direct the virus laboratory.

Schaeffer was born in Ukraine and came to New York City at 5 years of age (27). He did undergraduate work at the University of Alabama and earned a Ph.D. in microbiology from New York University in 1935 and an M.D. from the same institution in 1944. Schaeffer took over the virus and rickettsial lab at the CDC in 1949. Before undertaking that task, he had been on the faculty at Western Reserve University and directed the contagious diseases service at the municipal hospital in Cleveland. He was later to write about ". . . how well this laboratory and clinical experience had prepared me for the job at CDC" (M. Schaeffer, unpublished writing, provided through the courtesy of W. Dowdle). He is best remembered for his role in cracking down on irresponsible medical labs in his role as Director of the Bureau of Laboratories in New York City from 1959 to 1971 and for assisting Senator Jacob Javits in crafting the Clinical Laboratory Improvement Act (CLIA) of 1967 (26, 27).

The virology laboratory at the CDC was established on a temporary basis in the outskirts of Montgomery, AL, because the CDC buildings in Atlanta were considered unsafe for virology (Schaeffer, unpublished). The location was at the site of a rabies unit set up by the state with the assistance of the Rockefeller Foundation (28). That work had been led by Harald Johnson and Charles Leach. In yet another convergence in the early history of diagnostic virology, Reginald Reagan, who was to work with Harry Plotz in setting up the first diagnostic virology laboratory at Walter Reed, had helped open the Rabies Study Laboratory with Leach in 1936 (29). Reagan stayed on in Montgomery for 4 years before being assigned to

FIGURE 3 G.-D. Hsiung with the 1979 diagnostic virology class. Given annually or biannually for many years, the diagnostic virology class was an intensive 2-week course with lectures and hands-on laboratory sessions. The photograph shows Hsiung (front row, center) and Kenneth McIntosh to her left, with students, faculty, and staff at the VAMC, West Haven, CT. The authors of the first and second editions of this book are pictured: John Booss (second row, right), Marilyn J. August (third row, left), and Marie L. Landry (last row, third from left). (Personal collection, Marilyn J. August.)

Walter Reed. After the rabies work was completed, the state deeded the land and facilities to the U.S. Public Health Service (USPHS), which established the virology labs of the CDC. Prior to Schaeffer's arrival in Montgomery, Seward Miller had hired Beatrice Howitt "to take charge of the fledgling lab" (Schaeffer, unpublished). Howitt had had a productive career in virology to that point, including the first human isolation of western equine encephalitis while working at the Hooper Foundation of the University of California at San Francisco (30). She stayed on at the CDC after Schaeffer took leadership of the lab.

Schaeffer was confronted with major facility issues, including water, sewage, incineration, and air conditioning for animal quarters. "It took a lot of persuasion to convince the bureaucrats in those days that it was cheaper in the long run to provide proper housing and air conditioning for mice which otherwise died in large numbers" (Schaeffer, unpublished). He noted ironically that it took even longer to make the argument for air conditioning for the humans. One can envision that this early experience establishing the fundamental needs for laboratory operations prepared him well for his later work on laboratories in New York City and to counsel Senator Javits for the CLIA legislation. He was pleased with the collaborations between entomologists and veterinarians in his unit, which resulted in field and laboratory studies to define the arboviral life cycle. Collaborations with visiting scientists were

FIGURE 4 *Chen Pien Li and Morris Schaeffer (seated) of the CDC. They are shown conducting polio research at the CDC location in Montgomery, AL, during a 1953 study. Schaeffer was the first director of CDC's virology labs from 1949 until 1959, when they were still in Montgomery. In 1959 the labs were moved to Atlanta, GA. (Courtesy of CDC-PHIL [#2442], 1953.)*

productive and included the work by R. A. Goldwasser and R. E. Kissling in 1958 on fluorescent antibody detection of rabies, which became the standard method (31).

Schaeffer was particularly pleased to recruit Chen Pien Li (Fig. 4), who had worked with Thomas Rivers at the Rockefeller Institute on vaccinia virus and with Karl Habel at the NIH on adapting poliovirus type 3 to mice. At the CDC, Li produced a type 1 polio variant in mice which would ultimately be further passaged by Albert Sabin for use in an oral vaccine. The CDC was one of the collaborating laboratories in the Salk polio vaccine trial.

With rising international prominence based on continued contributions to public health virology, the CDC lab was designated the WHO Influenza Reference Laboratory for the western hemisphere during the 1957 influenza epidemic. Walter Dowdle (Fig. 5) joined the CDC as a virologist in 1961 and would later

FIGURE 5 *Walter Dowdle of the CDC. A distinguished virologist, Dowdle was with the CDC for 33 years and is a former deputy director. Among Dowdle's scientific interests are influenza, polio, HIV, and malaria. (Courtesy of CDC-PHIL [#8374], 1986.)*

lead the influenza collaborating center from 1968 to 1979 and also serve as Deputy Director of the CDC from 1987 to 1994 (32, 33).

Somewhat surprisingly, Schaeffer had a series of chiefs of the viral diagnostic unit: Tom Hughes, Michael Sigel, Seymour Kalter, and Andrew Fodor. Schaeffer attributed the high turnover to the large number of diagnostic specimens, which reduced the opportunities for research (Schaeffer, unpublished). In 1950, the CDC inaugurated what appeared to be the first diagnostic virology course offered anywhere (34). The course was attended by scientists from all over the United States, and the outcome was an increased number of states offering diagnostic virology services with individuals exposed to the rigors and constraints of a diagnostic laboratory practice.

Schaeffer admonished his staff that a diagnostic lab had not "arrived" until taken notice of by Smadel at Walter Reed and by Lennette in California (Schaeffer, unpublished). Apparently, Schaeffer succeeded; Smadel was impressed with the polio and arboviral work, and Lennette sent his staff to observe the CDC's work on rabies and encephalitis. The diagnostic work at the CDC must have been a highly informative experience, for Schaeffer, some years later, was to write a masterful, state-of-the-art review of the laboratory diagnosis of viral and rickettsial diseases (35).

Schaeffer left the CDC in 1959 to assume the position of Assistant Commissioner of Health and Director of the Bureau of Laboratories in the City of New York. He had started in the New York laboratories, and he returned to undertake the challenge of rebuilding their public laboratories. Yet he called the years at the CDC ". . . the most productive and gratifying of my career; the happiest time of my life" (Schaeffer, unpublished). When the virology labs moved into permanent quarters in Atlanta in 1960, leadership of the Virology Division was taken over by Telford H. Work (26). Work was an outstanding arbovirologist from the Rockefeller Foundation who was noted for describing Kyasanur Forest disease and its virus. He continued arboviral studies at the CDC, including Venezuelan equine encephalitis, established the CDC Arbovirus Unit as a WHO Collaborating Center, and recruited several scientists who would make significant contributions to the field. He remained at the CDC until 1967, when he was appointed the head of the Division of Infectious and Tropical Diseases at the UCLA School of Public Health (36).

THE LID AT THE NIH

From the latter 1940s, scientists at the Division of Infectious Diseases, which became the National Microbiological Institute and then the National Institute of Allergy and Infectious Diseases (NIAID), produced a remarkable series of studies of rickettsial and viral diseases. Within the NIAID, the Laboratory of Infectious Diseases (LID) played a vital role in defining disease entities (R. Chanock, 1996, NIAID, overview, LID, report to the Board of Scientific Counselors). Its first chief in 1948, Charles Armstrong (Fig. 6),

FIGURE 6 *Charles Armstrong of the NIH. Armstrong made numerous contributions to virology, including the understanding of polio, St. Louis encephalitis, and lymphocytic choriomeningitis. He was the first chief of the Division of Infectious Diseases at the NIH. Here he established the philosophy of the LID of fully working out an infectious disease process from agent isolation through prevention. (Courtesy of the National Library of Medicine.)*

had earlier been part of the team which had isolated St. Louis encephalitis virus during the outbreak in St. Louis in 1933 (37, 38). In studies subsequent to that epidemic, Armstrong isolated lymphocytic choriomeningitis virus, defined its association with the syndrome of aseptic meningitis, and demonstrated the source of the infection in mice (39).

Armstrong established the LID philosophy of working out an infectious disease fully, from the isolation of the agent to definition of the epidemiology and the clinical syndrome(s) and to preventive measures, such as development of vaccines. Armstrong was followed as LID Chief by Karl Habel, Dorland Davis, and Robert Huebner. Habel, an international expert on rabies (40), made fundamental contributions to the understanding of mumps (41) and rubella (42) before moving into studies of viral carcinogenesis (43). Davis ultimately served as the Director of the NIAID from 1964 to 1975. However, it is the work of Huebner and the array of brilliant virologists that he attracted to the burgeoning field of medical virology on which we focus here.

Huebner's first scientific success, and it was a major one, was immortalized by *The New Yorker* magazine writer Berton Roueché in "A Reporter at Large: the Alerting of Mr. Pomerantz" (44), later collected in *Eleven Blue Men and Other Narratives of Medical Detection* (45). *The New Yorker* essay concerned an outbreak in 1946 of a hitherto-unknown rickettsial disease, rickettsialpox, in Kew Gardens, Queens, in New York City. The infectious agent, *Rickettsia akari*, was transmitted by mites from house mice (46–48). The story demonstrated a number of Huebner's qualities, including a capacity to untangle complex problems, the ability to organize a team of specialists with diverse talents, and unselfish willingness to apportion credit where it was due. The Mr. Pomerantz of the title was an exterminator and self-taught entomologist who had pointed out the proliferation of mites and the possibility that they might be associated with the disease. Huebner included him as a coauthor of the report in which the association of the infectious agent with a rodent mite was identified (47). In *The New Yorker* piece, Huebner is reported as saying, "Well Charlie, we've made it." Pomerantz related, "I was suddenly stricken dumb. . .That 'we' included *me!*" (44) At the end of Huebner's career in 1982, Robert Chanock, who had followed Huebner as Chief of the LID, said, ". . .you demonstrated that one could lead without compromising the self-esteem or dignity of your associates" (49). This was high praise, indeed, for Huebner had gathered around him outstanding virologists, including Chanock, Wallace Rowe, Janet Hartley, Leon Rosen, Albert Kapikian, Robert Purcell, Robert Parrott, and others, as well as the outstanding epidemiologist Joseph Bell.

Huebner (Fig. 7) was born and raised in Ohio and endured the hard times of the Depression. In medical school he had to work outside of his studies, against

FIGURE 7 *Robert J. Huebner of the NIH. Although without formal scientific research training, Huebner was hired by Charles Armstrong at the NIH. Huebner quickly demonstrated a remarkable capacity to grasp the fundamental concepts of epidemiology and laboratory virology and their application to human viral diseases. He became the leader of the LID. A man of diverse interests, he is shown here on his farm with a prize Angus bull. (Courtesy of the Office of History, NIH.)*

the regulations of St. Louis University. According to Edward Beeman in his biography of Huebner, employment at one point included being a "bouncer" in a brothel (49). However, that information had been provided by Huebner himself, and Beeman elsewhere commented on Huebner's capacity as an entertaining raconteur. After medical school and internship, he was commissioned in the Public Health Service Reserve. Assigned to ship's duty in Alaska, he later applied for and won appointment in the Regular Commissioned Corps of the USPHS in the Ear, Nose and Throat Dispensary in Washington, DC. While the clinic was not to his liking, he managed to meet Charles Armstrong, who was the Chief of the Division of Infectious Diseases of the NIH. Remarkably, although Huebner had had no formal research training, Armstrong was to give him an appointment in the Division of Infectious Diseases in November 1944. He was assigned to a rickettsia laboratory to study Q fever.

The call to Kew Gardens, Queens, came shortly after Victory in Japan Day (V-J Day), and the end of World War II, 2 September 1945. According to the *New York Times*, Huebner was the only officer on duty in his institute because "senior officers were taking their first extended breaks in four years" (50). After defining the Kew Gardens outbreak as rickettsialpox, he and his colleagues investigated an outbreak of Q fever at the NIH. He was next assigned to evaluate a California outbreak of Q fever in which human cases were associated with dairy cattle. Finding the rickettsia in unpasteurized milk (51), Huebner apparently became *persona non grata* with the dairy farmers (50). Soon thereafter, his studies of viral diseases were to begin. Among the first were studies on coxsackieviruses in collaboration with his future biographer Edward Beeman.

Beeman came to the LID on the invitation of Charles Armstrong in 1948 after an internship in internal medicine (37) and was assigned to work with Huebner. In the same year, Gilbert Dalldorf and Grace Sickles reported the isolation of a virus from fecal specimens from two children with muscle weakness (52). This agent was isolated by inoculation of suckling mice, and neutralizing antibody was found in each patient. The virus came to be called coxsackievirus, after the small Hudson River village south of Albany, NY, where the children lived. Dalldorf and Sickles were circumspect in pathogenic claims for the virus, stating, "That it induces paralysis in man is unproven." Other reports associated the virus with a variety of illnesses.

In August and September 1949, Huebner and a team including Beeman investigated a small outbreak of a febrile illness in a Maryland suburb called Parkwood (53). Eight patients were involved, mostly children age 5 or younger. The illness was of brief duration, lasted an average of 3 days, and left no residua. Since a coxsackievirus was found repeatedly, a survey of the community was undertaken. Persistent viral shedding in stool was found for the eight patients originally studied as well as from three other persons living in their households. Five persons in 75 other households in the survey were found to be shedding virus, only one of whom was found to have been ill in the August-September time period. No cases of poliomyelitis or meningitis were found in the community at that time. The authors thus counseled caution in ascribing a causal relationship in sporadic associations of virus and illness. That cautionary note remains valid, yet another lesson was to be learned the following summer in Parkwood.

Six cases in households in proximity to each other in Parkwood involved symptoms similar to the cases of the previous year, yet a defining observation was made by a colleague. After his daughter reported a sore throat, small ulcers were observed on her tonsils and soft palate. This observation was repeated in subsequent cases and came to define the coxsackie A syndrome of herpangina (54). The roles of coxsackieviruses A and B in the genesis of herpangina and pleurodynia, respectively, were subsequently recognized (55, 56), yet the cautionary problem of assigning causal relationships remained. It was a topic to which Huebner would return in "The Virologist's Dilemma" (1). Much of Huebner's subsequent work in virology was undertaken with Wallace Rowe.

Rowe was born in Baltimore in 1926. During World War II, he completed an accelerated, combined undergraduate and medical education, receiving an M.D. from Johns Hopkins in 1948 (49). Following an internship and fellowship at Bowman Gray School of Medicine, he was assigned as a naval officer to the National Naval Medical Center. He worked with Eric Traub on the pathogenesis of lymphocytic choriomeningitis virus, which years later they would help classify as part of a taxonomic group, the arenaviruses (57). Rowe moved to the Huebner laboratory (Fig. 8) at the LID as a USPHS commissioned officer to take the place of Edward Beeman, who returned to training in internal medicine (49).

Karl Habel, a previous director at the LID, had sent Alexis Shelokov to John Enders' lab to learn tissue culture techniques for cultivation of viruses. Shelokov instructed Rowe in tissue culture systems which he then set up for Huebner's unit (reference 49; E. Beeman, Oral History Project, courtesy of the Office of History, NIH).

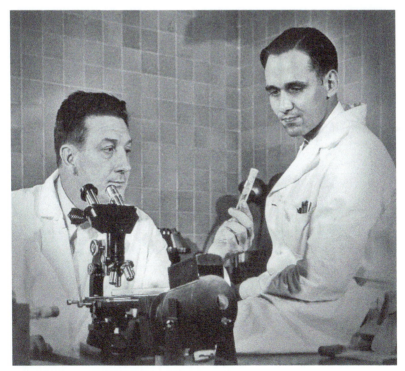

FIGURE 8 *Robert J. Huebner (left) and Wallace Rowe of the NIH. Rowe made numerous contributions to human virology and to experimental viral oncology. Among these contributions were the isolations of adenovirus and CMV. (Courtesy of the National Library of Medicine.)*

The application of tissue culture was to prove extremely productive. Rowe had arrived in Huebner's laboratory in August 1952 and by the following year had described the isolation of an "adenoid degeneration" agent, the first of the adenoviruses to be recognized (58). The origin of the observations was a study of human adenoidal cell growth in roller tubes. Adenoid tissues removed during routine tonsillectomies had been obtained from the Children's Hospital in Washington, DC, and the U.S. Naval Hospital in Bethesda, MD, during the winter of 1952–1953. Spontaneous rounding of cells was observed after 8 to 28 days of establishment of adenoid culture. The CPE was transmissible to other types of cell cultures. Very soon thereafter, Maurice Hilleman and J. H. Werner, investigating an outbreak of acute respiratory disease at Fort Leonard Wood in Missouri, isolated a cytopathogenic agent in HeLa cells (59). In an addendum to that report, they noted that tests in both labs found an immunological relationship between their isolated agent and that found by Rowe et al. Because of the anatomic associations that were found subsequently, these agents were originally called adenoidal-pharyngeal-conjunctival (APC) viruses and were later renamed the adenovirus group (60).

The spontaneous emergence of another cytopathology in cell cultures from surgically removed adenoids yielded yet another virus. Rowe, Janet Hartley, and their collaborators formed one of three groups to report the first isolations of another major human virus, CMV (61). Working with adenoidal tissue, Rowe and Hartley observed spontaneous degeneration of the monolayers after 22 to 51 days in epithelial cells in culture. In cultures which had not shown adenovirus CPE, a different type of CPE was observed in fibroblasts and could be passaged to other human fibroblast cultures but not to nonhuman cell lines or to mice or rabbits. This was to be later recognized as the species specificity of CMVs. In unfixed preparations, the changes were clearly localized foci of oval or rounded cells which progressed to a necrotic mass. In fixed preparations, large eosinophilic intranuclear inclusions, with a halo and chromatin margination, were found on hematoxylin and eosin staining. Giemsa-stained preparations occasionally demonstrated one or two basophilic bodies on the nuclear membrane or in the halo area.

Rowe and Hartley isolated three strains of CMV, including the prototype, AD 169. Comparison of the growth characteristics and antibody responses of AD 169 with viruses isolated separately by Margaret Smith and by Thomas Weller found them to be similar. Rowe et al. commented, "The value of growing tissue in culture for unmasking indigenous cytopathologic agents is exemplified here as in similar studies of APC viruses."

Janet Hartley joined the Huebner lab as a technician in 1953, when she was a graduate student and worked with Wallace Rowe (Janet Hartley, oral history interview, courtesy of the Office of History, NIH). She was to do her doctoral thesis on CMVs and stayed on at the NIH as a scientist with Rowe and Huebner. Their interests evolved to indigenous viruses of mice and oncogenic viruses, starting with polyomavirus, and the genetics of tumor viruses. Huebner's interest in viral oncogenesis and the viral oncogene hypothesis (62) would stimulate his move to the National Cancer Institute (NCI) in 1968 to direct the viral oncogenesis program (50). Rowe and Hartley would remain at NIAID, but their close collaboration with Huebner continued. Rowe passed away in July 1983 of cancer (63). Hartley would comment that Rowe's illness "was devastating for all of us" (Hartley, oral history), and she characterized him as ". . . tremendously knowledgeable, a very, very brilliant mind. He understood genetics and mathematics and statistics and all aspects of medicine and virology."

Robert Chanock (Fig. 9) was working at Johns Hopkins when Huebner invited him to join him at the NIH. Chanock had already isolated a number of respiratory viruses, the importance of which would emerge from his body of work at the LID. He would go on to become a highly productive virologist, take over the reins of leadership of the LID when Huebner moved on to the viral oncology program at the NCI, and collaborate with several scientists in his group who were to make major contributions to diagnostic virology. In retrospect, Chanock's early experiences as a pediatric house officer at the University of Chicago and his initial training in virology were to provide the motivation and the scientific foundation for his later successes.

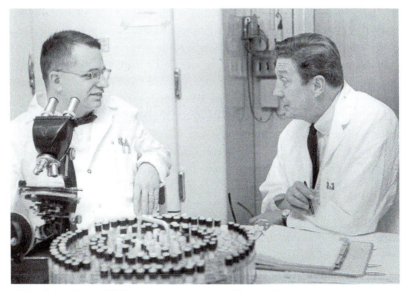

FIGURE 9 *Robert Chanock (left) and Robert J. Huebner of the NIH. Chanock made major contributions to the understanding of human viral respiratory disease, starting with the isolation of RSV. On Huebner's move to the NCI, Chanock took over as the leader of the LID. (Courtesy of the Office of History, NIH.)*

As Chanock described, ". . . the defining moment that set the direction of my career occurred during my residency" (Robert Chanock, oral history interview, courtesy of the Office of History, NIH). During his residency in pediatrics, he cared for the daughter of a colleague, who had severe croup, required tracheotomy, and almost succumbed. He realized that very little was known about viral croup, otherwise known as acute laryngotracheobronchitis, as well as about viral pneumonia and bronchiolitis. He characterized these three illnesses as ". . . the triad of serious lower respiratory tract disease in infants and children." He was to add considerably to the understanding of the causes of those illnesses within a few years. Chanock also noted the lack of knowledge of nonbacterial diarrheal disease: ". . . there was a total void in our understanding of the etiology of pediatric enteric and respiratory infectious diseases." He credited his residency director, Howell Wright, with observing that he needed more discipline and for sending him to Albert Sabin's lab in Cincinnati in 1950.

Sabin, known best for his determined development of the attenuated oral polio vaccine, was a highly productive virologist with a long list of accomplishments in addition to his work with poliovirus. Born in 1906 in Bialystok, Russia (now Poland), he came with his parents to the United States in 1921 (64). Apparently, an epidemic of polio in New York City the summer after graduation from the Medical School at NYU influenced his decision to study polio and other infectious diseases. He had already demonstrated precociousness, setting up a system for typing pneumococci while in medical school. During his internship at Bellevue Hospital in New York City, he isolated herpes B virus of monkeys from a colleague who had a fatal infection (65). Subsequent work included a stint at the Lister Institute in London and a return to the United States, where he worked on poliovirus at the Rockefeller Institute with Peter Olitsky. He moved to the University of Cincinnati in 1939, to which he returned after the war. During World War II, he served in the Army Medical Corps working on several infectious illnesses, including dengue fever, sandfly fever, Japanese encephalitis, and toxoplasmosis. Sabin was recognized for his wide range of contributions by election to the National Academy of Sciences in 1951, at the time Chanock joined his lab.

Chanock characterized Sabin as "severe and demanding," stating that he "monitored everything" but that they apparently "hit it off," and Chanock was able to survive Sabin's rigorous demands (Chanock, oral history). They studied the hemagglutinin of St. Louis encephalitis virus (66). Work in Sabin's laboratory was interrupted by service in the Army. Chanock was assigned to the 406 General Medical Laboratory in Tokyo and studied insect-borne encephalitis viruses. On his return to Cincinnati from the military, he was pressed by Sabin to choose a research field different from Sabin's. Chanock responded that he would study lower respiratory tract viruses, based on his experience during his internship. He was fortunate to have access to clinical material through the Children's Hospital Research Foundation (Chanock, oral history).

He was soon to have success, isolating two viruses associated with croup in 2 of 12 infants (67). The CPE in monkey kidney cells was first seen as focal syncytia and then a sponge-like appearance due to small vacuoles and increased granularity. Five of the twelve infants developed antibodies as measured by neutralization, hemagglutination inhibition, and complement fixation. The croup-associated (CA) viruses were noted to have properties of the myxovirus group, which shared hemadsorption (HA) as a diagnostic feature in tissue culture (68). HA had been first described by Vogel and Shelokov for the diagnosis of influenza in monkey kidney tissue culture (69) but had been extended to other viruses, including CA viruses (70).

In 1956, Chanock left Sabin's lab to take a position at Johns Hopkins, where he set up a study at the Baltimore City Hospital. As Chanock was to put it later, "it was a home run with the bases loaded" (Chanock, oral history). They isolated the first human strain of respiratory syncytial virus (RSV), "which proved to be the most important single cause of serious viral lower respiratory tract disease in infants and children worldwide." Throat swabs were obtained from sick infants and inoculated into three types of tissue culture. "An unusual cytopathogenic agent" was isolated in cell cultures from one infant with bronchopneumonia and one with laryngotracheobronchitis (71). The predominant cytopathology was characterized by syncytium formation. The virus would not agglutinate chicken or human O red blood cells, nor did it grow in embryonated eggs or 1-day-old mice. It was found to be the same as a cytopathogenic agent which had been recently isolated from a chimpanzee at the Walter Reed Army Institute of Research, the chimpanzee coryza agent, and associated with a respiratory illness in a laboratory worker (72). The agent was soon to be named RSV.

Robert Parrott, a pediatrician who had worked in the LID, left for Children's Hospital in Washington, DC, at the same time that Chanock arrived at the LID, in July 1957. They would form a productive collaboration. Surveillance for RSV was set up using complement fixation antibody testing in children with acute lower respiratory tract disease (73). Later in the surveillance effort, viral isolations were improved by immediate inoculation of clinical specimens into culture, rather than after a freeze-thaw step, as had been reported by others (74). Isolations were most common for infants under 7 months of age. RSV was found in 42% of patients with bronchiolitis, 24% of patients with pneumonia, and 12% of infants and children with febrile respiratory disease who did not merit hospitalization. One percent of controls were found to have the agent.

A study of an outbreak of febrile illness and pneumonia in the nursery group (6–50 months of age) at Junior Village, a children's home in Washington, DC, was undertaken by Albert Kapikian and colleagues at the LID (75). Eighty percent of young children in the infirmary developed pneumonia, as did 25% of children in the nursery group housed together in a cottage. Overall, RSV seroconversion was found in 91% of the study group. Thus, RSV was found to be highly contagious and a major cause of respiratory disease in young children. Like Chanock, Kapikian had arrived at the LID in July of 1957 at Huebner's invitation (49). He was to make major contributions, especially in immunoelectron microscopy of viral gastroenteritis, which is discussed in detail in the following chapter.

While at the LID, Chanock and his collaborators were able to demonstrate the association of HA viruses with croup, pneumonia, bronchiolitis, pharyngitis, and febrile illness in a large cohort of hospitalized children (76). Noting the proliferation of names for CA and HA and related viruses with similar properties, it was suggested by an international committee that they be called parainfluenza viruses (77). Hence, one finds the original CA virus of Chanock called an HA virus and then parainfluenza virus type 2 (49).

Chanock became the Head of the Respiratory Virus Section of the LID in 1959. While Chanock was to make numerous other contributions and receive many accolades (78), one further contribution to diagnostic virology in this era should be noted. For years, it had been assumed that an agent associated with cold agglutinin-positive pneumonia, called the Eaton agent, was a virus. Others had suggested that it was a pleuropneumonia-like agent, but isolation remained elusive. Chanock and his colleagues reported the growth of the agent on cell-free artificial media and characterized it as a pleuropneumonia-like organism (79). Success was achieved by growth studies on agar that incorporated yeast extract and horse serum. Characterized as a mycoplasma and found to differ from other members of the genus, it was named *Mycoplasma pneumoniae* (80) and was also found to be susceptible to antibiotic treatment with tetracycline (81).

Thus, in assessing Robert Huebner's accomplishments at the LID, it is apparent that he had a remarkable breadth and blend of strengths, including the ability to foster the work of others. While he did not spring from an experimental background, he grasped the key interrelationships of the emerging science of tissue culture virology and epidemiological characterization of disease outbreaks. At the LID, he cut his teeth on the outbreak of rickettsialpox in New York City, describing a newly isolated agent, its vector, and its reservoir in mice (48), and on studies of Q fever in Southern California (51). Expanding on the work on the association of coxsackie A virus and herpangina in a Maryland community, his group set up long-term viral monitoring studies in area hospitals. These included Children's Hospital in Washington, DC, in association with Robert Parrott and at Junior

Village, a children's welfare home where Kapikian had described an outbreak of RSV (82). Such collaborations allowed longitudinal and cross-sectional studies of viral infections and diseases. In certain instances, isolation of viruses was not clearly connected with human disease. Huebner chaired a conference in 1957 of the New York Academy of Sciences, "Viruses in Search of Disease." In that conference, he revisited Thomas Rivers's consideration of Koch's postulates, calling it "the virologist's dilemma" (1).

Huebner had a keen sense of people, selecting and supporting talented scientists who made landmark contributions to the flowering of clinical virology. Rowe and Hartley used fundamental tissue culture techniques to isolate adenoviruses and CMV. While Chanock had already isolated what was to be named a parainfluenza virus and what was to become known as RSV prior to arriving at the LID, his studies convincingly linking these viruses to acute respiratory diseases in children were performed during his tenure at the LID. Perhaps Huebner's most impressive quality was the capacity to recognize the broad outlines of where science was moving. Hence, he moved into molecular genetic studies of experimental cancer at the NCI and originated the concept of the oncogene in collaboration with George J. Todaro.

Following Huebner's move to the NCI in 1968, Chanock assumed the role of Chief at the LID. Important contributions to medical virology continued to flow. Agents of viral gastroenteritis and hepatitis, resistant to growth in tissue culture, were identified using immunoelectron microscopy. Innovative strategies of vaccine immunization using the techniques of molecular biology to create attenuated strains of agents followed the LID policy of a "beginning to end" study of infectious diseases (Chanock, report to the Board of Scientific Counselors).

REFINEMENTS IN CELL CULTURE METHODS AND DIFFERENTIAL SUSCEPTIBILITY FOR VIRAL DIAGNOSIS

In "Diagnostic Virology: from Animals to Automation," Hsiung characterized the years from 1950 to 1965 as the early period (83). On the heels of the demonstration by the Enders group that polio could be grown *in vitro* in nonneural tissue and that CPE could be used as a diagnostic marker, there was swift development of improvements in culture technique. Plaque assay under agar overlay was adapted by Renato Dulbecco from bacteriophage work to allow quantitation of viral inocula (15). Primary cell cultures and cell lines were tested for their capacity to isolate, grow, and characterize viruses. For example, kidney cells from different species of monkeys were found to be differentially susceptible to various enteric viruses. Plaque morphology could be used to distinguish isolates (84), and on occasion plaque formation under agar could be used to detect viruses when CPE failed to appear (85).

Continuous cell lines were developed, allowing serial propagation of cell cultures in diagnostic laboratories, in contrast to primary cells prepared directly from animal tissues that senesced after a limited number of passages. One of the most widely used continuous cell lines, HeLa cells, was derived from cervical carcinoma tissue propagated and distributed by George Otto Gey at the Johns Hopkins Hospital (86). The story of Henrietta Lacks, who succumbed to cervical cancer at age 31, and the immortality of her cells have been movingly recounted in *The Immortal Life of Henrietta Lacks* (87). Numerous other cell lines and HeLa cells were found to have various susceptibilities to a spectrum of viruses (12).

Further refinements included the addition of antibiotics to culture media to reduce bacterial contamination (88). Trypsin dispersion of single cell suspensions allowed quantitation and greater standardization of monolayer cultures (89). The use of defined media with serum supplements adjusted for cell growth or cell maintenance allowed the establishment and long-term maintenance of cultures for prolonged observation. Biological supply houses provided reliable sources of cell cultures, defined media, and sera, freeing laboratory time and ensuring standardization of reagents.

Despite the success of "dispersed" cell culture monolayers for the isolation of many important respiratory viruses, including rhinoviruses in 1956, the etiology of many cases of the common cold remained elusive. Rhinovirus (RV) was first isolated in 1956 by Dr. Winston Price at Johns Hopkins University and was quickly determined to be the most common cause of cold symptoms in adults (90). Then, in 1965, B. Hoorn and David Arthur John Tyrrell reported the infection of intact nasal and tracheal organ cultures *in vitro* by several known viruses and found a variety of tissue responses, including disruption of ciliary activity, focal changes, and complete destruction of the epithelium (91). In the same year, Tyrrell and M. L. Bynoe reported that an isolate, B814, obtained from a patient with a common cold grew in cultures of human fetal trachea but had not replicated in standard cell cultures or in eggs (92). The successful application of organ culture techniques to the isolation of agents from nasal washings of persons who had experienced spontaneous onset of the common cold was reported by the same authors early in the following year (93). June Almeida and Tyrrell soon applied electron microscopy to suspensions of organ cultures and discovered that B814 and another human virus had a morphological appearance similar to avian infectious bronchitis virus of chickens (94). In the same year, Kenneth McIntosh et al. detected eight agents in tracheal organ cultures from nasopharyngeal washings from 23 patients (95). The specimens were taken on or before the fourth day of an acute upper respiratory illness. Six of these agents had morphology similar to that of the avian infectious bronchitis virus. Because of the appearance "recalling the solar corona" and shared biochemical characteristics, these and similar viruses formed the basis of a new group of RNA viruses,

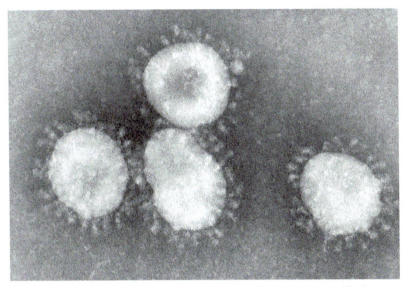

FIGURE 10 *Coronavirus, negative-contrast electron micrograph. The virus is named for the corona-like or crown spikes seen electron microscopically. This type of virus was first isolated from the human common cold using nasal and tracheal organ cultures. Magnification, approximately ×60,000. (Courtesy of CDC-PHIL [#10270], Dr. Fred Murphy; Sylvia Whitfield, 1975.)*

coronaviruses (96, 97) (Fig. 10). It would be from this group of viruses that severe acute respiratory syndrome coronavirus (SARS-CoV) would emerge in 2003. In 2019 SARS-CoV-2 emerged to devastate the world (see chapter 10).

In the 1960s, chronic diseases of the central nervous system were thought to be caused by viral agents, but here, too, isolations had remained elusive. One puzzling disease, subacute sclerosing panencephalitis, received intense scrutiny, and considerable evidence indicated that measles virus was involved. However, despite serological evidence of high levels of antibody in spinal fluid, electron microscopic evidence of paramyxovirus nucleocapsids, and measles antigen in the brain by immunohistology, measles virus had not been isolated in the laboratory. A breakthrough came in 1969 when two groups recovered measles virus when brain biopsy specimens were placed into culture with cell lines susceptible to measles virus (98, 99). This concept of propagating together cells of different types and origin in a single monolayer, termed cocultivation, has facilitated the detection and isolation of pathogens which otherwise would have gone undetected (100).

Fast-forward 20 years—in 1984, C. A. Gleaves et al. described a rapid viral diagnostic technique in which specimens suspected of containing CMV were centrifuged at a low speed in shell vials which contained cell monolayers of human diploid fibroblasts grown on coverslips (101). CMV was identified using a monoclonal antibody against a CMV early nuclear antigen and both immunofluorescent and immunohistological staining. All positive specimens were identified by 36 hours after specimen inoculation, whereas the development of standard diagnosis by CPE took an average of 9 days. This represented a significant improvement for CMV diagnosis, particularly for immunocompromised populations such as transplant recipients and persons with human immunodeficiency virus (HIV). Early therapeutic intervention was facilitated. This "shell vial technique" and the principle of centrifugation-enhanced inoculation were adapted to the diagnosis of many other viral infections, particularly for rapid respiratory virus diagnosis (100). Yet the practice of centrifugation inoculation was known as early as 1954. George Otto Gey et al. demonstrated that cells tolerated centrifugation and that viruses inoculated by centrifugation produced a greater destructive effect than did viruses inoculated without centrifugation (102).

PROFUSION OF ISOLATES, TAXONOMY, AND THE QUESTION OF DISEASE CAUSATION

As tissue culture systems were applied, a profusion of viruses infecting humans were identified. Inoculation of suckling mice contributed to the isolation of arboviruses as well as coxsackieviruses (52).

Whereas 20 truly human viruses were known in 1948, Huebner noted that the number had increased by 70 10 years later (103). The American Public Health Association's *Diagnostic Procedures for Viral and Rickettsial Diseases* had grown from 347 pages in its first edition in 1948 to 814 pages by the third edition in 1964 (104). The discovery of a great number of human viruses was a boon to the understanding of human infectious diseases, but it also created a number of challenges, not the least of these being the large number of cytopathic agents isolated from enteric specimens discovered in the course of polio studies. Often isolated from the stools of healthy individuals, a group of related viruses was called the enteric cytopathogenic human orphan (ECHO) group (105). Ironically, perhaps, despite the number of enteric viral isolates in tissue culture, the isolation of the most significant viral causes of gastroenteritis would await further technological advances such as immunoelectron microscopy. The same remained true for hepatitis. Despite the development of cell lines designed specifically to isolate the agents of hepatitis (12), isolation and identification of hepatitis viruses would await further developments in diagnostic technology.

The case was quite the opposite for viral causes of respiratory diseases. Isolations were achieved in tissue culture and disease associations established with cross-sectional and longitudinal studies (106). The associations of coxsackie A viruses with herpangina and coxsackie B viruses with pleurodynia were noted above, and the association of enteric viruses, subsequently known as enteroviruses, with the syndrome of aseptic meningitis was found by several investigators. Soon after the adenoviruses were isolated by the Rowe and Hilleman groups, the associations with pharyngitis, conjunctivitis, and acute respiratory diseases, particularly in military recruits, became apparent (107). Overall, one senses that disease associations were demonstrated to general satisfaction. The caveat held, of course, that any of several agents might cause a syndrome, such as an acute respiratory syndrome, and a single agent might cause more than one syndrome.

Arboviral studies were spurred by the Rockefeller Foundation, which initiated worldwide programs in the 1950s (108). The issues faced by arbovirologists were as complex as any in virology. Added to the mix of clinical syndromes, seroepidemiology, and viral isolations were patterns of transmission by vectors. Reflecting that transmission, in 1943 W. M. Hammon proposed the term "arthropod borne virus encephalitides," later contracted to arboviral (109). While the suckling mouse was to be the mainstay of arboviral isolations, tissue culture isolation systems were developed. Sonja M. Buckley, trained in cell culture by George Otto Gey, the developer of the HeLa cell line (see p. 152), joined the Rockefeller laboratories (108). Among many other contributions, including the later isolation of Lassa virus in cell culture (110), Buckley demonstrated the usefulness of tissue culture for propagating arboviruses and detecting antibodies (111). For certain arboviruses, isolation in cell culture under agar was more sensitive than CPE in fluid overlay culture or in infant mice (112). However, the major organizing principle for the arboviruses in this period was antigenic grouping based on careful studies of Jordi Casals and his colleagues, L. V. Brown and D. H. Clarke, at the Rockefeller Institute (113, 114).

While attention was focused on disease associations of viruses and clinical textbooks were organized by viral diseases of organ systems, the increasing amount of information about the biological and physical characteristics of viral isolates spurred taxonomic formulations. A report to the International Nomenclature Committee by the virus subcommittee proposed a non-Linnaean binomial system (11). Virus groups included in that report included poxviruses, herpesviruses, polioviruses, and myxoviruses. The enteroviruses, including poliovirus, coxsackie A virus, coxsackie B virus, and ECHO viruses, were taken up by the Committee on the Enteroviruses of the National Foundation for Infantile Paralysis (115). C. H. Andrewes et al., working on behalf of the International Nomenclature Committee, defined the myxovirus group for viruses related to influenza A virus (116). Later studies found closely related viruses which would be grouped as parainfluenza viruses (77). As more viruses were characterized, new groups were classified and further reclassified as additional information was obtained. For example, the virus originally classified as ECHO-10 was found to have distinctive biological characteristics (117). In light of biological differences and a significantly larger size, the virus was reclassified in a new group, the reoviruses (118).

The next era was marked by the explosion of viral isolations in tissue culture and would make use of immunological, chemical, and electron microscopic techniques. Hsiung defined this period as extending from 1965 to 1980 (83). It would bring with it the identification of agents that had been resistant to isolation in tissue culture, such as those associated with gastroenteritis and hepatitis. In those early days, viral diagnosis by isolation in tissue culture had been mostly confined to government and research laboratories, focusing on public health issues and surveillance of infectious diseases. A small number of hospitals associated with medical schools or research institutions performed limited diagnostic testing by viral isolation.

The most common approach to diagnosing viral infections was serologic testing performed in state and county health departments. Diagnostic studies of ill patients relied on the submission of paired sera, collected during the acute disease and the recovery phase, to demonstrate significant changes in antibody titers that might correlate with recent infection. Frequently, weeks to months would pass before results became available. At best, such studies provided a retrospective diagnosis after the individual had either recovered or died.

With advances in technology and the commercial availability of tissue culture cells, media, and reagents for virus identification, the opportunity to perform timely and accurate viral isolation became more widely feasible. Results could be available within a few days. The myth that virus laboratories were black holes where clinical specimens disappeared had to be dispelled to demonstrate that clinical diagnostic virology could be a valuable asset to clinicians for patient management.

W. Lawrence Drew, M.D., Ph.D. (Fig. 11), championed the proliferation of clinical virology laboratories in hospital microbiology laboratories, making the case that technologists trained in complex testing in a bacteriology laboratory could also be trained with the same rigor to perform viral cultures (119). With the development of testing with rapid turnaround times, the virus lab could make an impact on medical decisions. Out of 1,000 isolates in the early days of viral culture in his laboratory, 1972 to 1978, Drew reported an average of 4.1 days for virus detection (120). Further, most isolates of influenza A and herpes simplex viruses (HSVs) were reported within 2 to 3 days of specimen receipt.

As technology progressed, even more rapid results could be obtained by using immunofluorescent monoclonal antibody staining methods to detect viral antigens (121), for example, in skin smears for HSV or varicella-zoster virus or in throat swabs for influenza and respiratory syncytial viruses. Knowing that an infection was viral rather than bacterial could obviate the need for antibiotics and guide patient management and disease prognosis. Beyond this, the advent of targeted antiviral agents would require a definitive diagnosis for appropriate, effective treatment.

FIGURE 11 *W. Lawrence Drew championed the integration of virology into the mainstream with the other divisions of the microbiology laboratory. Here he is examining tissue culture tubes in the microbiology laboratory at Mt. Zion Hospital and Medical Center in San Francisco in 1982. (Courtesy of University of California, San Francisco Library, UCSF Medical Center at Mount Zion Archives.)*

Virology laboratories proliferated in the 1980s, and over the years the classical tissue culture laboratory was transformed in favor of very rapid immunologic and molecular techniques applied not only for diagnosis, but also to detect and track the evolution and spread of emerging viral agents. Such advances enhanced the value of viral laboratory diagnosis for patient care.

REFERENCES

1. **Huebner RJ.** 1957. Criteria for etiologic association of prevalent viruses with prevalent diseases; the virologist's dilemma. *Ann N Y Acad Sci* **67:**430–438 http://dx.doi.org/10.1111/j.1749-6632.1957.tb46066.x.

2. **Robbins FC, Enders JF, Weller TH.** 1950. Cytopathogenic effect of poliomyelitis viruses *in vitro* on, human embryonic tissues. *Proc Soc Exp Biol Med* **75:**370–374 http://dx.doi.org/10.3181/00379727-75-18202.

3. **Robbins FC, Enders JF, Weller TH, Florentino GL.** 1951. Studies on the cultivation of poliomyelitis viruses in tissue culture. V. The direct isolation and serologic identification of virus strains in tissue culture from patients with nonparalytic and paralytic poliomyelitis. *Am J Hyg* **54:**286–293.

4. **Lennette EH.**1988. *Pioneer of Diagnostic Virology with the California Department of Public Health*. Oral history conducted in 1982, 1983, and 1986 by S. Hughes. Regional Oral History Office, The Bancroft Library, University of California, Berkeley, Berkeley, CA.

5. **zur Hausen H.** 1980. Werner Henle 70 years. *Med Microbiol Immunol (Berl)* **168:**235–237 http://dx.doi.org/10.1007/BF02121806.

6. **Levy JA.** 2006. Gertrude S. Henle (1912–2006). *Virology* **358**:248–250 http://dx.doi.org/10.1016/j.virol.2006.11.001.
7. **Henle G, Henle W, Harris S.** 1947. The serological differentiation of mumps complement-fixation antigens. *Proc Soc Exp Biol Med* **64**:290–295 http://dx.doi.org/10.3181/00379727-64-15772.
8. **Henle W, Wiener M.** 1944. Complement fixation antigens of influenza viruses type A and B. *Proc Soc Exp Biol Med* **57**:176–179 http://dx.doi.org/10.3181/00379727-57-14744.
9. **Horsfall FL Jr.** 1949. Introduction, p 3–4. *In* Horsfall FL Jr (ed), *Diagnosis of Viral and Rickettsial Infections.* Columbia University Press, New York, NY. http://dx.doi.org/10.7312/hors90844-002.
10. **Henle G, Henle W.** 1949. The diagnosis of mumps, p 15–24. *In* Horsfall FL Jr (ed), *Diagnosis of Viral and Rickettsial Infection.* Columbia University Press, New York, NY. http://dx.doi.org/10.7312/hors90844-004.
11. **American Association of Immunologists.** 2002. In memoriam: Klaus Hummeler. *AAI Newsl* **August 2002**:15.
12. **Deinhardt F, Henle G.** 1957. Studies on the viral spectra of tissue culture lines of human cells. *J Immunol* **79**:60–67.
13. **Henle G, Deinhardt F.** 1957. The establishment of strains of human cells in tissue culture. *J Immunol* **79**:54–59.
14. **Landry ML.** 2006. Dr. Edith Hsiung remembered. *J Clin Virol* **37**:235–236 http://dx.doi.org/10.1016/j.jcv.2006.10.005.
15. **Dulbecco R.** 1952. Production of plaques in monolayer tissue cultures by single particles of an animal virus. *Proc Natl Acad Sci USA* **38**:747–752 http://dx.doi.org/10.1073/pnas.38.8.747.
16. **Hsiung GD, Melnick JL.** 1955. Plaque formation with poliomyelitis, Coxsackie, and orphan (echo) viruses in bottle cultures of monkey epithelial cells. *Virology* **1**:533–535 http://dx.doi.org/10.1016/0042-6822(55)90041-6.
17. **Hsiung GD.** 1961. Applications of primary cell cultures in the study of animal viruses. III. Biological and genetic studies of enteric viruses of man (enteroviruses). *Yale J Biol Med* **33**:359–371.
18. **Hsiung GD.** 1980. Progress in clinical virology—1960 to 1980: a recollection of twenty years. *Yale J Biol Med* **53**:1–4.
19. **Horstmann DM, Hsiung GD.** 1965. Principles of diagnostic virology, p 405–424. *In* Horsfall FL Jr, Tamm I (ed), *Viral and Rickettsial Infections of Man*, 4th ed. J B Lippincott Co, Philadelphia, PA.
20. **Hsiung GD, Gaylord WH Jr.** 1961. The vacuolating virus of monkeys. I. Isolation, growth characteristics, and inclusion body formation. *J Exp Med* **114**:975–986 http://dx.doi.org/10.1084/jem.114.6.975.
21. **Sweet BH, Hilleman MR.** 1960. The vacuolating virus, S.V. 40. *Proc Soc Exp Biol Med* **105**:420–427 http://dx.doi.org/10.3181/00379727-105-26128.
22. **Offit PA.** 2007. *Vaccinated. One Man's Quest To Defeat the World's Deadliest Diseases.* Harper Collins, New York, NY.
23. **Hsiung GD.** 1968. Latent virus infections in primate tissues with special reference to simian viruses. *Bacteriol Rev* **32**:185–205 http://dx.doi.org/10.1128/br.32.3.185-205.1968.
24. **Hsiung GD, Henderson JR.** 1964. *Diagnostic Virology.* Yale University Press, New Haven, CT.
25. **Hsiung GD, Fong CKY, Landry ML.** 1994. *Hsiung's Diagnostic Virology as Illustrated by Light and Electron Microscopy*, 4th ed. Yale University Press, New Haven, CT.
26. **Etheridge EW.** 1992. *Sentinel for Health: a History of the Centers for Disease Control.* University of California Press, Berkeley, CA. http://dx.doi.org/10.1525/9780520910416.
27. **Saxon W.** 25 March 1997. Morris Schaeffer, 89, virologist who tightened controls on labs. New York Times Online. http://www.nytimes.com/1997/03/25/nyregion/morris-schaeffer-89-virologist-who-tightened-controls-on-labs.html.
28. **Johnson HN.** 1991. *Virologist and Naturalist with the Rockefeller Foundation and the California Department of Public Health.* Oral history by S. Smith Hughes, Bancroft Library, University of California, Berkeley, Berkeley, CA.
29. **Reagan RL.** 1980. *One Man's Research: the Autobiography of Reginald L. Reagan.* Dan River Press, Stafford, VA.

30. **Howitt B.** 1938. Recovery of the virus of equine encephalomyelitis from the brain of a child. *Science* **88:**455–456 http://dx.doi.org/10.1126/science.88.2289.455.
31. **Goldwasser RA, Kissling RE.** 1958. Fluorescent antibody staining of street and fixed rabies virus antigens. *Proc Soc Exp Biol Med* **98:**219–223 http://dx.doi.org/10.3181/00379727-98-23996.
32. **Dowdle WR.** 1994. Walter R. Dowdle, Ph.D., in honor of 33 years' service at CDC. *MMWR Morb Mortal Wkly Rep* **43:**227.
33. **Task Force for Global Health.** Accessed 23 January 2010. Walter Dowdle—global polio eradication. Task Force for Global Health, Decatur, GA. http://task-force.citizenstudio.com/our-team/our-staff/walter-dowdle-global-polio-eradication.
34. **Andrewes CH.** 1954. NOMENCLATURE of viruses. *Nature* **173:**620–621 http://dx.doi.org/10.1038/173620a0.
35. **Schaeffer M.**1965. Diagnosis of viral and rickettsial diseases, p 2–37. *In* Sanders M, Lennette EH (ed), *The First Annual Symposium on Applied Virology.* Olympic Press, Sheboygan, WI.
36. **Calisher CH.** 1996. Telford H. Work—a tribute. *J Am Mosq Control Assoc* **12:**385–395.
37. **Beeman EA.** 2007. *Charles Armstrong, M.D.: a Biography.* National Institutes of Health, Washington, DC. Accessed 9 September 2011. http://history.nih.gov/research/downloads/ArmstrongBiography.pdf.
38. **Muckenfuss RS, Armstrong C, McCordock HA.** 1933. Encephalitis: studies on experimental transmission. *Public Health Rep* **48:**1341–1343 http://dx.doi.org/10.2307/4580968.
39. **Armstrong CR, Lillie D.** 1934. Experimental lymphocytic choriomeningitis of monkeys and mice produced by a virus encountered in studies of the 1933 St. Louis epidemic. *Public Health Rep* **49:**1019–1027 http://dx.doi.org/10.2307/4581290.
40. **Habel K.** 1957. Rabies prophylaxis in man. *Public Health Rep* **72:**667–673 http://dx.doi.org/10.2307/4589865.
41. **Habel K.** 1945. Cultivation of mumps virus in the developing chick embryo and its application to studies of immunity to mumps in man. *Public Health Rep* **60:**201–212 http://dx.doi.org/10.2307/4585189.
42. **Habel K.** 1942. Transmission of rubella to *Macacus mulatta* monkeys. *Public Health Rep* **57:**1126–1139 http://dx.doi.org/10.2307/4584176.
43. **Habel K.** 1968. The biology of viral carcinogenesis. *Cancer Res* **28:**1825–1831.
44. **Roueché B.**1947. A reporter at large: the alerting of Mr. Pomerantz. *The New Yorker* **23(28):**28–37.
45. **Roueché B.** 1967. *Eleven Blue Men and Other Narratives of Medical Detection.* Little Brown, Boston, MA.
46. **Huebner RJ, Stamps P, Armstrong C.** 1946. Rickettsialpox; a newly recognized rickettsial disease; isolation of the etiological agent. *Public Health Rep* **61:**1605–1614 http://dx.doi.org/10.2307/4585895.
47. **Huebner RJ, Jellison WL, Pomerantz C.** 1946. Rickettsialpox, a newly recognized rickettsial disease; isolation of a Rickettsia apparently identical with the causative agent of rickettsialpox from Allodermanyssus sanguineus, a rodent mite. *Public Health Rep* **61:**1677–1682 http://dx.doi.org/10.2307/4585913.
48. **Huebner RJ, Jellison WL, Armstrong C.** 1947. Rickettsialpox; a newly recognized rickettsial disease; recovery of *Rickettsia akari* from a house mouse (*Mus musculus*). *Public Health Rep* **62:**777–780 http://dx.doi.org/10.2307/4586142.
49. **Beeman EA.** 2005. *Robert J. Huebner, M.D.: a Virologist's Odyssey.* National Institutes of Health, Washington, DC. Accessed 9 September 2011. http://history.nih.gov/research/downloads/HuebnerBiography.pdf.
50. **Noble HB.** 5 September 1998. Robert Huebner, 84, dies; found virus-cancer connections. New York Times Online. http://www.nytimes.com/1998/09/05.
51. **Huebner RJ, Jellison WL, Beck MD, Parker RR, Shepard CC.** 1948. Q fever studies in southern California; recovery of *Rickettsia burneti* from raw milk. *Public Health Rep* **63:**214–222 http://dx.doi.org/10.2307/4586445.
52. **Dalldorf G, Sickles GM.** 1948. An unidentified, filterable agent isolated from the feces of children with paralysis. *Science* **108:**61–62 http://dx.doi.org/10.1126/science.108.2794.61.

53. **Huebner RJ, Armstrong C, Beeman EA, Cole RM.** 1950. Studies of coxsackie viruses; preliminary report on occurrence of Coxsackie virus in a Southern Maryland community. *J Am Med Assoc* **144:**609–613 http://dx.doi.org/10.1001/jama.1950.02920080011004.

54. **Huebner RJ, Cole RM, Beeman EA, Bell JA, Peers JH.** 1951. Herpangina; etiological studies of a specific infectious disease. *J Am Med Assoc* **145:**628–633 http://dx.doi.org/10.1001/jama.1951.02920270022005.

55. **Huebner RJ, Beeman EA, Cole RM, Beigelman PM, Bell JA.** 1952. The importance of Coxsackie viruses in human disease, particularly herpangina and epidemic pleurodynia. *N Engl J Med* **247:**249–256 http://dx.doi.org/10.1056/NEJM195208142470705.

56. **Huebner RJ, Beeman EA, Cole RM, Beigelman PM, Bell JA.** 1952. The importance of Coxsackie viruses in human disease, particularly herpangina and epidemic pleurodynia. *N Engl J Med* **247:**285–289 http://dx.doi.org/10.1056/NEJM195208212470805.

57. **Rowe WP, Murphy FA, Bergold GH, Casals J, Hotchin J, Johnson KM, Lehmann-Grube F, Mims CA, Traub E, Webb PA.** 1970. Arenoviruses: proposed name for a newly defined virus group. *J Virol* **5:**651–652 http://dx.doi.org/10.1128/jvi.5.5.651-652.1970.

58. **Rowe WP, Huebner RJ, Gilmore LK, Parrott RH, Ward TG.** 1953. Isolation of a cytopathogenic agent from human adenoids undergoing spontaneous degeneration in tissue culture. *Proc Soc Exp Biol Med* **84:**570–573 http://dx.doi.org/10.3181/00379727-84-20714.

59. **Hilleman MR, Werner JH.** 1954. Recovery of new agent from patients with acute respiratory illness. *Proc Soc Exp Biol Med* **85:**183–188 http://dx.doi.org/10.3181/00379727-85-20825.

60. **Rowe WP, Huebner RJ, Bell JA.** 1957. Definition and outline of contemporary information on the adenovirus group. *Ann N Y Acad Sci* **67:**255–261 http://dx.doi.org/10.1111/j.1749-6632.1957.tb46048.x.

61. **Rowe WP, Hartley JW, Waterman S, Turner HC, Huebner RJ.** 1956. Cytopathogenic agent resembling human salivary gland virus recovered from tissue cultures of human adenoids. *Proc Soc Exp Biol Med* **92:**418–424 http://dx.doi.org/10.3181/00379727-92-22497.

62. **Huebner RJ, Todaro GJ.** 1969. Oncogenes of RNA tumor viruses as determinants of cancer. *Proc Natl Acad Sci USA* **64:**1087–1094 http://dx.doi.org/10.1073/pnas.64.3.1087.

63. **Rowe WP.** 9 July 1983. Dr. Wallace P. Rowe, 57, dies; a leader in cancer research. Obituary, *New York Times.*

64. **Schmeck HM Jr.** 4 March 1993. Albert Sabin, polio researcher, 86, dies. New York Times Online. http://www.nytimes.com/1993/03/04/us.

65. **Gale Group.** 2010. Albert Sabin. Notable scientists: from 1900 to the present. Online. Gale Group, 2008. (Reproduced in *Biography Resource Center*. Gale, Farmington Hills, MI, 2010.) http://galenet.galegroup.com/serviet/BioRC.

66. **Chanock RM, Sabin AB.** 1953. The hemagglutinin of St. Louis encephalitis virus. I. Recovery of stable hemagglutinin from the brains of infected mice. *J Immunol* **70:**271–285.

67. **Chanock RM.** 1956. Association of a new type of cytopathogenic myxovirus with infantile croup. *J Exp Med* **104:**555–576 http://dx.doi.org/10.1084/jem.104.4.555.

68. **Chanock RM, Parrott RH, Cook K, Andrews BE, Bell JA, Reichelderfer T, Kapikian AZ, Mastrota FM, Huebner RJ.** 1958. Newly recognized myxoviruses from children with respiratory disease. *N Engl J Med* **258:**207–213 http://dx.doi.org/10.1056/NEJM195801302580502.

69. **Vogel J, Shelokov A.** 1957. Adsorption-hemagglutination test for influenza virus in monkey kidney tissue culture. *Science* **126:**358–359 http://dx.doi.org/10.1126/science.126.3269.358.b.

70. **Shelokov A, Vogel JE, Chi L.** 1958. Hemadsorption (adsorption-hemagglutination) test for viral agents in tissue culture with special reference to influenza. *Proc Soc Exp Biol Med* **97:**802–809 http://dx.doi.org/10.3181/00379727-97-23884.

71. **Chanock R, Roizman B, Myers R.** 1957. Recovery from infants with respiratory illness of a virus related to chimpanzee coryza agent (CCA). I. Isolation, properties and characterization. *Am J Hyg* **66:**281–290.

72. **Blount RE Jr, Morris JA, Savage RE.** 1956. Recovery of cytopathogenic agent from chimpanzees with coryza. *Proc Soc Exp Biol Med* **92:**544–549 http://dx.doi.org/10.3181/00379727-92-22538.

73. **Chanock RM, Kim HW, Vargosko AJ, Deleva A, Johnson KM, Cumming C, Parrott RH.** 1961. Respiratory syncytial virus. I. Virus recovery and other observations during 1960 outbreak of bronchiolitis, pneumonia, and minor respiratory diseases in children. *JAMA* **176:**647–653.

74. **Beem M, Wright FH, Hamre D, Egerer R, Oehme M.** 1960. Association of the chimpanzee coryza agent with acute respiratory disease in children. *N Engl J Med* **263:**523–530 http://dx.doi.org/10.1056/NEJM196009152631101.

75. **Kapikian AZ, Bell JA, Mastrota FM, Johnson KM, Huebner RJ, Chanock RM.** 1961. An outbreak of febrile illness and pneumonia associated with respiratory syncytial virus infection. *Am J Hyg* **74:**234–248.

76. **Chanock RM, Vargosko A, Luckey A, Cook MK, Kapikian AZ, Reicelderfer T, Parrott RH.** 1959. Association of hemadsorption viruses with respiratory illness in childhood. *J Am Med Assoc* **169:**548–553 http://dx.doi.org/10.1001/jama.1959.03000230004002.

77. **Andrewes CH, Bang FB, Chanock RM, Zhdanov VM.** 1959. Para-influenza viruses 1, 2, and 3: suggested names for recently described myxoviruses. *Virology* **8:**129–130 http://dx.doi.org/10.1016/0042-6822(59)90027-3.

78. **Ligon BL.** 1998. Robert M. Chanock, MD: a living legend in the war against viruses. *Semin Pediatr Infect Dis* **9:**258–269 http://dx.doi.org/10.1016/S1045-1870(98)80040-X.

79. **Chanock RM, Hayflick L, Barile MF.** 1962. Growth on artificial medium of an agent associated with atypical pneumonia and its identification as a PPLO. *Proc Natl Acad Sci USA* **48:**41–49 http://dx.doi.org/10.1073/pnas.48.1.41.

80. **Chanock RM, Dienes L, Eaton MD, Edward DG, Freundt EA, Hayflick L, Hers JF, Jensen KE, Liu C, Marmion BP, Morton HE, Mufson MA, Smith PF, Somerson NL, Taylor-Robinson D.** 1963. *Mycoplasma pneumoniae:* proposed nomenclature for atypical pneumonia organism (Eaton agent). *Science* **140:**662 http://dx.doi.org/10.1126/science.140.3567.662.a.

81. **Kingston JR, Chanock RM, Mufson MA, Hellman LP, James WD, Fox HH, Manko MA, Boyers J.** 1961. Eaton agent pneumonia. *JAMA* **176:**118–123 http://dx.doi.org/10.1001/jama.1961.03040150034009.

82. **Bell JA, Huebner RJ, Rosen L, Rowe WP, Cole RM, Mastrota FM, Floyd TM, Chanock RM, Shvedoff RA.** 1961. Illness and microbial experiences of nursery children at Junior Village. *Am J Hyg* **74:**267–292.

83. **Hsiung GD.** 1984. Diagnostic virology: from animals to automation. *Yale J Biol Med* **57:**727–733.

84. **Hsiung GD, Melnick JL.** 1957. Morphologic characteristics of plaques produced on monkey kidney monolayer cultures by enteric viruses (poliomyelitis, Coxsackie, and echo groups). *J Immunol* **78:**128–135.

85. **Hsiung GD.** 1959. The use of agar overlay cultures for detection of new virus isolates. *Virology* **9:**717–719 http://dx.doi.org/10.1016/0042-6822(59)90167-9.

86. **Scherer WF, Syverton JT, Gey GO.** 1953. Studies on the propagation *in vitro* of poliomyelitis viruses. IV. Viral multiplication in a stable strain of human malignant epithelial cells (strain HeLa) derived from an epidermoid carcinoma of the cervix. *J Exp Med* **97:**695–710 http://dx.doi.org/10.1084/jem.97.5.695.

87. **Skloot R.** 2010. *The Immortal Life of Henrietta Lacks.* Crown Publishers, New York, NY.

88. **Mortimer P, Weller TH, Robbins FC.** 2009. Classic paper: How monolayer cell culture transformed diagnostic virology: a review of a classic paper and the developments that stemmed from it. (*Science,* New Series, Vol. 109, No. 2822 (Jan. 28, 1949), p 85–87). *Rev Med Virol* **19:**241–249.

89. **Youngner JS.** 1954. Monolayer tissue cultures. I. Preparation and standardization of suspensions of trypsin-dispersed monkey kidney cells. *Proc Soc Exp Biol Med* **85:**202–205 http://dx.doi.org/10.3181/00379727-85-20830.

90. **Price WH.** 1956. The isolation of a new virus associated with respiratory clinical disease in humans. *Proc Natl Acad Sci U S A* **42:**892–896.

91. **Hoorn B, Tyrrell DAJ.** 1965. On the growth of certain "newer" respiratory viruses in organ cultures. *Br J Exp Pathol* **46:**109–118.

92. **Tyrrell DAJ, Bynoe ML.** 1965. Cultivation of a novel type of common-cold virus in organ cultures. *BMJ* **1:**1467–1470 http://dx.doi.org/10.1136/bmj.1.5448.1467.

93. **Tyrrell DAJ, Bynoe ML.** 1966. Cultivation of viruses from a high proportion of patients with colds. *Lancet* **1:**76–77 http://dx.doi.org/10.1016/S0140-6736(66)92364-6.

94. **Almeida JD, Tyrrell DAJ.** 1967. The morphology of three previously uncharacterized human respiratory viruses that grow in organ culture. *J Gen Virol* **1:**175–178 http://dx.doi.org/10.1099/0022-1317-1-2-175.

95. **McIntosh K, Dees JH, Becker WB, Kapikian AZ, Chanock RM.** 1967. Recovery in tracheal organ cultures of novel viruses from patients with respiratory disease. *Proc Natl Acad Sci USA* **57:**933–940 http://dx.doi.org/10.1073/pnas.57.4.933.

96. **Almeida JD, Berry DM, Cunningham CH, Hamre D, Hofstad MS, Mallucci L, McIntosh K, Tyrrell DAJ.** 1968. Coronaviruses. *Nature* **220:**650 http://dx.doi.org/10.1038/220650b0.

97. **Estola T.** 1970. Coronaviruses, a new group of animal RNA viruses. *Avian Dis* **14:**330–336 http://dx.doi.org/10.2307/1588476.

98. **Horta-Barbosa L, Fuccillo DA, Sever JL, Zeman W.** 1969. Subacute sclerosing panencephalitis: isolation of measles virus from a brain biopsy. *Nature* **221:**974 http://dx.doi.org/10.1038/221974a0.

99. **Payne FE, Baublis JV, Itabashi HH.** 1969. Isolation of measles virus from cell cultures of brain from a patient with subacute sclerosing panencephalitis. *N Engl J Med* **281:**585–589 http://dx.doi.org/10.1056/NEJM196909112811103.

100. **Leland DS, Ginocchio CC.** 2007. Role of cell culture for virus detection in the age of technology. *Clin Microbiol Rev* **20:**49–78 http://dx.doi.org/10.1128/CMR.00002-06.

101. **Gleaves CA, Smith TF, Shuster EA, Pearson GR.** 1984. Rapid detection of cytomegalovirus in MRC-5 cells inoculated with urine specimens by using low-speed centrifugation and monoclonal antibody to an early antigen. *J Clin Microbiol* **19:**917–919 http://dx.doi.org/10.1128/jcm.19.6.917-919.1984.

102. **Gey GO, Bang FB, Gey MK.** 1954. Responses of a variety of normal and malignant cells to continuous cultivation, and some practical applications of these responses to problems in the biology of disease. *Ann N Y Acad Sci* **58:**976–999 http://dx.doi.org/10.1111/j.1749-6632.1954.tb45886.x.

103. **Huebner RJ.** 1959. 70 Newly recognized viruses in man. *Public Health Rep* **74:**6–12 http://dx.doi.org/10.2307/4590355.

104. **Lennette EH, Schmidt NJ (ed).** 1964. *Diagnostic Procedures for Viral and Rickettsial Diseases*, 3rd ed. American Public Health Association, Inc, New York, NY.

105. **Committee on the ECHO Viruses.** 1955. ENTERIC cytopathogenic human orphan (ECHO) viruses. *Science* **122:**1187–1188 http://dx.doi.org/10.1126/science.122.3181.1187.

106. **Chanock RM, Parrott RH.** 1965. Acute respiratory disease in infancy and childhood: present understanding and prospects for prevention: E. Mead Johnson Address, October, 1964. *Pediatrics* **36:**21–39 http://dx.doi.org/10.1542/peds.36.1.21.

107. **Huebner RJ, Rowe WP, Ward TG, Parrott RH, Bell JA.** 1954. Adenoidal-pharyngeal-conjunctival agents: a newly recognized group of common viruses of the respiratory system. *N Engl J Med* **251:**1077–1086 http://dx.doi.org/10.1056/NEJM195412302512701.

108. **Calisher CH.** 2005. A very brief history of arbovirology, focusing on contributions by workers of the Rockefeller foundation. *Vector Borne Zoonotic Dis* **5:**202–211 http://dx.doi.org/10.1089/vbz.2005.5.202.

109. **Hammon WM.** 1943. Encephalitis. Eastern and Western Equine and St. Louis Types, as observed in 1941 in Washington, Arizona, New Mexico, and Texas. *JAMA* **121:**560–566 http://dx.doi.org/10.1001/jama.1943.02840080008002.

110. **Buckley SM, Casals J, Downs WG.** 1970. Isolation and antigenic characterization of Lassa virus. *Nature* **227:**174 http://dx.doi.org/10.1038/227174a0.

111. **Buckley SM.** 1959. Propagation, cytopathogenicity, and hemagglutination-hemadsorption of some arthropod-borne viruses in tissue culture. *Ann N Y Acad Sci* **81:**172–187 http://dx.doi.org/10.1111/j.1749-6632.1959.tb49305.x.

112. **Henderson JR, Taylor RM.** 1959. Arthropod-borne virus plaques in agar overlaid tube cultures. *Proc Soc Exp Biol Med* **101:**257–259 http://dx.doi.org/10.3181/00379727-101-24902.

113. **Casals J, Brown LV.** 1954. Hemagglutination with arthropod-borne viruses. *J Exp Med* **99:**429–449 http://dx.doi.org/10.1084/jem.99.5.429.

114. **Clarke DH, Casals J.** 1958. Techniques for hemagglutination and hemagglutination-inhibition with arthropod-borne viruses. *Am J Trop Med Hyg* **7:**561–573 http://dx.doi.org/10.4269/ajtmh.1958.7.561.

115. **Committee on the Enteroviruses, National Foundation for Infantile Paralysis.** 1957. The enteroviruses. *Am J Public Health Nations Health* **47:**1556–1566 http://dx.doi.org/10.2105/AJPH.47.12.1556.

116. **Andrewes CH, Bang FB, Burnet FM.** 1955. A short description of the *Myxovirus* group (influenza and related viruses). *Virology* **1:**176–184 http://dx.doi.org/10.1016/0042-6822(55)90014-3.

117. **Hsiung GD.** 1958. Some distinctive biological characteristics of ECHO-10 virus. *Proc Soc Exp Biol Med* **99:**387–390 http://dx.doi.org/10.3181/00379727-99-24359.

118. **Sabin AB.** 1959. Reoviruses. A new group of respiratory and enteric viruses formerly classified as ECHO type 10 is described. *Science* **130:**1387–1389 http://dx.doi.org/10.1126/science.130.3386.1387.

119. **Drew WL, Stevens GR.** 1979. Should your laboratory perform viral studies? *Lab Med* **10:**663–667 http://dx.doi.org/10.1093/labmed/10.11.663.

120. **Drew WL, Stevens GR.** 1979. How your laboratory should perform viral studies: laboratory equipment, specimen types, cell culture techniques. *Lab Med* **10:**741–746 http://dx.doi.org/10.1093/labmed/10.12.741.

121. **Drew WL, Stevens GR.** 1980. How your laboratory should perform viral studies (continued): isolation and identification of commonly encountered viruses. *Lab Med* **11:**14–23 http://dx.doi.org/10.1093/labmed/11.1.14.

7 Imaging Viruses and Tagging Their Antigens

Rapid diagnosis of a virus infection is a relative term but in this context should be interpreted in hours.

P. S. Gardner and J. McQuillin (1)

INTRODUCTION

Undefined viral illnesses

Following the application of tissue culture to viral isolation, a veritable cornucopia of viruses was found, particularly agents associated with acute respiratory diseases in children. However, acute, nonbacterial gastroenteritis remained one of the most important disease puzzles to be solved. Despite the successes of growing polioviruses in tissue culture and isolating the coxsackieviruses and enteric cytopathogenic human orphan (ECHO) viruses from stool, they were not the etiological agents of acute gastroenteritis.

Although yellow fever had been successfully isolated in human subjects and in laboratory animals and a vaccine had been developed (chapter 1), the agent of hepatitis still eluded detection in tissue culture. A clue to the etiology of hepatitis, however, was discovered during World War II when an outbreak of hepatitis in troops followed yellow fever vaccination. Subsequent investigations revealed that a hepatitis-causing agent had contaminated the serum blended with the yellow fever vaccine, but the agent of hepatitis remained undiscovered in tissue culture. Subsequent understanding

To Catch a Virus, Second Edition. Authored by John Booss and Marie Louise Landry.
© 2023 American Society for Microbiology. DOI: 10.1128/9781683673828.ch07

of hepatitis discerned two types of illness, serum and infectious hepatitis. Yet neither type had yielded to tissue culture investigations.

Timeliness of diagnosis

While isolation and characterization in cell culture, in animal systems, and by serology could give definitive answers in some cases of viral illness, the results often came well after the acute illnesses resolved. Thus, without further refinement, the usefulness of the new tissue culture systems was in the discovery and definition of viral diseases and for public health applications of serological findings, not for immediate patient care.

Early on, rapid viral diagnosis could be achieved through histological observation of intracellular inclusion bodies. Within hours of specimen receipt, for example, observation of Negri bodies allowed the prompt diagnosis of rabies virus infection. The application of electron microscopy (EM) could morphologically distinguish poxvirus from varicella virus in skin scrapings within hours of specimen receipt. However, these were exceptions, and the bulk of clinical diseases of suspected viral origin had to await the outcome of viral isolations using incubation in cell cultures or in animals or the development of an antibody response as the patient's acute illness ran its course.

There were other sound clinical reasons to wish that the viral diagnostic sequence could be accelerated. If a distinction could not be made between bacterial and viral disease, a course of unneeded antibiotics would be given, with the attendant risk of selecting for resistant organisms. The situation became even more urgent as antiviral medications were developed. Ward management, too, was a matter of importance, as Phillip S. Gardner pointed out (2). In the absence of a rapid viral diagnosis, a child admitted to hospital with a single febrile seizure in the context of an otherwise benign influenza virus infection might roam the ward spreading flu to more vulnerable children.

In what follows, we first examine the refinements in EM that greatly expanded the understanding of the structure and replication of viruses and facilitated the application of EM to viral diagnosis. The refinements included thin sectioning, negative staining, and immunoelectron microscopy (IEM) developed for clinical diagnostic work. The use of EM in discovering the viral causes of acute gastroenteritis and infectious hepatitis are then considered. We conclude with a section considering the use of fluorescence microscopy for rapid viral diagnosis (3).

REFINEMENTS OF EM

Early diagnostic EM

In the early days of tissue culture virology in the 1950s, the role of EM was minimal in defining human viral diseases. However, diagnosis by EM was evident in 1952, when Joseph L. Melnick and his colleagues used EM to investigate human viral skin lesions (4). During outbreaks of smallpox, EM was useful to distinguish atypical smallpox from chicken pox (5, 6). Smallpox viruses appeared as relatively large, brick-like structures, while the agent of chicken pox, a herpesvirus, was characterized by concentric light and dark circles around a usually electron-lucent central core.

Thin sectioning

Advances were made in thin-sectioning techniques which allowed preserved tissue to be cut in such a way that tissue relationships and submicroscopic structures could be examined (7). Progress was made in the design of microtome cutting devices, in development of glass and diamond knives, in improvements to fixation and embedding of specimens, and in application of novel stains (8). Yet for EM, the advances in producing thin sections came slowly. Cecil E. Hall, a professor of biophysics at MIT in Cambridge, MA, declared, "Microtomy is almost entirely an empirical art . . . what will be tried is likely to be restricted by custom or prejudice" (9). Hall noted that the three most important elements were the cutting edge, the cutting machine, and the embedding matrix. Representative advances included modification of the microtome for consistent production of thin sections by D. C. Pease and R. F. Barker (10), introduction of a methacrylate resin for embedding of biological specimens by S. B. Newman et al. (11), and introduction of a glass edge for cutting sections by H. Latta and J. F. Hartmann (12).

Shadow casting

To determine the thickness of objects under the electron microscope, R. C. Williams and R. W. G. Wyckoff developed a method called shadow casting (13). A thin metallic film deposited on a specimen at known angles produced shadows of lengths proportional to the thickness of the object studied. The cast shadows were photographed and measured to estimate the thickness of objects. Williams and Wyckoff soon applied this method to the study of viruses, using tobacco mosaic virus (TMV) and influenza virus (14). A marked improvement of visualization resulted from what they termed the "three-dimensional effect," yet the details of viral morphology remained obscured.

Negative staining

The principle underlying negative staining, defining an object by staining the background and its interstices, was recognized by Sydney Brenner (Fig. 1) and Robert Horne (Fig. 2). They refined the technique as applied to viruses in a trio of handsomely illustrated papers published in 1959 (15–17).

Prior to the work of Brenner and Horne, Cecil E. Hall and, separately, H. E. Huxley reported novel observations on negative staining; however, the technique

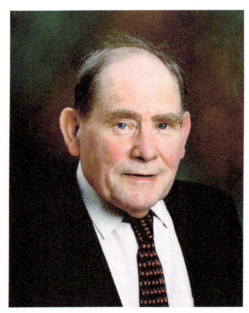

FIGURE 1 *Sydney Brenner. With Robert Horne, Brenner developed negative staining to rapidly screen a large number of T-even bacteriophage fractions by EM. It became a crucial tool in the investigation and classification of viruses. Brenner continued fundamental work in molecular biology, including the triplet nature of the genetic code and the demonstration of mRNA. He won a Nobel Prize in 2002 for the development of a unique model system with which to study organ development. (Courtesy of the Salk Institute for Biological Studies.)*

appeared not to have been followed up in detail. Hall was interested in EM of small molecules and had published a text, *Introduction to Electron Microscopy* (9). He published the results of his investigations into the amount of heavy metal stains, such as phosphotungstic acid (PTA), which could be absorbed by viruses and measured by quantitative electron densitometry (18). Hall sought to measure the amount of stain absorbed by the viruses, positive staining, under various pH conditions, studying tomato bushy stunt virus (BSV) and TMV.

Hall found that greater stain absorption occurred with a lowering of the pH but that the integrity of the viral structure was compromised at those low pHs. At the end of his report he described "Anomalous Images" and included illustrative figures. One image showed BSV with stain between the particles which were lightly stained. Another image showed BSV in which the stain had been incompletely washed away. Both figures demonstrate what would somewhat later be called negative staining. Hall himself recognized that ". . . the visibility of particles of low scattering power can be enhanced as well, if not better, by surrounding them with dense material rather than impregnating them with dense material."

FIGURE 2 *Robert Horne standing beside a poster showing the atomic lattice of gold after optical linear integration of an electron micrograph of gold foil. Photograph taken by Alec Bangham, 1973. With Sydney Brenner, Robert Horne developed the technique of negative staining in a study of the components of a T-even bacteriophage. This technique was to revolutionize the morphological study of all types of viruses. (Reprinted with permission from Harris JR, Munn EA. 2011. Micron 42:528–530 [reference 136] © Elsevier, 2011.)*

Some 2 years later, H. E. Huxley at the University College London addressed the question of whether high-resolution EM might demonstrate structural features of TMV as had been shown with X-ray diffraction studies (19). Using the technique of Hall in sections in which stain had not been fully washed away, the viral particles were "outlined in a very distinctive manner." Also found was a stained line running within the particle. Huxley concluded that "The 'outlining' technique would appear to be a quite useful one for this type of specimen, particularly as it is so simple and gives excellent contrast and resolution." Prophetic words, indeed.

The *raison d'être* for negative staining arose from Sydney Brenner's need to screen a large number of T-even bacteriophage fractions ". . . to be monitored morphologically at relatively short intervals" (20). In his retrospective review, Horne detailed the steps needed to establish the conditions of the technique. The result was a simple, rapid procedure that allowed the combination of chemical techniques with negative-stain EM to analyze the components of a T-even bacteriophage (16). The investigators disassociated the bacteriophage into its component parts of head, tail sheath, core, and tail fibers. In addition, the authors analyzed the chemical composition and produced striking photomicrographs by negatively staining the individual parts (Fig. 3). They demonstrated filled and empty heads, extended and contracted tail sheaths, tail cores, and tail fibers. By chemical means, they evaluated the protein patterns of the head, sheath, and tail fibers and found that the primary protein structures differed. The investigators contrasted the complexity of the bacteriophage with plant viruses such as TMV. While the individual components of the bacteriophage seemed to adhere to a simple plan, they concluded that "the assembly of the parts would seem to pose a formidable problem." Read more than five decades after its publication, the paper of Brenner et al. is richly satisfying in its combination of biochemical and morphological analyses.

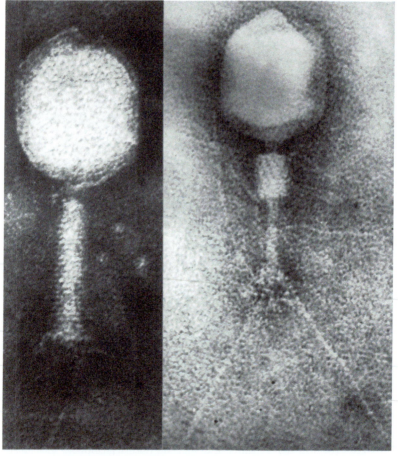

FIGURE 3 *"The first electron micrographs of negatively stained bacteriophages, prepared by Bob Horne." (Reprinted with permission from Harris JR, Munn EA. 2011. Micron 42:528–530 [reference 136] © Elsevier, 2011.)*

EM photomicrographs published by Brenner and Horne in 1959 using negative staining show a remarkable increase in structural detail compared with contemporary EM photomicrographs using techniques such as shadow casting. In a study of adenovirus, Horne et al. noted previous conflicting findings produced by other methods (17). A polyhedral shape had been suggested by standard staining and carbon replica techniques, yet a more spherical form was suggested by the shadow technique. Using their recently perfected method of embedding the viral suspension in PTA, which was electron dense, they demonstrated that the adenovirus was an icosahedron. They found it to be composed of a defined number of subunits "arranged not at random" but in a specific pattern, arranged with fivefold symmetry (Fig. 4).

The clear demonstration of subunit construction with cubic symmetry of adenovirus allowed Horne and Brenner to consider and support the theoretical suggestions for the construction of small viruses. F. H. C. Crick and J. D. Watson had hypothesized that small viruses were made up of "identical subunits, packed together in a regular manner" and were made up of identical protein molecules (21). For round viruses, they speculated on the need for a spherical shell and for rod-shaped viruses a cylindrical shell. Each would surround and protect a ribonucleic acid core. They reasoned that "the virus, when in the cell, finds it easier to control the production of a large number of identical small protein molecules rather than that of one or two very large molecules to act as its shell." The subunits

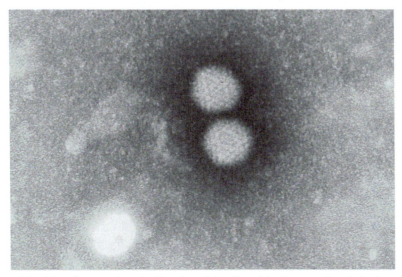

FIGURE 4 *Adenovirus, negative stain. Negative staining allowed the demonstration of subunit construction of viruses. With nucleic acid type, the architecture revealed by negative staining served as a basis for classification of viruses. (Photo by C. K. Y. Fong. Collection of Marilyn J. August.)*

would aggregate around the nucleic acid using a regular packing pattern. It would have a symmetry that "... in favorable cases" could "be discovered experimentally." Horne and Brenner appeared to have demonstrated just that.

The morphological findings revealed by negative staining allowed a proposal by André Lwoff and others, including Horne, of a general classification of viruses based on the nucleic acid type and the architecture of the capsid (22). A first division was made based on whether the virus contained RNA or DNA; a second division was made on whether the capsid symmetry was helical, cubical, or mixed; and a third division was made on whether the capsids were naked or enveloped. Other considerations were the number of capsomeres in viruses with cubic symmetry and the diameter of the capsid in viruses with helical symmetry. While such a straightforward classification based on criteria that are "molecular and structural" appears in retrospect to have been logical and intuitive, objection was raised in the day by a discussant who claimed that the criteria were "arbitrarily weighted." Yet the importance of the observations provided by negative staining on viral architecture and on capsid symmetry was to prove fundamental to other virologists in considering classifications of viruses. For example, June Almeida included capsid symmetry among the principal classification criteria (23). Once again, Crick and Watson had hit upon a fundamental biological principle. Horne and Brenner's study using negative staining of adenovirus had provided crucial supporting data. While from a clinical perspective, criteria based on organ symptomatology and epidemiological patterns would remain important, the negative-staining technique would also make major contributions to diagnostic virology (24–26).

Of the studies on negative-staining EM published by Brenner and Horne in 1959, the one commonly referred to in the viral diagnostic literature is "A Negative Staining Method for High Resolution Electron Microscopy of Viruses" (15). The paper starts by noting the possibility that the EM might give direct information about the appearance of viruses, but such information could be compromised by the problems of then-current methods, including obscuration of structural details or insufficient contrast. Negative staining was described as a technique which "extends the range of electron microscopical study of virus structures" by "embedding" viruses in electron-dense PTA. The authors meticulously detailed the conditions of mixing PTA with the virus suspension, the preparation of carbon-coated grids, and adjusting the concentration of each virus preparation. Images of TMV were compared with findings reported from X-ray diffraction studies. Negative-stained images of turnip yellow mosaic virus revealed "ghost" particles, that is, empty protein shells devoid of nucleic acid. The previous work of Hall and of Huxley, noted above, was credited with recognizing the technique's potentialities; but in history's judgment, the paper by Brenner and Horne became the seminal work for advancing the use of negative staining.

Horne, taken with the beauty of viral architecture, continued to make basic contributions to the understanding of viral form and construction (Fig. 5). In a review, "The Structure of Viruses," published 4 years later, he would write, "Viewing the micrographs one has the impression of being shown how the inanimate world of atoms and molecules shades imperceptibly into the world of forms possessing some of the attributes of life" (27). He was to go on to lead units in ultrastructural studies. Retiring from the John Innes Institute in 1983, he took up marine painting and was to write that it "... came about as a result of his studies of the development and form of sailing ships, with special reference to early shipwright's trades" (John Innes Foundation Historical Collections, Norwich, United Kingdom). One is tempted to replace "sailing ships" with "viruses" as a description of his professional work.

Brenner became one of the giants of molecular biology. Making pivotal contributions, he established the triplet nature of the genetic code and, working with others, the discovery of mRNA. Were those accomplishments not enough, he set about the study of the genetics of behavior (28).

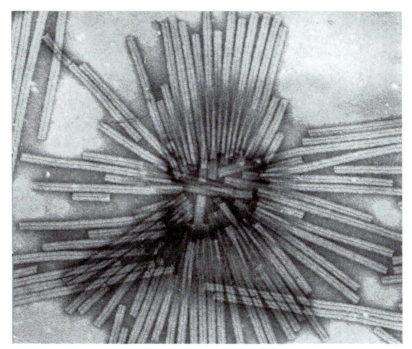

FIGURE 5 *TMV prepared by Robert Horne using the Horne and Pasquali-Ronchetti "Negative Staining-Carbon Film technique" showing two-dimensional paracrystalline/crystalline arrays of viruses. (Reprinted with permission from Harris JR, Munn EA. 2011.* Micron *42:528–530 [reference 136] © Elsevier, 2011.)*

Immunoelectron microscopy

EM demonstrations of specific antibody aggregating viral particles were published in 1941 from Berlin (29) and from Camden and Princeton, NJ (30). The intent was to examine the morphology of the precipitin reaction of antigen and antibody, rather than the discovery of new agents associated with medical illnesses. That would come later, and notably with acute nonbacterial gastroenteritis. In 1941, immunological theory suggested that the reaction of antibody with antigens in a precipitin reaction resulted from a lattice of antigens. Both the Berlin and New Jersey groups published micrographs demonstrating just that. In both laboratories, TMV was reacted with specific rabbit antiserum. The distinctive rod shape of TMV facilitated the findings of virus aggregates (29) and irregular frameworks of viruses and antibodies (30), respectively. The New Jersey group demonstrated antigenic specificity in that no aggregates were observed when anti-TMV antibody was mixed with BSV.

Thus, by 1941, a proof of principle, aggregation of virus by virus-specific antibody observable by EM, had been established independently in American and German laboratories. Yet with the hindsight of history, little advance occurred for two decades. The progressive development of ultrathin sectioning aided the observation of viruses in cells, but there was insufficient detail in most cases to allow specific identification. The introduction of shadow casting improved the three-dimensional understanding of viral morphology, yet its technical aspects precluded observation of virus-antibody interactions. These technical factors included the granularity of the metal used, which obscured detail, and the drying process for the metal application, which disrupted the virus-antibody structures under study (31). That rapidly changed with the advent of negative staining for the EM of viruses (15). The technique was rapid and simple, provided excellent detailed observations of viral morphology, and, in the words of Almeida and Anthony P. Waterson, ". . . overnight revolutionized the electron microscopy of viruses" (31).

Negative staining allowed visualization of the attachment of antibody to virus, and studies of the immune interaction of viruses and antibody were off and running in the early 1960s. T. F. Anderson et al. found that the attachment of antibody to virus embedded in phosphotungstate did not interfere with the observation of viral details (32). This feature allowed K. J. Lafferty and S. J. Oertelis to examine the kinetics of neutralization of influenza virus (33). They observed two site attachments of antibody when the antibody was not in great excess. Almeida and her colleagues in Toronto studied the interactions of polyomavirus and wart virus with antibodies raised in goats or rabbits (34). They had previously exploited negative staining to examine the fine structure of polyomavirus (35) and the morphology of wart virus (13). In the studies of viral and antibody interactions, Almeida et al.

reported that antibodies had the appearance of cylindrical rods, that the combining sites were on the short ends of the antibody, and that the specificities of most of the two combining sites were identical. They documented that viral-antibody aggregate formation was dependent on the concentrations of antibody used.

June Almeida (Fig. 6) played a crucial role in adapting the electron microscope to clinical diagnostic virology work. Born in 1930, she was brought up in Glasgow and left school at 16 (36). Without funds to go to university, she trained as a technician in histopathology at the Glasgow Infirmary. After a move to St. Bartholomew's Hospital, London, she married and immigrated to Canada. There she became an EM technician at the Ontario Cancer Institute in Toronto. Remarkable to consider, at Toronto with her colleagues she produced a series of studies applying negative staining to clinical problems as well as to basic studies, without the usual foundation of formal training in a degree-granting program. Clinical studies focused on skin lesions (37, 38). Vesicular lesions and benign cellular hyperplasia were found to yield the three morphological types of viruses: herpesviruses, poxviruses, and wart viruses (38). The conclusion was reached that negative staining would be of use in the diagnosis of "difficult or atypical viral lesions of the skin." An investigation of clinical varicella by negative staining revealed the morphological identity with herpes simplex virus and no morphological relationship to the poxvirus group (39).

At the invitation of Anthony Waterson, Almeida, who received an Sc.D. for her published contributions, returned to London in 1964 to continue her electron micrographic studies of antibodies (40). With David Arthur John Tyrrell, she demonstrated the usefulness of direct EM examination of organ cultures for viral isolation and characterization (41) and contributed to the study of respiratory viruses grown in organ culture (42). Notably, she developed IEM for clinical work. Early achievements with IEM included the first conclusive EM demonstration of rubella virus (43) and the demonstration of two components of the Dane particle (44).

An additional contribution was the training of others, both in the lab and by her writings on the techniques of clinical diagnostic virology by EM, for example, "Uses and Abuses of Diagnostic Electron Microscopy" (45). One of the most productive of those who worked with her was Albert Kapikian from the National Institutes of Health (NIH), who spent 6 months in her lab learning IEM (46, 47). Kapikian returned to the United States and used IEM to crack a conundrum that had baffled science for decades, the agent of acute nonbacterial gastroenteritis. That story is told next.

June Almeida was remarried in 1979 to Phillip Gardner, the pioneer developer of fluorescent-antibody (FA) studies for rapid viral diagnosis. In their retirement they had an antiques business, which seems nicely ironic for a couple who had each

FIGURE 6 *June Almeida. With training at the technical level, Almeida went on to receive a doctorate based on the body of her work. She pioneered the application of EM to clinical diagnosis, including IEM. (Courtesy of Joyce Almeida.)*

been at the forefront of modern virological methods. Gardner predeceased his wife in 1994 (48). June Almeida would pass away in 2007. Gardner's role in the development of FA studies for rapid viral diagnosis is recounted later in this chapter.

Winter vomiting disease

In the face of the marked success of tissue culture to isolate viruses associated with acute respiratory disease, the failure of such techniques to identify an etiological agent in acute nonbacterial gastroenteritis stood out. One particularly frustrating disease entity, included among a number of acute infectious illnesses of the intestinal tract, was winter vomiting disease (49, 50). It was separated from "the multitudinous forms of vomiting which occur in winter months" by John Zahorsky in 1929 (51) (Fig. 7).

Zahorsky, a St. Louis pediatrician, was born in Hungary and came to America with his parents at 6 months of age. He graduated from the Missouri Medical College in 1895 (52). Zahorsky was careful to separate the illness of winter vomiting disease from the intestinal symptoms which were sometimes associated with respiratory infections. He identified four diagnostic considerations. First, he noted disease occurrence in epidemics, reporting that a severe epidemic had occurred in 1925. In a later paper, reemphasizing its existence as a specific illness, he focused on an epidemic in St. Louis in February 1940, in which 3,000 children were afflicted (51). Second, he commented on the usually afebrile nature of the illness but did note a temperature of 99 to 101°F in some uncomplicated cases. Third, persistent vomiting was associated with "peculiar offensive light-colored stools." Fourth, he observed "the absence of an acute respiratory infection." Winter vomiting disease would be studied for decades, including prospective community surveys and transmission of symptoms by stool filtrates to volunteers, yet it was only at the start of the 1970s that IEM would reveal a viral etiology (53).

While Zahorsky relied on personal experience with hundreds of cases and observations of outbreaks of acute gastrointestinal illness, John Dingle and his colleagues prospectively observed families in Cleveland, OH, for signs of illness in the period from 1948 to 1950 (54). Among 1,466 cases of gastrointestinal symptoms, specific causes were found for 362 cases. As in Zahorsky's description, cases associated with respiratory symptoms were excluded from further study, leaving 683 cases of acute gastroenteritis for analysis. Symptoms selected for case inclusion were vomiting, diarrhea, and abdominal pain. The illness was characterized by abrupt onset and brief duration, and patients were frequently afebrile. Outbreaks were investigated epidemiologically and etiological agents sought. The CDC studied two Ohio outbreaks in 1968 in Norwalk and in Columbus (55). In the Norwalk outbreak, 50% of teachers and students in an elementary school became acutely ill over a 2-day period with nausea, vomiting, and abdominal pain of brief

FIGURE 7 *John Zahorsky, a pediatrician who first described "winter vomiting disease." He character-ized the clinical characteristics of outbreaks as early as 1925. Later, the illness was called acute non-bacterial gastroenteritis. A characteristic outbreak in Norwalk, OH, in 1968, investigated by the CDC, resulted in the isolation of the Norwalk agent, soon imaged by Kapikian and colleagues. (Reprinted with permission from* From the Hills: an Autobiography of a Pediatrician *[52].)*

duration. A common source exposure was suspected, and person-to-person spread was thought to account for secondary cases. No bacterial cause was demonstrated, and it was concluded that ". . . the association of specific viruses with epidemic winter vomiting disease awaits laboratory confirmation." Bacterium-free stool filtrates from the Norwalk outbreak were used in human volunteer studies of oral transmission (56). Confirming previous reports, it was found that experimental transmission occurred and that serial transmission was achieved, suggesting an infectious etiology. The investigators also induced gastrointestinal symptoms in a volunteer administered fluids passaged three times in human fetal intestinal culture.

At the start of the 1970s, acute infectious nonbacterial gastroenteritis was characterized as ". . . the commonest of the remaining acute infectious diseases whose etiologic agents are still to be cultivated in the laboratory" (49). The Norwalk agent had been transmitted to human volunteers by bacterium-free stool filtrates, and certain of its characteristics had been defined. It was found to be small, probably smaller than 36 nm, without a lipid coat, and acid and heat stable. The virus failed to grow in standard tissue culture systems and in infant mice and did not produce disease in rhesus monkeys, rabbits, mice, or guinea pigs. It could be passaged in fetal intestinal organ culture as noted above, but no morphological changes were detected by light microscopy. Thus, the identification of the causative agent remained stubbornly elusive.

That would change in November 1972 with the publication of the visualization by IEM of the Norwalk agent as a 27-nm particle (Fig. 8). The work was reported by Albert Kapikian and his colleagues at the Laboratory of Infectious Diseases (LID) at the NIH, the same laboratory that had been so successful in exploiting tissue culture technology in elucidating viral causes of other types of illness, particularly respiratory illness. Kapikian has termed the approach "direct virology" or "particle virology," in which the virus is sought without the intervention of tissue culture or animals (57). Human stool filtrates were mixed with convalescent-phase serum, centrifuged, and mixed with distilled water and PTA. The filtrate was from a volunteer who had developed diarrhea after ingesting a passaged Norwalk filtrate. The inactivated convalescent-phase serum was obtained from volunteers who had been experimentally infected.

The initial experiment revealed negatively stained aggregates of viral particles that were heavily coated with antibody. One can only imagine the excitement on first viewing the viral aggregates after decades of others failing to identify a cause of acute nonbacterial gastroenteritis. Yet there were hurdles to be overcome. Specifically, was the observed agent a mere passenger, unrelated to disease causation? Hence, further studies were undertaken to use IEM to demonstrate an immune response comparing acute- and convalescent-phase sera. Scales were devised

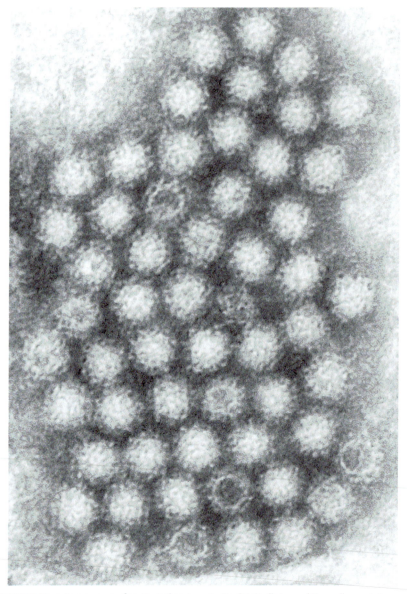

FIGURE 8 *Norovirus, a cause of acute viral gastroenteritis. Originally termed Norwalk agent, it was found by Albert Kapikian and colleagues at the NIH using antibodies to aggregate the virus by IEM. Kapikian had studied in the laboratory of June Almeida, where he learned the technique. (Courtesy of CDC-PHIL [#10704], Charles D. Humphrey.)*

which quantified both the amount of antibody according to the glistening appearance of aggregates and the number of aggregates. This allowed the demonstration of a serological response to the visualized virus. A serological response was found both in volunteers and in persons infected naturally. Furthermore, specificity of the serological response was demonstrated using isolates from different outbreaks. Subsequent buoyant-density gradient studies with the virus, in combination with its appearance and its acid, ether, and heat stabilities, allowed the investigators to suggest that it was "parvovirus-like" (58).

Albert Kapikian (Fig. 9), the first to demonstrate a significant viral cause of human gastroenteritis, was a legacy of Robert Huebner's tenure at the LID at the NIH. Born in Brooklyn, NY, Kapikian (1930–2014) went to Queens College, from which he graduated *cum laude* and where he was a star baseball pitcher. Notably, he had an 11-and-0 won-lost record as the starting pitcher on the Queens College baseball team. In medical school at Cornell, he came under the influence of Edwin D. Kilbourne and Walsh McDermott in public health and infectious diseases. As a medical student, Kapikian visited the NIH and had a personal tour of the LID arranged by McDermott and conducted by Huebner (59) (http://www.sabin.org/updates/pressreleases/vaccine-developer-albert-z-kapikian-md-awarded-sabin-gold-medal). Offered a position by Huebner, Kapikian joined the

FIGURE 9 *Albert Kapikian. One of the members of the LID of the NIH recruited by Robert Huebner, Kapikian was to make seminal contributions to the understanding of viral gastroenteritis. He was the first to demonstrate the Norwalk agent to be a virus (norovirus), assisted in the demonstration of hepatitis A virus, and was one of the first investigators to demonstrate rotavirus associated with acute infantile diarrhea. Kapikian is shown seated (center) with Robert Chanock, seated on the left. (Courtesy of the NIH.)*

LID in July 1957 following an internship at the Meadowbrook Hospital on Long Island. Apparently, Kapikian remained drawn toward clinical medicine, keeping a stethoscope with him for several years while working in the lab until teased by Huebner (59). Coincidentally, Robert Chanock and Al Kapikian both arrived at the NIH on 1 July 1957. It is remarkable that Chanock was to break open the study of viral respiratory diseases with cell culture techniques and Kapikian, after cell cultures had failed, broke open the study of gastroenteritis viruses with IEM. Kapikian was assigned to the Epidemiology Unit under Joseph Bell and worked on the Children's Village project. He became the head of the Epidemiology Unit after Bell's retirement.

According to Kapikian, the IEM work in Tony Waterson's lab in London with June Almeida was originally aimed at the study of coronaviruses, most of which would not replicate in cell culture (http://www.sabin.org/files/gold medalspeech2005.pdf). That goal was reached on his return to the LID with the publication of immunoelectron micrographs of a coronavirus harvested from human embryonic tracheal organ culture (60). In an additional groundbreaking application of IEM with Stephen Feinstone and Robert Purcell, the agent of hepatitis A was demonstrated (61). However, it was to studies of viral gastroenteritis and studies of rotavirus vaccine development that Kapikian devoted his greatest energies in coming years.

Following his initial demonstration of the Norwalk agent, Kapikian continued his studies in search of the agent at Children's Hospital in Washington, DC, but he found instead a high prevalence of rotaviruses among infants hospitalized for diarrhea (http://www.sabin.org/updates/pressreleases/vaccine-developer-albert-z-kapikian-md-awarded-sabin-gold-medal). Rotavirus disease, in contrast to the benign course of illness induced by the Norwalk agent, can be a cause of lethal diarrhea in infants and young children. Kapikian's lab developed vaccines employing a Jennerian strategy using an animal strain of rotavirus and inserting several human rotavirus antigens by molecular techniques. Detailing that fascinating vaccine work is not our aim here, and we turn next to the discovery of rotaviruses by EM.

Acute infantile diarrhea

It is paradoxical that IEM successfully defined the viral etiology of one of the acute nonbacterial gastroenteritis syndromes, winter vomiting disease, before basic EM defined the viral etiology of another highly prevalent gastroenteritis, acute infantile diarrhea. However, when the EM definition of acute infantile diarrhea was demonstrated to be viral in nature, it was reported virtually simultaneously from research groups on several continents within a matter of months. Various names were proposed by investigators before the name was settled as rotavirus. That name was suggested based on the wheel-like appearance of the virus, with *rota* derived from the

Latin for "wheel." It was also recognized that calf diarrhea virus had the same appearance (62). Remarkably, as early as 1943, stool filtrates from newborn human infants with epidemic diarrhea were demonstrated to induce diarrhea in calves (63). That early insight was apparently lost in later attempts to isolate an agent. Then, a lyophilized sample prepared in 1943 was examined by IEM in 1976, revealing characteristic virus particles (Wyatt et al., cited in reference 64).

In 1968, Hugh Moffet and his colleagues at the Children's Memorial Hospital in Chicago reported a prospective 2-year study of the infectious etiology of severe, acute infantile diarrhea (65). The peak annual onset of disease was in December, and the median age of onset was 6 months. A broad range of media was used to isolate pathogenic microbes, including shigella, salmonella, and candida species and enteropathogenic *Escherichia coli*. Three tissue cultures were used to isolate viruses: primary human embryonic kidney, HEp-2, and primary rhesus monkey kidney cells. They found adenovirus in 17% of patients and in only 3% of healthy subjects. This was a new finding which was attributed to the use of human embryonic kidney cells, which readily support the growth of adenoviruses. Yet no etiological agent of diarrhea was found in the majority of cases.

In 1970, M. D. Yow et al., working with Joseph L. Melnick, one of the leaders in virology, reported etiological studies on infantile diarrhea in Houston (66). Two types of cell cultures were used, rhesus monkey kidney and human embryonic kidney. While viruses were isolated from 27% of study participants, that number did not differ significantly from isolations detected in 19% of control subjects. Hence, even in the hands of a leader in virology, cell cultures yielded disappointing results. The concluding sentence of the report was prescient: ". . . there may be agents which do not produce cytopathogenic effect and their presence may have to be demonstrated by other methods."

The successful method, soon to be applied, was negative-stain EM. The initial link was an anatomic investigation, not microbiological. In November 1973, Ruth Bishop (Fig. 10), a bacteriologist, and her colleagues in Melbourne, Australia, reported the observation by EM of virus-like particles in duodenal epithelial cells from two of four patients with gastroenteritis (67). The tissue had been obtained by capsule biopsy from children under 3 years of age with gastroenteritis. The capsule biopsy procedure had previously been utilized to study children with malabsorption (68). In a study by G. L. Barnes and R. R. W. Townley using capsule biopsies, viral cultures done on the first 10 of the 31 patients studied were reported to be "unrewarding" (69). Thus, in following the publication record, there appears to be a bit of serendipity. In the article describing the discovery of the viral particles, Bishop et al. pointed out the "marked resemblance" to the agent of epizootic diarrhea of infant mice. Hence, from the start it was recognized that the agent that would later become identified as rotavirus could be found in numerous species.

FIGURE 10 *Discoverers of rotavirus as the cause of acute infantile diarrhea. Ruth Bishop (left) of Melbourne, Australia, was the first to report the virus of infantile diarrhea by EM in biopsy samples of the duodenum. Thomas Flewett (middle) of Birmingham, England, established the presence of rotavirus in stool specimens by EM. Albert Kapikian of the NIH, Bethesda, MD (right), identified the virus by using IEM and conventional EM (see Fig. 9). (Courtesy of Graham Beards, under license CC BY-SA 4.0.)*

The following month, in December 1973, not knowing the fate of the submission of the letter to the *New England Journal of Medicine* because of a postal strike (70), Bishop and her colleagues published electron microscopic findings in *The Lancet*, describing duodenal biopsies from nine children less than 3 years of age with acute gastroenteritis (71). Samples from six of the nine patients revealed viral particles described as having an electron-dense core 33 nm in diameter, with a capsid of 67 nm; 10% appeared to be enveloped, with an overall diameter of 87 nm. Examination of repeat biopsy samples 4 to 8 weeks later from three of the virus-positive infants failed to disclose the virus. The authors regarded the viral particles as important in the etiology of the disease because they were present during the illness but not after recovery. Based on the morphology, size, and presence of enveloped and nonenveloped particles, they concluded that these viruses belonged to the orbivirus group, which had been recently described. They again pointed out the resemblance to the virus of epizootic diarrhea of infant mice and noted the difference from the Norwalk agent as found by Kapikian et al. by IEM (53). Years later, Bishop would be quoted as saying that the report brought a worldwide response of others finding the virus (68). She went on to say that they were concerned about people with less skill doing the biopsies, but noted ". . . that vets had developed a staining technique for rotavirus in cow feces and we developed this into a clinical diagnostic test."

Acknowledging Andrew Turner and his colleagues at the Department of Veterinary Research Laboratory, Melbourne, for the purification technique for fecal extracts in February 1974, Bishop and her colleagues reported the successful identification of viral particles from samples of stool from children with acute nonbacterial gastroenteritis (72). No such viral particles were seen in fecal samples from control children. The samples were extracted and subjected to a series of centrifugation steps prior to negative staining for EM. The authors concluded that differential centrifugation for extraction of feces was more sensitive than biopsy for detection of the particles and that duodenal biopsy was no longer necessary. They expressed the expectation that this method would allow global epidemiological studies of viral distribution. As it turned out, laboratories on other continents were demonstrating the virus in feces and in intestinal juices. Prominent among the investigators were Thomas Flewett and his colleagues in Birmingham, England (Fig. 10).

Reporting just 3 weeks after the first *Lancet* article of Bishop et al., Flewett and his coworkers found virus particles in stools of children with gastroenteritis (73). The procedure used was a two-step centrifugation, the first at low speed to remove sediment and bacteria and the second to pellet the viral particles; negative staining followed. The particles were 75 to 84 nm in diameter and were observed to resemble reoviruses, although the investigators felt that it was premature to classify them in a subgroup. Hence, it appears that the Melbourne group of Bishop established priority in finding the virus of acute infantile diarrhea in tissues, but the Birmingham group of Flewett was the first to demonstrate it in fecal specimens.

In 1974, the Flewett group published a pair of papers discussing the viral flora of feces by EM (74) and gastroenteritis associated with reovirus-like particles (75). In the first of those papers, the authors observed that electron microscopic examination of feces for viruses was "—a technique long neglected—," a point that is quite striking in that negative staining was first applied to viral studies in 1959. One can speculate that diagnostic virology was focused on tissue culture isolation and note that the tissue culture revolution started with the successful growth of an enteric virus, poliovirus.

Thomas Flewett had been appointed Director of the Regional Virus Laboratory when it was established at East Birmingham Hospital. "Here his interests in virus diseases and electron microscopy led him to provide a comprehensive service and to investigate the viral causes of childhood diarrhea..." (76). Flewett was described as a "born tinkerer, and electron microscopy with its mystique and its machinery suited his temperament exactly." It was from the Regional Virus Laboratory, in collaboration with colleagues at the Institute for Research on Animal Diseases at Compton, that Flewett and colleagues suggested the name rotavirus based on its appearance by EM (62) (Fig. 11). The human and newborn calf diarrhea viruses were described as differing from orbiviruses and reoviruses, hence deserving of their own name. "Rotavirus" was a name that stuck and served to identify a group of viruses of similar morphology causing gastroenteritis in the young in several species of animals (77).

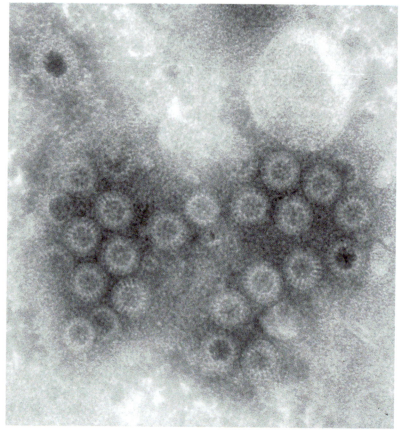

FIGURE 11 *Rotavirus in stool. This negative-stained preparation shows the characteristic wheel-like appearance of rotavirus. The name was derived from the Latin* rota, *for "wheel." (Collection of Marilyn J. August.)*

Meanwhile, in Toronto, Peter Middleton's group at the Hospital for Sick Children had embarked on a broad-ranging study to identify etiological agents in acute nonbacterial gastroenteritis of infants (78, 79). Stool samples for negative-stain EM were prepared by direct examination or following a two-stage centrifugation process. An indirect immunofluorescence assay was also developed. Feces were found positive by both techniques for what the investigators termed an orbivirus. It was positive in 48 cases but negative in 86 controls, positive in duodenal aspirates from 10 patients but negative in 3 controls, and positive in 7 autopsy samples of upper jejunum of patients who had not been hospitalized but negative in 14 control autopsy samples. Virus was not found in the convalescent period in a more limited number of patients.

The observed virus had a mean diameter of 64 nm, and pseudo-enveloped forms measured just over 79 nm in diameter. Parallel studies failed to reveal a viral isolate in an extensive array of cell cultures. Virus was not demonstrated in human embryo organ cultures of upper jejunum or trachea or from suckling mice or embryonated hens' eggs. A preponderance of virus-positive samples was found in the colder months of the year, and the age distribution was predominantly under 2 years.

Thus, the work of Middleton et al. in Canada identifying a virus resembling orbiviruses confirmed the observations of Bishop et al. in Australia and of Flewett et al. in England. Their extensive sampling procedures found the virus in duodenum, in upper jejunum, and in stool. They confirmed that the presence of virus was limited to the acute phase of the illness and was not found in controls or after convalescence.

In the United States, Kapikian, the discoverer of the Norwalk agent by IEM, applied the technique to studies of infants and young children hospitalized for acute gastroenteritis at Children's Hospital in Washington, DC, from January to March 1974 (80). Stool filtrates were incubated with serum from a convalescent patient. The investigators applied the IEM assay for antibody as had been developed for the Norwalk agent, in which scoring is based on the amount of antibody coating viral particles. A complement fixation (CF) assay was also developed which used a diluted stool filtrate as the antigen. Stool samples from 13 of 21 patients were positive by IEM and conventional EM for what the investigators termed reovirus-like particles. Each of five patients tested showed the development of a serological response by the IEM method. The CF assay was positive for these patients and five additional patients. Using the CF assay, the investigators were able to show an antigenic relationship of the human virus to the virus of epizootic diarrhea of infant mice as well as a relationship with the Nebraska calf diarrhea virus. C. A. Mebus et al. found that the agent of infantile gastroenteritis could induce diarrhea in gnotobiotic calves (81). In another study, investigators were able to find evidence by indirect immunofluorescence of growth of the agent in human fetal intestinal organ culture (82).

Kapikian et al. noted that the CF assay would facilitate laboratory and epidemiological studies of the stool agent (80). In preliminary studies, they found that the CF antibody was "quite common" in adults in the United States, implying widespread distribution of the virus. They speculated that it was possible that the virus would "emerge as a major etiologic agent of diarrhea in infants and young children." These were again prophetic words, for global epidemiological studies would find that rotavirus was the leading cause of death in children hospitalized with diarrhea (83). It has been estimated that rotavirus was responsible for over 600,000 diarrhea-related deaths in children annually worldwide. There was a steep gradient of deaths related to income in the countries surveyed, with 85% of deaths occurring in countries classified as low income. The importance of effective vaccines, particularly for low-income countries, could not be clearer (84–87).

TIMELINESS OF DIAGNOSIS: THE DEVELOPMENT OF FA

The advent of tissue culture techniques opened up diagnostic virology in the 1950s and 1960s, defining dozens of viruses with or without disease associations. Yet the application of these techniques required several days to weeks for isolation and characterization of disease-associated viruses. The observations were of great importance to further define the spectrum of clinical infectious diseases, to provide a starting point for the understanding of human host-virus interactions, and to support epidemiological surveillance of diseases in communities and regions. On the other hand, EM could, on occasion, provide a definitive diagnosis within hours. This was particularly important for patient care and public health management in distinguishing a poxvirus or a herpesvirus as the cause of an ambiguous rash with potential lethal risk. Yet the use of diagnostic EM was not widely practiced, requiring special, costly equipment and highly trained personnel. The clinical sample needed to have a high concentration of viral particles in the original material or be enriched by centrifugation to yield a prompt diagnosis without going through an amplification step in tissue culture, eggs, or animals.

With limited laboratory resources for a rapid diagnosis, the clinician responsible for a child acutely ill with bronchiolitis, for example, would have to rely entirely on clinical acumen for management of the acute phase of the illness. The situation was made more urgent as potential antiviral treatments started to be investigated for specific viral illnesses. In sporadic encephalitis, for example, antiviral agents for herpes simplex encephalitis began to be investigated. However, the first compounds investigated had significant morbidity. Thus, there were several clinical circumstances of patient management that would benefit from rapid viral diagnosis. In practical terms, this would mean an answer with virus identification within a few hours to 24 hours of admission to the hospital (see epigraph). A technique that would meet those needs was FA staining, developed thanks to the imagination and perseverance of and research support for Albert Hewlett Coons.

Coons and the development of the FA technique

Albert Hewlett Coons (Fig. 12) was born in 1912 in a small upstate New York community, Gloversville, which was named after its principal industry. It is located about fifty miles west and north of Albany at the southern edge of the Adirondacks region. The city's glove industry blossomed in the second half of the 19th century after the invention of the sewing machine and the stimulus of demand during the Civil War.

FIGURE 12 *Albert Coons, the originator of the FA staining technique. The technique identified antigens in tissues using antibodies tagged with compounds which would fluoresce under illumination by ultraviolet light. It allowed the development of rapid viral diagnosis. (Courtesy of the Lasker Foundation.)*

After attending local schools, Albert Coons graduated from Williams College in 1933 and went on to medical school at Harvard (88). Here, in his second year, he encountered the charismatic teaching of Hans Zinsser, who gave a course in immunology. "Zinsser was in his prime, full of vitality, intensely curious and completely informed about medical bacteriology" (89). Coons was hooked, and the following summer he worked in the laboratory of John Enders on a problem relating to anaphylactic shock in guinea pigs. While the project itself was not completed, it stimulated a lifelong interest in microbial pathogenesis and in Coons's words "bent the twig."

According to Hugh McDevitt, Coons had originally intended to become a clinician (90). However, as Coons himself told the story, a trip to Berlin as a tourist in the summer of 1939, before World War II, would be decisive for his thinking (89). He spent his mornings watching autopsies at a pathological institute and his afternoons wandering about the city or sitting in his room "reading or brooding." In thinking about the Aschoff nodule, a finding in rheumatic fever, and the relationship of the antigens of the streptococcus and antibodies, he realized the importance of devising a way to demonstrate the immunological components. "The notion of labeling an antibody molecule with a visible label was perfectly obvious in such a context (89)." McDevitt thought that the comment was too modest and that as of 1939 "the concept of putting a visible label on an antibody seemed both bold and original, even if technically naïve" (90). Coons, in a fitting image, would later characterize the effort as "putting tail lights on antibodies" (88). The perfection of the technique to attach a signal visible by microscopy was to consume several years, interrupted by World War II, and for which he was greatly aided by the scientific resources available at Harvard.

For the young Coons, the project would require a change in career direction from clinical work to a research focus which would require support, mentoring, and collaboration with chemists. Fortunately, all were available (89). The Director of the Thorndike Memorial Laboratory of the Boston City Hospital, George Minot, a Nobel Prize winner for work on pernicious anemia, encouraged Coons to pursue research on his idea, and he provided funds for reagents. John Enders took Coons into his lab as a research fellow the following year. The first issue was whether antibodies could be chemically manipulated to attach a visible marker while retaining specific immunological activity. It turned out that there had been prior encouraging studies. L. Reiner, whose goal was to "increase the destructive effect of antibodies on pathogenic antigens" by attaching diazo compounds, found that modified antipneumococcal antibodies retained their immunological specificity (91). Interestingly, the product was called an "antibody-dye."

In 1934, John Marrack reported that typhoid or *Vibrio cholerae* antibodies, which had been coupled to a red dye, gave a visible reaction against their immunospecific target (92). While this was a putative proof of concept, when Coons repeated the experiments, he found that there was insufficient intensity of color to be diagnostically useful in tissue. Hence, it became necessary to couple some other more visible type of marker to the antibodies.

It turned out that the laboratory of an organic chemist at Harvard, Louis Fieser, was investigating the conjugation of carcinogenic hydrocarbons to proteins for the purpose of immunization against cancer (93). Fieser volunteered two chemists in his lab, Hugh Creech and Norman Jones, to assist Coons. According to Fieser, they were "already busy hooking fluorescent compounds to proteins" (89). The pair of chemists reported monitoring the degree of conjugation of horse serum albumin to an isocyanate by ultraviolet absorption spectrophotometry (94). Together with Coons, they produced an anthracene-coupled antiserum against pneumococci, which agglutinated pneumococci and produced a bright blue fluorescence when viewed under ultraviolet light (95). However, since mammalian connective tissue also produced a blue fluorescence, they concluded that the conjugate was "inadequate for the demonstration of antigen in tissues."

Coons turned again to Fieser to request assistance in making another compound, fluorescein isocyanate. Fieser assigned Ernst Berliner, a graduate student, to the task; he quickly succeeded. Additionally, Berliner instructed Coons in the technique. According to Coons, Berliner produced the compound, Creech coupled it to the antibodies, and Jones measured the degree of attachment of the fluorescent label to the specific antibody (89).

The next piece of the puzzle was a means to visualize the fluorescein-labeled organisms in prepared specimens. To seek pneumococci in fixed tissues, Coons needed a fluorescence microscope. In "another stroke of luck," a former laboratory instructor of his, Allan Grafflin, in the Department of Anatomy, was at the time piecing together such an instrument. Together the investigators succeeded in showing specific, green fluorescence of pneumococcal antigen in tissues of heavily infected mice (96). The choice of fluorescein was to prove a "happy one," for as Coons later reported, its emission wavelength was close to the maximum sensitivity of the retina (97). Coons has told a charming story about the publication of what is, after all, one of the seminal papers for the use of FA staining to find antigens in tissue (89). On leaving for the Army on a cross-country trip, he wrote the paper describing his success. He sent the manuscript back to Enders for editing and submission but heard nothing. Ultimately, Coons saw that his paper had been published when he opened his newly subscribed journals on the day he embarked by ship for New Guinea.

Returning to Harvard after the war, with the chairman Howard Mueller's encouragement to work independently, Coons undertook a series of investigations which refined the technique of fluorescein staining and demonstrated its applications. In the January 1950 issue of the *Journal of Experimental Medicine*, Coons and his colleagues published three reports advancing the technique. With Melvin Kaplan he tackled the problem of nonspecific staining. They purified the conjugates and empirically discovered that absorption with suspensions of ground liver removed staining of normal tissue (98). Also studied was the application of FA staining to localize the antigens of the rickettsiae of epidemic typhus and Rocky Mountain spotted fever and mumps virus in experimental animals (97). The point was made that the identification of rickett-siae was considerably accelerated, from at least 2 weeks for inoculations of animals and serological studies to under 2 hours, by FA staining. The time advantage would play out in years to come as a first-line test for rapid viral diagnosis in clinical virology.

Meanwhile, fluorescence studies were applied in other virus-host systems, and reciprocally, viral studies were also used to advance the technique's usefulness. B. K. Watson used immunofluorescence to localize mumps viral antigen in tissue culture (99). This was a forerunner of the use of FA staining in clinical situations to detect viral antigens in tissue culture before the appearance of or in the absence of cytopathic effect, thereby speeding the diagnostic process. The indirect FA tech-nique was first described in studies of varicella and herpes zoster by Thomas Weller and Coons (100). With the direct technique, each primary antibody directed against the antigens of interest was conjugated to a fluorescent label. In contrast, in the indirect technique, a secondary antibody directed against the animal species of the primary antibody is conjugated to the fluorescein label. The indirect proce-dure was considerably simplified, since one fluorescein-conjugated antibody stock could be used against numerous specific antibodies as long as they were made in the same species. For example, Weller and Coons used as a secondary antibody an anti-human gamma globulin made in rabbits and conjugated it with fluorescein; thus, it could be used to detect any virus-specific antibody in human serum.

Technical variations soon were reported to enlarge the scope of the assay. R. A. Goldwasser and C. C. Shepard introduced a method which utilized the fixation of complement in an antibody-antigen reaction. Only one conjugate is, therefore, needed to react with the complement for the visualization of antigens, reducing the number of reagents that required labeling (101). The capacity to describe the anatomic relationship of two different antigens in cells or the presence of a dual infection was opened up when A. M. Silverstein described a two-color system for fluorescent labels to distinguish different antibodies in the same biological sam-ple (102). Application of the indirect FA method to measure antibody develop-ment over time was reported by Ch'ien Liu et al. for patients with primary atypical pneumonia (103). They demonstrated the development of antibody in the second and third weeks of infection which lasted for more than a year.

A considerable advance was reported by J. L. Riggs et al. in 1958 in the prepara-tion of conjugated antibodies (104). Until this time, fluorescein isocyanate con-jugates were prepared with phosgene, a dangerous substance which had a limited half-life. Riggs and his colleagues substituted isothiocyanate for isocyanate, thereby eliminating the danger of phosgene gas and producing a more stable compound. They demonstrated the usefulness of isothiocyanate for both green-fluorescing fluorescein and red-fluorescing rhodamine as well as for several antigens, including adenovirus. Commercial preparation of such stains extended the use to diagnostic labs that would not otherwise have had the resources to produce the reagents.

Early viral diagnostic studies with FA techniques

"Time lost" is a crucial factor in the management of viral diseases, both from the medical perspective of an individual's care and from the public health perspective for the management of the risk of disease spreading in populations. Medically, for example, the decision to start postexposure rabies prophylaxis was not to be taken lightly for an individual bitten by an animal suspected of being rabid. The risk of an adverse immune response to the vaccine itself and the pain and discomfort of a series of inoculations required with earlier vaccine formulations made difficult the decision to start treatment.

Likewise, in public health, particularly as rapid international air travel flourished, smallpox posed a risk before its global elimination. The decisions were complex concerning a traveler with a suspicious rash arriving from an area where smallpox was endemic. For example, was quarantine required for the individual and contacts? Was mass immunization in crowded urban centers needed, with the risk of ensuing panic (105)? Decisions in these cases could be accelerated and made more accurate by the application of FA staining techniques for diagnosis with clinical specimens. The need for urgent rapid viral diagnosis and the elimination of time lost in clinical treatment became more evident as specific antiviral medications were developed. For example, while antiviral therapy trials would depend primarily on viral isolation (106, 107), FA studies on brain tissue helped to define the specific diagnosis of herpes simplex virus and distinguish it from other conditions presenting in a similar manner (108, 109). The first demonstration of the usefulness of FA studies for diagnosis of herpesvirus infections was by J. Z. Biegeleisen et al. in 1959 in vesicular lesions (110), yet the application of FA techniques to clinical virological diagnosis had begun 3 years earlier.

Ch'ien Liu, a colleague of Coons, is credited with the first study using FA for clinical diagnosis (111). Liu noted that the laboratory diagnosis of influenza was usually made retrospectively by serological studies or viral isolations. The conditions seemed ripe for Liu to undertake a clinical diagnostic study of influenza in humans; he had studied the pathogenesis of influenza in the ferret model with FA (112). Working in cooperation with Coons, whom he credited for "encouragement and interest" and for the FA preparations (111), Liu obtained specimens from students with a clinical diagnosis of influenza or respiratory infection at the Harvard University and Phillips Exeter Academy infirmaries. Collection of specimens was thorough, if a bit arduous for the subjects. Nasal washings were collected from fluids instilled into the nose and allowed to drip out into a collection dish. An aliquot of washings was stored for virus isolation, and the remainder was prepared for FA examination. For some patients, swabs from nasal turbinates were also examined. Sera were drawn for hemagglutination inhibition antibody studies. According to the tabulated data for influenza A, FA on nasal washings was more sensitive than viral isolation but less so than hemagglutination inhibition serology. Swabs from turbinates were unproductive for viral diagnosis. Sensitivity for diagnosing influenza by FA was 71% for type A and 38% for type B. It was suggested that the difference could be due either to a less potent conjugate or to a milder infection. There was one false positive in 23 patients without influenza virus infection. The author observed that whereas the serological test was more sensitive, it required 10 to 14 days for convalescent-phase serum collection. In contrast, FA allowed diagnosis on the day of specimen collection. Presciently, it was observed that "if therapeutic agents for viral infections ever become available, retrospective diagnostic tests might become impractical and a rapid diagnosis of disease becomes desirable."

The demonstration of FA diagnosis of rabies virus infection in 1958 by Goldwasser and R. E. Kissling of the CDC in Atlanta had a major impact on the standards of practice for rabies diagnosis (113). Previously, the diagnosis relied on the observation of Negri bodies in tissues; however, there was a rare but consistent failure to diagnose rabid animals. Noting the risk inherent in antirabies vaccines, the authors pointed out the need for a rapid and accurate assay. A new test was developed using impression smears of brains from experimentally infected mice. They confirmed that Negri bodies contained viral antigens throughout, not only in the inner granules. Staining was observed against so-called fixed virus, that is, virus which had been stabilized during passage through laboratory animals so that the incubation period had become fixed. Since fixed virus did not form Negri bodies but stained with specific antibodies, the finding raised the likelihood that fluorescence could be used for rapid diagnosis in suspected rabid animals in which Negri bodies had not been observed. In addition, assay of sera of persons vaccinated against rabies demonstrated good fluorescence, suggesting a role for documenting immunity using this technique.

Preliminary studies with specimens from naturally infected animals demonstrated rabies virus antigen in the salivary glands. The usefulness of salivary gland examination for viral antigen by FAs was confirmed in a study of specimens submitted for rabies determination in the state of Alabama (114). Confirmation of the accuracy and sensitivity of the assay emerged from other labs (115), and as Riggs reported, FA staining became the standard procedure for the diagnosis of rabies in the California State Department of Public Health (116).

Another report from the CDC, which had become a leader in the application of FAs to clinical diagnosis, demonstrated the usefulness of immunofluorescence in the presumptive diagnosis of smallpox (117). The initial study involved a boy with a vesicular rash returning to Canada from Brazil. His travel connections were through a New York airport, and it was of crucial importance to determine if contacts needed to be located and vaccinated. Vesicular fluid was positive for poxvirus by FA, the result being compatible with either variola or vaccinia. Two days later, inoculated eggs determined that the agent was variola virus. In a postscript, the authors reported that FA accurately assisted diagnosis in 10 further cases, 3 of which were variola.

Concerning the CDC, Coons modestly observed that ". . . although my colleagues and I had a hand in introducing fluorescent antibodies as useful immunological reagents, the largest and most active group now working with them is at the Communicable Disease Center " (118). Coons and his colleagues at Harvard had introduced and refined the technique and made initial studies of viral agents, and the CDC had amplified its public health applications (119). The prominence of FA as a rapid virus diagnostic technique would be most fully exploited by Phillip S. Gardner and Joyce McQuillin in the United Kingdom.

Rapid viral diagnosis: the role of Phillip S. Gardner and Joyce McQuillin at Newcastle-upon-Tyne

With the pioneering efforts of Phillip S. Gardner and Joyce McQuillin, fluorescence microscopy came to be a broadly useful rapid virus diagnostic technique. Although the procedure had been developed in the early 1940s by Coons and his colleagues, it had taken a very long time for the technique to mature as a clinical tool (96). One wonders if it took the press of clinical urgency to overcome the previously encountered problems with the assay. Gardner defined rapid viral diagnosis as an etiological answer within a few hours of admission of the patient to the hospital (2).

As of the first decade of the 21st century, acute infection of the lower respiratory tract was still the leading cause of mortality in children worldwide and respiratory syncytial virus (RSV) was the most common cause of infection at that site (120). Decades earlier, in the latter 1950s and the early 1960s, soon after the virus had been described, its extremely important role in respiratory disease in children was still to be defined. Phillip Gardner would play a major role in that definition. He began his studies of childhood respiratory pathogens in 1959 (121) and progressed to the study of the rapid diagnosis of RSV by the FA technique with Joyce McQuillin (122). The focus on FA techniques led Gardner in 1975 to push for the formation of the European Group for Rapid Viral Diagnosis (48), an organization of virologists committed to the dissemination of advances in the field (123).

In 1959, Gardner and his colleagues set out to correlate clinical and microbiological findings in children with acute respiratory disease (121). Clinically, 146 children at three hospitals were divided into five disease categories: upper respiratory infections, croup, acute bronchitis, acute bronchiolitis, and pneumonia. Serological assays and bacteriological isolation cultures were undertaken. Viral isolation studies on "cough swabs," which had been frozen at the bedside, included inoculation of hens' eggs, HeLa cells, and monkey kidney cells, with follow-up hemadsorption assays on cell cultures. No correlation was demonstrated with any particular bacterium, and isolates were frequently normal flora. Several different viruses were found in association with each clinical category, except among 27 individuals with acute bronchiolitis. In fact, four of the six deaths observed in the study group were due to acute bronchiolitis, yet no virus was identified. The authors speculated that the lack of virus isolation may have been due to the inability to sample the lower respiratory tract or, presciently, the presence of an unidentified virus. This negative finding was to lead to studies focused on RSV.

The puzzling absence of isolates in bronchiolitis was explained by the work of M. Beem et al. (124) published in the same year as the survey by Phillip Gardner et al. These investigators found that RSV was a significant cause of lower respiratory tract infection in children and in infants less than 6 months of age. Beem and his colleagues found that isolates peaked in a brief period in the winter. Importantly, they had found the crucial piece of the puzzle: RSV was inactivated by freeze-thaw specimen preparation, and specimens had to be immediately inoculated into culture for successful isolation. Robert Chanock, the discoverer of RSV in children, and his colleagues at the NIH and the Children's Hospital in Washington, DC, soon confirmed these observations (125). They had done surveillance on hospitalized children with lower tract respiratory disease explicitly for RSV infection yet had had a paucity of isolates. When they adopted the strategy of inoculating tissue culture fresh, within hours of collection, they found that 42% of children with bronchiolitis and 21% with pneumonia yielded RSV in culture. The virus was most commonly found in children in the first half-year of life. They found, too, that RSV occurred in outbreaks of 3 to 5 months in duration and that inter-outbreak periods were virtually bereft of RSV isolates.

Tucked away in the "Virus Recovery" section of the paper by Chanock et al. was an observation that would be a harbinger of the future work of Phillip Gardner and Joyce McQuillin. Chanock et al. reported that viral cytopathic effect in tissue culture was not evident before 6 days, and although the majority of isolations occurred by 2 weeks, the maximum time for isolation was 22 days (125). This delay for an answer that was only suggestive, not definitive, was clearly insufficient for decisive clinical intervention in an illness that could be lethal to young children.

The lost-time problem in RSV infections was emblematic of viral diagnostic studies in general and would lead to the movement for rapid virus diagnosis. That effort would be spearheaded by Gardner and McQuillin in their perfection and application of FA techniques.

Gardner, working with a senior registrar physician in Newcastle-upon-Tyne, reconfirmed the need for unfrozen material to isolate RSV (126). In the winter of 1962–1963, they studied exactly the same number of cases of bronchiolitis as in the earlier negative study, 27. They were able to isolate RSV from 6, or 22%. They found that some of the viral isolates did not emerge until 4 weeks in culture, even longer than the period found by the Chanock group. Gardner speculated that while calf serum may have preserved viral viability, it may also have slowed viral growth.

Further studies were performed examining the serological response to RSV (127). Over the course of three winters, they found that RSV was the most common cause of severe lower respiratory tract infection in infants and children. They also found that the serological response to the virus increased as maternal antibody waned.

The seriousness of RSV infection was found by Gardner and his colleagues in a 27-month study of deaths in 22 children with respiratory tract infections (1). Among the victims, six deaths were associated with bacterial infections, and in eight RSV was the only pathogen isolated. Nine of the children had major congenital abnormalities which were felt to have contributed to the fatality of the infection. Hence, this study underlined the urgency of management decisions in caring for infants and young children suspected of having acute RSV infection of the lower respiratory tract.

The studies in the 1960s of acute lower respiratory tract RSV infection by Phillip Gardner and his colleagues had been brought to the point that "... it was felt that little progress could be made in the precise management and development of specific treatment unless an accurate diagnosis of infection with this virus could be made speedily" (128). Precedent had been established in the diagnosis of various other viral infections such as influenza and rabies, as already described, and some work had previously been done on the diagnosis of RSV by FA. J. H. Schieble et al. at the California State Public Health Laboratory had developed an FA method for the identification of RSV in cell culture but had not applied it directly to clinical specimens (129). The challenges faced in developing a reliable FA diagnostic test on clinical samples were considerable, not the least of which was reduction of nonspecific staining. It was recognized that "[g]rave damage may be done to diagnostic virology by the indiscriminate use of inadequately controlled tests and reagents" (130). However, in Joyce McQuillin, Gardner had a senior lab associate who, in the words of Dick Madeley, was capable of "... gold standard laboratory work in preparing antisera from which all unwanted cross reactions had been removed by adsorption" and who ". . . had great practical knowledge of infection patterns by viruses. . ." (131).

The introduction to the paper by Gardner and McQuillin on the rapid diagnosis of RSV infection by the FA technique contains a comment that is a crucial warning: "A test that gives equivocal results or readings which vary with different observers is useless as a routine procedure" (122). Critical to their success was a two-step absorption of the antisera and testing to confirm that nonspecific staining against components of the cells used for virus detection was removed. Figures were reproduced in the report demonstrating removal of nonspecific staining. Three types of specimens were tested with these optimized reagents: inoculated tissue culture cells, direct smears from throat swabs, and aspirated nasopharyngeal secretions. A progressively higher percentage of FA positivity was found during the first week of inoculated tissue culture. Direct smears proved unsatisfactory, but three of six nasopharyngeal aspirates gave a positive reaction on the day of admission to the hospital. In a response to a report of the FA diagnosis of RSV in another laboratory, Gardner and McQuillin emphasized the need for ". . . the most fastidious and tedious absorption to remove non-specific reactions," and they also noted ". . . a characteristic distribution and appearance" of intracellular RSV fluorescence (132) (Fig. 13).

Later in 1968, Gardner and McQuillin published the paper which established the clinical usefulness of the FA technique for the rapid diagnosis of RSV respiratory infection of children (130). Seventy-eight children under 2 years of age who were admitted to the hospital with acute respiratory infection were evaluated by isolation in cell cultures, by the development of complement-fixing antibody in paired sera, and by FA staining directly on cells derived from clinical specimens. They demonstrated the patterns of positivity in tissue culture cells and in cells from nasopharyngeal secretions. Because of the paucity of target cells in throat swabs, they were abandoned as a source of diagnosis by FA. The most dramatic results were obtained with nasopharyngeal secretions from children with bronchiolitis. RSV infection was demonstrated in 69% of children with bronchiolitis on the day of admission. Fifteen of sixteen cases that were positive by culture techniques were positive by FA, and there was complete concurrence of negativity. However, diagnosis by cell culture took up to a month, with results mostly retrospective by then and of no clinical use for patient management. Diagnosis by serology was in general agreement with the results by FA, but the authors noted that the antibody response in very young children might be variable. They concluded that FA study of nasopharyngeal secretions was the method of choice for rapid diagnosis of suspected RSV infections. They noted that the technique ". . . has revolutionized the time in which a positive report can be expected."

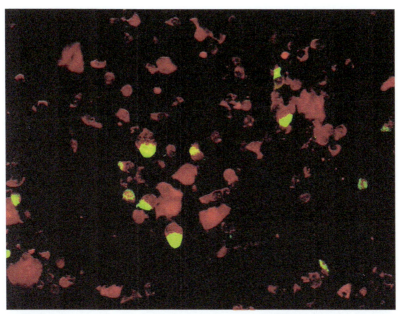

FIGURE 13 *Clinical specimen diagnosed as RSV by immunofluorescence. Rapid viral diagnosis was revolutionized with the implementation of immunofluorescence staining techniques. A sample collected on a swab from the nasopharynx was processed in the laboratory, and harvested cells were used to prepare a smear for staining. This is a direct immunofluorescent stain for RSV showing typical apple-green staining in the cytoplasm of infected cells against a background of negative cells counterstained with Evans blue. (Courtesy of Indiana Pathology Images.)*

In 1971, Gardner (Fig. 14) would write a review of acute viral respiratory infections of childhood. He set the topic in a historical context and discussed the categories of respiratory illnesses, agents, and diagnostic techniques (133). In comparison with conventional techniques, he advanced the argument for rapid diagnosis in the context of antiviral drugs which ". . . seem to be specific for particular groups of viruses and they must be given as early as possible in the disease." The experience gained in the evaluation of acute respiratory tract infections in children would serve as the stimulus and core of his book with McQuillin, *Rapid Virus Diagnosis: Application of Immunofluorescence* (128). Devoted to the application of immunofluorescence, Gardner and McQuillin noted in the introduction, "We have now evolved a system where rapid diagnosis by immunofluorescence can be used as a routine diagnostic tool." Speaking of acute respiratory infections of childhood, they noted that ". . . the great majority of causal agents are identified by immunofluorescence within hours of the patient's admission to the hospital." Their aim was a critical review of the literature to allow the reader to decide ". . . which part of the subject has useful application in his own laboratory. . ."

FIGURE 14 *Phillip Gardner. With Joyce McQuillin, Gardner pioneered the use of fluorescent-antibody staining techniques for the rapid diagnosis of viral infections. He also played a crucial role in the creation of a group for the advancement of rapid viral diagnosis in Europe and encouraged a similar group in North America. He is shown here reviewing the book he coauthored with Joyce McQuillin. (Courtesy of Dick Madeley and June Almeida.)*

Central to the advancement of rapid viral diagnostic techniques and accurate dissemination of those advances was the organization of groups of virologists committed to rapid methods. Phillip Gardner was a driving force behind the formation of the European Group for Rapid Laboratory Viral Diagnosis. The first symposium of the group was held in September 1977 in Amsterdam (123). Three sessions were held that emphasized immunofluorescence, EM, IEM, and detection of immunoglobulin M antibodies. Among the techniques to be covered in the last session were radioimmunoassay and enzyme-linked immunosorbent assay.

Gardner also played a crucial role in the formation of the North American Provisional Group for Rapid Viral Diagnosis. The idea germinated during a sabbatical of the American virologist Kenneth McIntosh with Gardner and McQuillin in 1976–1977 (134). In working with the Newcastle group, McIntosh applied the FA technique to clinical studies of coronavirus infection (64). With very strong encouragement from Gardner, McIntosh and Philadelphia virologist Stanley Plotkin initiated plans for an organizational meeting for the North American group at the NIH in Bethesda, MD, in January 1977. Key aims of the group included development of new techniques, quality control of reagents, dissemination of important information, development of training programs, and collaborative research. Because of interest from Mexico and Central and South America, the name of the group was changed to the Pan American Group for Rapid Viral Diagnosis in 1978. The organization further evolved to encompass all areas of clinical virology: viral pathogenesis, manifestations of disease, laboratory diagnosis, prevention, and therapy. With the expansion of purpose, the name was again changed in 1995 to the Pan American Society for Clinical Virology.

The WHO published *Manual for Rapid Viral Diagnosis* in 1979 (135). Four techniques were emphasized: EM, immunofluorescence, enzyme immunoassay, and radioimmunoassay. For developing countries, techniques for the rapid diagnosis of rabies, viral hepatitis, and rotavirus were encouraged. The role of the European and Pan American groups in ensuring reagent quality was noted. As the movement for rapid viral diagnostics gained momentum, numerous methods for antigen detection were implemented (64). However, Gardner had noted that "[d]iagnostic techniques are not in competition with one another, and ultimately, each technique will have its optimal place in diagnosis..." (2). Despite its promise, FA staining in clinical laboratories was limited until the development of highly specific and reproducible monoclonal antibodies.

REFERENCES

1. **Gardner PS, Turk DC, Aherne WA, Bird T, Holdaway MD, Court SDM.** 1967. Deaths associated with respiratory tract infection in childhood. *BMJ* **4:**316–320 http://dx.doi.org/10.1136/bmj.4.5575.316.
2. **Gardner PS.** 1977. Rapid virus diagnosis. *J Gen Virol* **36:**1–29 http://dx.doi.org/10.1099/0022-1317-36-1-1.
3. **Madeley CR, Peiris JSM.** 2002. Methods in virus diagnosis: immunofluorescence revisited. *J Clin Virol* **25:**121–134 http://dx.doi.org/10.1016/S1386-6532(02)00039-2.
4. **Melnick JL, Bunting H, Banfield WG, Strauss MJ, Gaylord WH.** 1952. Electron microscopy of viruses of human papilloma, molluscum contagiosum, and vaccinia, including observations on the formation of virus within the cell. *Ann N Y Acad Sci* **54:**1214–1225 http://dx.doi.org/10.1111/j.1749-6632.1952.tb39990.x.
5. **Nagler FPO, Rake G.** 1948. The use of the electron microscope in diagnosis of variola, vaccinia, and varicella. *J Bacteriol* **55:**45–51 http://dx.doi.org/10.1128/jb.55.1.45-51.1948.
6. **Van Rooyen CE, Scott GD.** 1948. Smallpox diagnosis with special reference to electron microscopy. *Can J Public Health* **39:**467–477.
7. **Williams RC, Kallman F.** 1955. Interpretation of electron micrographs of single and serial sections. *J Biophys Biochem Cytol* **1:**301–314 http://dx.doi.org/10.1083/jcb.1.4.301.
8. **Williams RC.** 1954. Electron microscopy of viruses. *Adv Virus Res* **2:**183–239 http://dx.doi.org/10.1016/S0065-3527(08)60533-3.
9. **Hall CE.** 1966. *Introduction to Electron Microscopy*, 2nd ed. McGraw-Hill, New York, NY.

10. **Pease DC, Baker RF.** 1948. Sectioning techniques for electron microscopy using a conventional microtome. *Proc Soc Exp Biol Med* **67:**470–474 http://dx.doi.org/10.3181/00379727-67-16344.

11. **Newman SB, Borysko E, Swerdlow M.** 1949. New sectioning techniques for light and electron microscopy. *Science* **110:**66–68 http://dx.doi.org/10.1126/science.110.2846.66.

12. **Latta H, Hartmann JF.** 1950. Use of a glass edge in thin sectioning for electron microscopy. *Proc Soc Exp Biol Med* **74:**436–439 http://dx.doi.org/10.3181/00379727-74-17931.

13. **Williams RC, Wyckoff RWG.** 1944. The thickness of electron microscopic objects. *J Appl Phys* **15:**712–716 http://dx.doi.org/10.1063/1.1707376.

14. **Williams RC, Wyckoff RWG.** 1945. Electron shadow- micrography of virus particles. *Proc Soc Exp Biol Med* **58:**265–270 http://dx.doi.org/10.3181/00379727-58-14918.

15. **Brenner S, Horne RW.** 1959. A negative staining method for high resolution electron microscopy of viruses. *Biochim Biophys Acta* **34:**103–110 http://dx.doi.org/10.1016/0006-3002(59)90237-9.

16. **Brenner S, Streisinger G, Horne RW, Champe SP, Barnett L, Benzer S, Rees MW.** 1959. Structural components of bacteriophage. *J Mol Biol* **1:**281–292 http://dx.doi.org/10.1016/S0022-2836(59)80035-8.

17. **Horne RW, Brenner S, Waterson AP, Wildy P.**1959. Letters to the editor: the icosahedral form of an adenovirus. *J Mol Biol* **1:**84–86

18. **Hall CE.** 1955. Electron densitometry of stained virus particles. *J Biophys Biochem Cytol* **1:**1–12 http://dx.doi.org/10.1083/jcb.1.1.1.

19. **Huxley HE.**1956. Some observations on the structure of tobacco mosaic virus, p 260–261. *Proc 1st Eur Regional Conf Electron Microsc, Stockholm*. Academic Press, New York, NY

20. **Horne RW.** 1991. Early developments in the negative staining technique for electron microscopy. *Micron Microsc Acta* **22:**321–326 http://dx.doi.org/10.1016/0739-6260(91)90051-Z.

21. **Crick FHC, Watson JD.** 1956. Structure of small viruses. *Nature* **177:**473–475 http://dx.doi.org/10.1038/177473a0.

22. **Lwoff A, Horne R, Tournier P.** 1962. A system of viruses. *Cold Spring Harb Symp Quant Biol* **27:**51–55 http://dx.doi.org/10.1101/SQB.1962.027.001.008.

23. **Almeida JD.** 1963. A classification of virus particles based on morphology. *Can Med Assoc J* **89:** 787–798.

24. **Almeida JD.** 1980. Practical aspects of diagnostic electron microscopy. *Yale J Biol Med* **53:** 5–18.

25. **Gelderblom HR, Renz H, Ozel M.** 1991. Negative staining in diagnostic virology. *Micron Microsc Acta* **22:**435–447 http://dx.doi.org/10.1016/0739-6260(91)90061-4.

26. **Hsiung G-D, Fong CKY, August MJ.** 1979. The use of electron microscopy for diagnosis of virus infections: an overview. *Prog Med Virol* **25:**133–159.

27. **Horne RW.** 1963. The structure of viruses. *Sci Am* **208:**48–56 http://dx.doi.org/10.1038/scientificamerican0163-48.

28. **Wolpert S.** 2001. *Sydney Brenner: My Life in Science*. Science Archive Limited, London, United Kingdom.

29. **Ardenne MV, Fredrich-Freksa H, Schramm G.** 1941. Elektronenmikroskopische Untersuchung der Pracipitinreaktion von Tabak-mosaikvirus mit Kaninchenantiserum. *Arch Virol* **2:**80–86.

30. **Anderson TF, Stanley WM.** 1941. A study of means of the electron microscope of the reaction between tobacco mosaic virus and its antiserum. *J Biol Chem* **139:**339–344 http://dx.doi.org/10.1016/S0021-9258(19)51390-4.

31. **Almeida JD, Waterson AP.** 1969. The morphology of virus-antibody interaction. *Adv Virus Res* **15:**307–338 http://dx.doi.org/10.1016/S0065-3527(08)60878-7.

32. **Anderson TF, Yamamoto N, Hummeler K.** 1961. Specific agglutination of phages and poliovirus by antibody molecules as seen in the electron microscope. *J Appl Phys* **32:**1639.

33. **Lafferty KJ, Oertelis SJ.** 1961. Attachment of antibody to influenza virus. *Nature* **192:**764–765 http://dx.doi.org/10.1038/192764a0.

34. **Almeida J, Cinader B, Howatson A.** 1963. The structure of antigen-antibody complexes: a study by electron microscopy. *J Exp Med* **118:**327–340 http://dx.doi.org/10.1084/jem.118.3.327.

35. **Howatson AF, Almeida JD.** 1960. Observations on the fine structure of polyoma virus. *J Biophys Biochem Cytol* **8:**828–834 http://dx.doi.org/10.1083/jcb.8.3.828.

36. **Almeida J.** 2008. Obituaries: June Almeida (nee Hart). *BMJ* **336:**1511 http://dx.doi.org/10.1136/bmj.a434.

37. **Williams MG, Howatson AF, Almeida JD.** 1961. Morphological characterization of the virus of the human common wart (*verruca vulgaris*). *Nature* **189:**895–897 http://dx.doi.org/10.1038/189895a0.

38. **Williams MG, Almeida JD, Howatson AF.** 1962. Electron microscope studies on viral skin lesions. A simple and rapid method of identifying virus particles. *Arch Dermatol* **86:**290–297 http://dx.doi.org/10.1001/archderm.1962.01590090032010.

39. **Almeida JD, Howatson AF, Williams MG.** 1962. Morphology of varicella (chicken pox) virus. *Virology* **16:**353–355 http://dx.doi.org/10.1016/0042-6822(62)90261-1.

40. **Almeida JD, Brown F, Waterson AP.** 1967. The morphologic characteristics of 19S antibody. *J Immunol* **98:**186–193.

41. **Tyrrell DAJ, Almeida JD.** 1967. Direct electron-microscopy of organ culture for the detection and characterization of viruses. *Arch Gesamte Virusforsch* **22:**417–425 http://dx.doi.org/10.1007/BF01242962.

42. **Almeida JD, Tyrrell DA.** 1967. The morphology of three previously uncharacterized human respiratory viruses that grow in organ culture. *J Gen Virol* **1:**175–178 http://dx.doi.org/10.1099/0022-1317-1-2-175.

43. **Best JM, Banatvala JE, Almeida JD, Waterson AP.** 1967. Morphological characteristics of rubella virus. *Lancet* **290:**237–239 http://dx.doi.org/10.1016/S0140-6736(67)92302-1.

44. **Almeida JD, Rubenstein D, Stott EJ.** 1971. New antigen-antibody system in Australia-antigen-positive hepatitis. *Lancet* **2:**1225–1227 http://dx.doi.org/10.1016/S0140-6736(71)90543-5.

45. **Almeida JD.** 1983. Uses and abuses of diagnostic electron microscopy, p 147–158. *In* Bachmann PA (ed), *New Developments in Diagnostic Virology.* Springer-Verlag, New York, NY. http://dx.doi.org/10.1007/978-3-642-68949-9_9.

46. **Kapikian AZ, Almeida JD, Stott EJ.** 1972. Immune electron microscopy of rhinoviruses. *J Virol* **10:**142–146 http://dx.doi.org/10.1128/jvi.10.1.142-146.1972.

47. **Kapikian AZ, Feinstone SM, Purcell RH, Wyatt RG, Thornhill TS, Kalica AR, Chanock RM.** 1975. Detection and identification by immune electron microscopy of fastidious agents associated with respiratory illness, acute nonbacterial gastroenteritis, and hepatitis A, p 9–48. *In* Pollard M (ed), *Antiviral Mechanisms: Perspectives in Virology IX.* Academic Press, New York, NY. http://dx.doi.org/10.1016/B978-0-12-560565-6.50010-3.

48. **British Medical Journal.** 1994. Obituary: gardner PS. *BMJ* **309:**950–951.

49. **Blacklow NR, Dolin R, Fedson DS, Dupont H, Northrup RS, Hornick RB, Chanock RM.** 1972. Acute infectious nonbacterial gastroenteritis: etiology and pathogenesis. *Ann Intern Med* **76:**993–1008 http://dx.doi.org/10.7326/0003-4819-76-6-993.

50. **Zahorsky J.** 1929. Hyperemesis hiemis or the winter vomiting disease. *Arch Pediatr* **49:**391–395.

51. **Zahorsky J.** 1940. The winter vomiting disease (hyperemesis hiemis). *Arch Pediatr* **57:**666–671.

52. **Zahorsky J.** 1949. *From the Hills: an Autobiography of a Pediatrician.* C V Mosby, St Louis, MO.

53. **Kapikian AZ, Wyatt RG, Dolin R, Thornhill TS, Kalica AR, Chanock RM.** 1972. Visualization by immune electron microscopy of a 27-nm particle associated with acute infectious nonbacterial gastroenteritis. *J Virol* **10:**1075–1081 http://dx.doi.org/10.1128/jvi.10.5.1075-1081.1972.

54. **Badger GF, Curtiss C, Dingle JH, Hodges RG, Jordan WS Jr, McCorkle LP.** 1956. A study of illness in a group of Cleveland families. XIII. Clinical description of acute nonbacterial gastroenteritis. *Am J Hyg* **64:**368–375.

55. **Adler JL, Zickl R.** 1969. Winter vomiting disease. *J Infect Dis* **119:**668–673 http://dx.doi.org/10.1093/infdis/119.6.668.

56. **Dolin R, Blacklow NR, DuPont H, Formal S, Buscho RF, Kasel JA, Chames RP, Hornick R, Chanock RM.** 1971. Transmission of acute infectious nonbacterial gastroenteritis to volunteers by oral administration of stool filtrates. *J Infect Dis* **123:**307–312 http://dx.doi.org/10.1093/infdis/123.3.307.

57. **Kapikian AZ.** 2000. The discovery of the 27-nm Norwalk virus: an historic perspective. *J Infect Dis* **181**(Suppl 2):S295–S302 http://dx.doi.org/10.1086/315584.

58. **Kapikian AZ, Gerin JL, Wyatt RG, Thornhill TS, Chanock RM.** 1973. Density in cesium chloride of the 27 nm "8FIIa" particle associated with acute infectious nonbacterial gastroenteritis: determination by ultra-centrifugation and immune electron microscopy. *Proc Soc Exp Biol Med* **142**:874–877 http://dx.doi.org/10.3181/00379727-142-37135.

59. **Beeman EA.** 2005. Robert J. Huebner, M.D.: a Virologist's Odyssey. Office of NIH History at the NIH. Accessed 9 August 2011.http://history.nih.gov/research/downloads/HuebnerBiography.pdf.

60. **Kapikian AZ, James HD Jr, Kelly SJ, Vaughn AL.** 1973. Detection of coronavirus strain 692 by immune electron microscopy. *Infect Immun* **7**:111–116 http://dx.doi.org/10.1128/iai.7.1.111-116.1973.

61. **Feinstone SM, Kapikian AZ, Purceli RH.** 1973. Hepatitis A: detection by immune electron microscopy of a viruslike antigen associated with acute illness. *Science* **182**:1026–1028 http://dx.doi.org/10.1126/science.182.4116.1026.

62. **Flewett TH, Bryden AS, Davies H, Woode GN, Bridger JC, Derrick JM.** 1974. Relation between viruses from acute gastroenteritis of children and newborn calves. *Lancet* **2**:61–63 http://dx.doi.org/10.1016/S0140-6736(74)91631-6.

63. **Light JS, Hodes HL.** 1943. Studies on epidemic diarrhea of the new-born: isolation of a filterable agent causing diarrhea in calves. *Am J Public Health Nations Health* **33**:1451–1454 http://dx.doi.org/10.2105/AJPH.33.12.1451.

64. **McIntosh K.** 1978. Recent advances in viral diagnosis. *Am J Dis Child* **132**:849–850.

65. **Moffet HL, Shulenberger HK, Burkholder ER.** 1968. Epidemiology and etiology of severe infantile diarrhea. *J Pediatr* **72**:1–14 http://dx.doi.org/10.1016/S0022-3476(68)80394-4.

66. **Yow MD, Melnick JL, Blattner RJ, Stephenson WBG, Robinson NM, Burkhardt MA.** 1970. The association of viruses and bacteria with infantile diarrhea. *Am J Epidemiol* **92**:33–39 http://dx.doi.org/10.1093/oxfordjournals.aje.a121177.

67. **Bishop RF, Davidson GP, Holmes IH, Ruck BJ.** 1973. Letter: evidence for viral gastroenteritis. *N Engl J Med* **289**:1096–1097 http://dx.doi.org/10.1056/NEJM197311152892025.

68. **Townley RRW, Barnes GL.** 1973. Intestinal biopsy in childhood. *Arch Dis Child* **48**:480–482 http://dx.doi.org/10.1136/adc.48.6.480.

69. **Barnes GL, Townley RRW.** 1973. Duodenal mucosal damage in 31 infants with gastroenteritis. *Arch Dis Child* **48**:343–349 http://dx.doi.org/10.1136/adc.48.5.343.

70. **Bishop R.** 1999. Ruth Bishop: rotaviruses and vaccines. Interview by Amanda Tattam. *Lancet* **353**:1860 http://dx.doi.org/10.1016/S0140-6736(05)75068-6.

71. **Bishop RF, Davidson GP, Holmes IH, Ruck BJ.** 1973. Virus particles in epithelial cells of duodenal mucosa from children with acute non-bacterial gastroenteritis. *Lancet* **2**:1281–1283 http://dx.doi.org/10.1016/S0140-6736(73)92867-5.

72. **Bishop RF, Davidson GP, Holmes IH, Ruck BJ.** 1974. Detection of a new virus by electron microscopy of faecal extracts from children with acute gastroenteritis. *Lancet* **1**:149–151 http://dx.doi.org/10.1016/S0140-6736(74)92440-4.

73. **Flewett TH, Bryden AS, Davies H.** 1973. Letter: virus particles in gastroenteritis. *Lancet* **2**:1497 http://dx.doi.org/10.1016/S0140-6736(73)92760-8.

74. **Flewett TH, Bryden AS, Davies H.** 1974. Diagnostic electron microscopy of faeces. I. The viral flora of the faeces as seen by electron microscopy. *J Clin Pathol* **27**:603–608 http://dx.doi.org/10.1136/jcp.27.8.603.

75. **Flewett TH, Davies H, Bryden AS, Robertson MJ.** 1974. Diagnostic electron microscopy of faeces. II. Acute gastroenteritis associated with reovirus-like particles. *J Clin Pathol* **27**:608–614 http://dx.doi.org/10.1136/jcp.27.8.608.

76. **Geddes A, Madeley D.** 2007. Thomas Henry Flewett. *BMJ* **334**:753.

77. **Flewett TH, Woode GN.** 1978. The rotaviruses. *Arch Virol* **57**:1–23 http://dx.doi.org/10.1007/BF01315633.

78. **Bortolussi R, Szymanski M, Hamilton R, Middleton P.** 1974. Studies on the etiology of acute infantile diarrhea. *Pediatr Res* **8:**379 http://dx.doi.org/10.1203/00006450-197404000-00234.

79. **Middleton PJ, Szymanski MR, Abbott GD, Bortolussi R, Hamilton JR.** 1974. Orbivirus acute gastroenteritis of infancy. *Lancet* **303:**1241–1244 http://dx.doi.org/10.1016/S0140-6736(74)90001-4.

80. **Kapikian AZ, Kim HW, Wyatt RG, Rodriguez WJ, Ross S, Cline WL, Parrott RH, Chanock RM.** 1974. Reoviruslike agent in stools: association with infantile diarrhea and development of serologic tests. *Science* **185:**1049–1053 http://dx.doi.org/10.1126/science.185.4156.1049.

81. **Mebus CA, Wyatt RG, Sharpee RL, Sereno MM, Kalica AR, Kapikian AZ, Twiehaus MJ.** 1976. Diarrhea in gnotobiotic calves caused by the reovirus-like agent of human infantile gastroenteritis. *Infect Immun* **14:**471–474 http://dx.doi.org/10.1128/iai.14.2.471-474.1976.

82. **Wyatt RG, Kapikian AZ, Thornhill TS, Sereno MM, Kim HW, Chanock RM.** 1974. In vitro cultivation in human fetal intestinal organ culture of a reovirus-like agent associated with nonbacterial gastroenteritis in infants and children. *J Infect Dis* **130:**523–528 http://dx.doi.org/10.1093/infdis/130.5.523.

83. **Parashar UD, Gibson CJ, Bresee JS, Glass RI.** 2006. Rotavirus and severe childhood diarrhea. *Emerg Infect Dis* **12:**304–306 http://dx.doi.org/10.3201/eid1202.050006.

84. **Madhi SA, Cunliffe NA, Steele D, Witte D, Kirsten M, Louw C, Ngwira B, Victor JC, Gillard PH, Cheuvart BB, Han HH, Neuzil KM.** 2010. Effect of human rotavirus vaccine on severe diarrhea in African infants. *N Engl J Med* **362:**289–298 http://dx.doi.org/10.1056/NEJMoa0904797.

85. **Patel NC, Hertel PM, Estes MK, de la Morena M, Petru AM, Noroski LM, Revell PA, Hanson IC, Paul ME, Rosenblatt HM, Abramson SL.** 2010. Vaccine-acquired rotavirus in infants with severe combined immunodeficiency. *N Engl J Med* **362:**314–319 http://dx.doi.org/10.1056/NEJMoa0904485.

86. **Richardson V, Hernandez-Pichardo J, Quintanar-Solares M, Esparza-Aguilar M, Johnson B, Gomez-Altamirano CM, Parashar U, Patel M.** 2010. Effect of rotavirus vaccination on death from childhood diarrhea in Mexico. *N Engl J Med* **362:**299–305 http://dx.doi.org/10.1056/NEJMoa0905211.

87. **Santosham M.** 2010. Rotavirus vaccine—a powerful tool to combat deaths from diarrhea. *N Engl J Med* **362:**358–360 http://dx.doi.org/10.1056/NEJMe0912141.

88. **Karnovsky MJ.** 1979. Dedication to Albert H. Coons 1912–1978. *J Histochem Cytochem* **27:**1117–1118 http://dx.doi.org/10.1177/27.8.383820.

89. **Coons AH.** 1961. The beginnings of immunofluorescence. *J Immunol* **87:**499–503.

90. **McDevitt H.** 1996. p 27–34. *In Albert Coons. Biographical Memoir.* National Academy of Science Press, Washington, DC.

91. **Reiner L.** 1930. On the chemical alteration of purified antibody-proteins. *Science* **72:**483–484 http://dx.doi.org/10.1126/science.72.1871.483.b.

92. **Marrack J.** 1934. Nature of antibodies. *Nature* **133:**292–293 http://dx.doi.org/10.1038/133292b0.

93. **Fieser LF, Creech HJ.** 1939. The conjugation of amino acids with isocyanates of the anthracene and 1,2-benzanthracene series. *J Am Chem Soc* **61:**3502–3506 http://dx.doi.org/10.1021/ja01267a082.

94. **Creech HJ, Jones RN.** 1940. The conjugation of horse serum albumin with 1,2-benzanthryl isocyanates. *J Am Chem Soc* **62:**1970–1975 http://dx.doi.org/10.1021/ja01865a020.

95 **Coons AH, Creech HJ, Jones RN.** 1941. Immunological properties of an antibody containing a fluorescent group. *Proc Soc Exp Biol Med* **47:**200–202 http://dx.doi.org/10.3181/00379727-47-13084P.

96. **Coons AH, Creech HJ, Jones RN, Berliner E.** 1942. The demonstration of pneumococcal antigen in tissues by the use of fluorescent antibody. *J Immunol* **45:**159–170.

97. **Coons AH, Snyder JC, Cheever FS, Murray ES.** 1950. Localization of antigen in tissue cells; antigens of rickettsiae and mumps virus. *J Exp Med* **91:**31–38 http://dx.doi.org/10.1084/jem.91.1.31.

98. **Coons AH, Kaplan MH.** 1950. Localization of antigen in tissue cells; improvements in a method for the detection of antigen by means of fluorescent antibody. *J Exp Med* **91:**1–13 http://dx.doi.org/10.1084/jem.91.1.1.

99. **Watson BK.** 1952. Distribution of mumps virus in tissue cultures as determined by fluorescein-labeled antiserum. *Proc Soc Exp Biol Med* **79:**222–224 http://dx.doi.org/10.3181/00379727-79-19329.

100. **Weller TH, Coons AH.** 1954. Fluorescent antibody studies with agents of varicella and herpes zoster propagated *in vitro. Proc Soc Exp Biol Med* **86:**789–794 http://dx.doi.org/10.3181/003797 27-86-21235.

101. **Goldwasser RA, Shepard CC.** 1958. Staining of complement and modification of fluorescent antibody procedures. *J Immunol* **80:**122–131.

102. **Silverstein AM.** 1957. Contrasting fluorescent labels for two antibodies. *J Histochem Cytochem* **5:**94–95 http://dx.doi.org/10.1177/5.1.94-b.

103. **Liu C, Eaton MD, Heyl JT.** 1959. Studies on primary atypical pneumonia. II. Observations concerning the development and immunological characteristics of antibody in patients. *J Exp Med* **109:**545–556 http://dx.doi.org/10.1084/jem.109.6.545.

104. **Riggs JL, Seiwald RJ, Burckhalter JH, Downs CM, Metcalf TG.** 1958. Isothiocyanate compounds as fluorescent labeling agents for immune serum. *Am J Pathol* **34:**1081–1097.

105. **Schaeffer M, Orsi EV, Widelock D.** 1964. Applications of immunofluorescence in public health virology. *Bacteriol Rev* **28:**402–408 http://dx.doi.org/10.1128/br.28.4.402-408.1964.

106. **Boston Interhospital Virus Study Group NIAID-Sponsored Cooperative Antiviral Clinical Study.** 1975. Failure of high dose 5-iodo-2′-deoxyuridine in the therapy of herpes simplex virus encephalitis. Evidence of unacceptable toxicity. *N Engl J Med* **292:**599–603 http://dx.doi.org/10.1056/NEJM197503202921201.

107. **Whitley RJ, Soong S-J, Dolin R, Galasso GJ, Ch'ien LT, Alford CA, Collaborative Antiviral Study.** 1977. Adenine arabinoside therapy of biopsy-proved herpes simplex encephalitis. National Institute of Allergy and Infectious Diseases collaborative antiviral study. *N Engl J Med* **297:**289–294 http://dx.doi.org/10.1056/NEJM197708112970601.

108. **Johnson KP, Rosenthal MS, Lerner PI.** 1972. Herpes simplex encephalitis. The course in five virologically proven cases. *Arch Neurol* **27:**103–108 http://dx.doi.org/10.1001/archneur.1972.00490140007003.

109. **Johnson RT, Olson LC, Buescher EL.** 1968. Herpes simplex virus infections of the nervous system. Problems in laboratory diagnosis. *Arch Neurol* **18:**260–264 http://dx.doi.org/10.1001/archneur.1968.00470330050004.

110. **Biegeleisen JZ Jr, Scott LV, Lewis V Jr.** 1959. Rapid diagnosis of herpes simplex virus infections with fluorescent antibody. *Science* **129:**640–641 http://dx.doi.org/10.1126/science.129.3349.640.

111. **Liu C.** 1956. Rapid diagnosis of human influenza infection from nasal smears by means of fluorescein-labeled antibody. *Proc Soc Exp Biol Med* **92:**883–887 http://dx.doi.org/10.3181/00379727-92-22642.

112. **Liu C.** 1955. Studies of influenza infection in ferrets by means of fluorescein-labelled antibody. I. The pathogenesis and diagnosis of the disease. *J Exp Med* **101:**665–676 http://dx.doi.org/10.1084/jem.101.6.665.

113. **Goldwasser RA, Kissling RE.** 1958. Fluorescent antibody staining of street and fixed rabies virus antigens. *Proc Soc Exp Biol Med* **98:**219–223 http://dx.doi.org/10.3181/00379727-98-23996.

114. **Goldwasser RA, Kissling RE, Carski TR, Hosty TS.** 1959. Fluorescent antibody staining of rabies virus antigens in the salivary glands of rabid animals. *Bull World Health Organ* **20:**579–588.

115. **McQueen JL, Lewis AL, Schneider NJ.** 1960. Rabies diagnosis by fluorescent antibody. I. Its evaluation in a public health laboratory. *Am J Public Health Nations Health* **50:**1743–1752 http://dx.doi.org/10.2105/AJPH.50.11.1743.

116. **Riggs JL.**1965. Application of fluorescent antibody techniques to viral infections, p 43–53. *In* Sanders M, Lennette EH (ed), *The First Annual Symposium on Applied Virology.* Olympic Press, Sheboygan, WI.

117. **Kirsh D, Kissling R.** 1963. The use of immunofluorescence in the rapid presumptive diagnosis of variola. *Bull World Health Organ* **29:**126–128.

118. **Coons AH.** 1960. Immunofluorescence. *Public Health Rep* **75:**937–943 http://dx.doi.org/10.2307/4590963.

119. **Cherry WB, Goldman M, Carski T, Moody MD.** 1960. *Fluorescent Antibody Techniques in the Diagnosis of Communicable Diseases.* US Department of Health, Education, and Welfare, Atlanta, GA.

120. **Nair H, Nokes DJ, Gessner BD, Dherani M, Madhi SA, Singleton RJ, O'Brien KL, Roca A, Wright PF, Bruce N, Chandran A, Theodoratou E, Sutanto A, Sedyaningsih ER, Ngama M, Munywoki PK, Kartasasmita C, Simões EAF, Rudan I, Weber MW, Campbell H.** 2010. Global burden of acute lower respiratory infections due to respiratory syncytial virus in young children: a systematic review and meta-analysis. *Lancet* **375:**1545–1555 http://dx.doi.org/10.1016/ S0140-6736(10)60206-1.

121. **Gardner PS, Stanfield JP, Wright AE, Court SDM, Green CA.** 1960. Viruses, bacteria, and respiratory disease in children. *BMJ* **1:**1077–1081 http://dx.doi.org/10.1136/bmj.1.5179.1077.

122. **McQuillin J, Gardner PS.** 1968. Rapid diagnosis of respiratory syncytial virus infection by immunofluorescent antibody techniques. *BMJ* **1:**602–605 http://dx.doi.org/10.1136/bmj.1.5592.602.

123. **Gardner PS.** 1978. European group for rapid laboratory viral diagnosis. Amsterdam symposium on rapid diagnosis. *J Gen Virol* **39:**201–203 http://dx.doi.org/10.1099/0022-1317-39-1-201.

124. **Beem M, Wright FH, Hamre D, Egerer R, Oehme M.** 1960. Association of the chimpanzee coryza agent with acute respiratory disease in children. *N Engl J Med* **263:**523–530 http://dx.doi.org/10.1056/NEJM196009152631101.

125. **Chanock RM, Kim HW, Vargosko AJ, Deleva A, Johnson KM, Cumming C, Parrott RH.** 1961. Respiratory syncytial virus. I. Virus recovery and other observations during 1960 outbreak of bronchiolitis, pneumonia, and minor respiratory diseases in children. *JAMA* **176:**647–653.

126. **Andrew JD, Gardner PS.** 1963. Occurrence of respiratory syncytial virus in acute respiratory diseases in infancy. *BMJ* **2:**1447–1448 http://dx.doi.org/10.1136/bmj.2.5370.1447.

127. **Gardner PS, Elderkin FM, Wall AH.** 1964. Serological study of respiratory syncytial virus infections in infancy and childhood. *BMJ* **2:**1570–1573 http://dx.doi.org/10.1136/bmj.2.5424.1570.

128. **Gardner PS, McQuillin J.** 1980. *Rapid Virus Diagnosis: Application of Immunofluorescence*, 2nd ed. Butterworths, Boston, MA.

129. **Schieble JH, Lennette EH, Kase A.** 1965. An immunofluorescent staining method for rapid identification of respiratory syncytial virus. *Proc Soc Exp Biol Med* **120:**203–208 http://dx.doi.org/10.3 181/00379727-120-30486.

130. **Gardner PS, McQuillin J.** 1968. Application of immunofluorescent antibody technique in rapid diagnosis of respiratory syncytial virus infection. *BMJ* **3:**340–343 http://dx.doi.org/10.1136/ bmj.3.5614.340.

131. **Madeley D.** 2001. Obituaries: Joyce McQuillin. *BMJ* **323:**1191.

132. **Gardner PS, McQuilin J.** 1968. Viral diagnosis by immunofluorescence. *Lancet* **1:**597–598 http:// dx.doi.org/10.1016/S0140-6736(68)92876-6.

133. **Gardner PS.** 1971. Acute respiratory virus infections of childhood, p 1–32. *In* Banatvala JE (ed), *Current Problems in Clinical Virology*. Williams & Wilkins, Baltimore, MD.

134. **Chernesky M.** 2000. 2. The organizational meeting for the Pan American Group for Rapid Viral Diagnosis (PAGRVD) and the first ten years (1977–1987). *J Clin Virol* **16:**S3–S8.

135. **Almeida JD, Atanasiu P, Bradley DW, Gardner PS, Maynard JE, Shuurs AW, Voller A, Yolken RH.** 1979. *Manual for Rapid Laboratory Viral Diagnosis.* World Health Organization, Geneva, Switzerland.

136. **Harris JR, Munn EA.** 2011. An appreciation: Robert (Bob) W. Horne (21st January 1923–13th November 2010). *Micron* **42:**528–530 http://dx.doi.org/10.1016/j.micron.2010.12.009.

8 Immunological Memory
INGENUITY AND SERENDIPITY

Kohler's and Milstein's development of the hybridoma technique for production of monoclonal antibodies have in less than a decade revolutionized the use of antibodies in healthcare and research.

Hans Wigzell (1)

INTRODUCTION

The continuing expansion of the diagnostic virologist's armamentarium by electron and fluorescence microscopy resulted in a progressive unmasking of the virological causes of disease and more rapid diagnosis. The application of tissue culture revealed great numbers of human pathogens. Even so, certain classes of suspected viral disease had remained recalcitrant to discovery, in particular, the cause of the class of hepatitis formerly known as serum hepatitis.

The distinction between hepatitis A and B was made in the 1940s, yet identification of a cause of serum hepatitis, hepatitis B, awaited a serendipitous finding by serology in the 1960s. The observation by Baruch Blumberg and colleagues of the "Australia antigen" led to major breakthroughs which identified hepatitis B virus (HBV). The diagnosis of HBV in particular, and viruses in general, took a great leap forward with adaptation of a diagnostic technique known as radioimmunoassay (RIA), developed by Solomon Berson and Rosalyn Yalow to measure very small amounts of insulin. Both achievements, the discovery of the Australia antigen and the development of RIA, yielded Nobel Prizes. Subsequently, enzyme immunoassays (EIA), also called

To Catch a Virus, Second Edition. Authored by John Booss and Marie Louise Landry.
© 2023 American Society for Microbiology. DOI: 10.1128/9781683673828.ch08

enzyme-linked immunosorbent assays (ELISA), and other immunological assays had major impacts on virological diagnosis (2). But first, we tell the hepatitis story.

HISTORICAL ORIGINS OF HEPATITIS IN CATARRHAL JAUNDICE AND HOMOLOGOUS SERUM JAUNDICE

Yellowing of skin and conjunctivae must have been among the most observable signs of illness for millennia. Following Rudolf Ludwig Virchow, that which later came to be known as infectious hepatitis was termed "catarrhal jaundice" (3) (Fig. 1). This was based on the notion that a mucous or catarrhal plug blocked the common bile duct. As F. O. MacCallum put it, it was a "stumbling block to an understanding of the basic pathology of the disease up to nearly 1940" (4). E. A. Cockayne's review in 1912 noted that catarrhal jaundice was familiar to the Greeks and Romans but that "the first undoubted reference" to the epidemic form was in 1745 in Minorca (5). The review noted several epidemics in the latter half of the 19th century with a broad global distribution, and it distinguished catarrhal jaundice from Weil's disease (leptospirosis). Leptospiral illness, yellow fever, and louse-borne relapsing fever were described by W. MacArthur as other causes of epidemic diseases with jaundice (6).

A new form of jaundice, homologous serum jaundice, later termed serum hepatitis, was recognized as a complication of yellow fever vaccination in the 1930s.

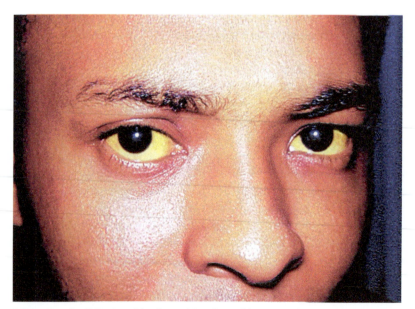

FIGURE 1 *Jaundice. Yellowing of the skin and the whites of the eyes can be an indication of disease of the liver. Viral hepatitis, due to any of several agents, has been progressively defined as newer diagnostic techniques have been developed. In this photograph, yellowing of the sclerae (whites of the eyes) is evident. (Courtesy of CDC-PHIL [#2860], Dr. Thomas F. Sellers, Emory University, 1963.)*

The term homologous was used to indicate serum from the same species, that is, of human origin, as distinct from therapeutic sera raised in animals for human treatment. It was noted that cases of jaundice following inoculations of serum

FIGURE 2 *F. O. MacCallum. A pioneer in diagnostic virology, MacCallum set up the first general virus reference laboratory in Britain at Colindale in 1946 and coauthored the textbook* Virus and Rickettsial Diseases of Man *in 1950. With G. M. Findlay, he made essential observations on homologous serum hepatitis and later devised the nomenclature which divided hepatitis into virus A (infectious) and virus B (homologous serum) hepatitis. (© Crown copyright. Reproduced with permission from the U.K. Health Security Agency, UKHSA.)*

had been recognized as early as 1885 in Bremen, Germany (7). As reviewed by F. O. MacCallum (Fig. 2), shipyard workers developed icterus several months following inoculation with human lymph against smallpox (4). Therapeutic serum inoculations against measles and mumps were also reported as sources of outbreaks of homologous serum jaundice. In 1937, a single lot of measles convalescent-phase serum produced jaundice in 41 of 109 recipients, with 8 deaths, in the south of England (8). Paul Beeson, later to become a major figure in academic medicine in the United States (9), reported the occurrence of hepatitis in almost 45% of men in a United Kingdom training regiment inoculated with convalescent-phase mumps serum (10). What linked all of these outbreaks were the injected human sera that were supposed to confer immunity to measles and mumps but in addition carried a further infectious complication. Paradoxically, it was hepatitis with a long incubation period following yellow fever immunization that focused the spotlight on homologous serum hepatitis.

The development of Theiler's yellow fever vaccine was a triumph (11). When coupled with the clearance of the mosquito vector, it promised to significantly reduce a scourge that had infected whole swaths of populations, killed uncountable numbers of people, and interfered with international commerce. Hence, as the world launched into World War II, the vaccine promised to eliminate one of the major noncombat threats to the military forces in tropical climates. Yet even before the onset of the war, trouble was brewing. In the makeup of the vaccine, human serum included for stabilization turned out to be a source of hepatitis. The added serum carried a yet-to-be-identified agent or agents. Fate was to deal a brutally cruel blow, for the very vaccine which was to prevent a virulent, acute form of jaundice, yellow fever, in some cases transmitted hepatitis of long incubation.

The first observation of hepatitis following vaccination against yellow fever was reported by G. Marshall Findlay and F. O. MacCallum from the Wellcome Bureau of Scientific Research in London (12). Careful records of 2,200 individuals who had been immunized with yellow fever vaccine over a 4.5-year period were kept to document immediate results and to monitor illnesses developing at a later date. Forty-eight cases of hepatitis were found that could not be explained by other causes such as gallstones or food poisoning. The symptoms were markedly similar: malaise, anorexia, nausea and occasional vomiting, jaundice of skin and conjunctivae, dark urine, and occasional pale-colored stools. Among those 48 affected, over 50% developed hepatitis at 2 months and 25% at between 2 and 3 months, and the remainder developed symptoms at between 3 and 7 months. Several explanations were considered for the delayed onset of hepatitis. With more data 2 years later, Findlay et al. concluded that the normal human serum was the source of the extraneous agent (13). Other investigators reached a similar conclusion, recommending "the complete elimination of the use of human serum" (14).

G. Marshall Findlay and F. O. MacCallum were two of the earliest medical virologists in the United Kingdom. Findlay was credited with having great energy, a wide range of interests, and a "facile pen and photographic mind" (15). His early work included thorough studies of viral inclusions, resulting later in an exhaustive scholarly review in R. Doerr and C. Hallauer's 1938 *Handbuch der Virusforschung* (16). He studied numerous viruses, but it was his work and observations on yellow fever immunization at the Wellcome Bureau of Scientific Research which had the greatest impact. MacCallum was credited with a wide variety of sophisticated cultural interests in addition to his accomplishments in virology (17). He was born in Canada, where he trained in medicine, and went to London in 1934 to work with S. P. Bedson on psittacosis. He moved to the Wellcome Research Institute in 1936 to work with Findlay on yellow fever, Rift Valley fever, and lymphogranuloma venereum. Following World War II, MacCallum was appointed as the first director of the Virus Reference Laboratory at Colindale in London. That lab was set up in 1946 as part of the Public Health Laboratory Service, which was the successor to the Emergency Public Health Service during World War II (18). The organization of the Public Health Laboratory Service was national in scope, with regional and area labs. According to N. R. Grist, those were "exciting, stimulating times, since everything was so new, new viruses to be found, studied and matched with their clinical manifestations, major syndromes to have the role of viruses investigated . . ." (17). MacCallum was a coauthor in 1950 and in subsequent editions of *Virus and Rickettsial Diseases of Man*, an early medical virology text (19). Forced to retire in 1960 from Colindale because of a coronary thrombosis, he later started diagnostic work at Oxford, where he made important contributions to the study of herpes simplex encephalitis. However, in the long run, his seminal contributions to the understanding of hepatitis were the most enduring. Before modification of the yellow fever vaccine, serum hepatitis following vaccination had a significant impact on the military. According to the United States secretary of war, Henry L. Stimson, in the first 6 months of 1943 28,585 cases of jaundice had developed in army personnel, with 62 deaths as a result of yellow fever vaccination (20). Apparently, Britain narrowly missed having its charismatic leader, Winston Churchill, put out of action by the vaccine during the war years. MacCallum reported that in 1942 he was summoned to Whitehall to discuss the potential immunization of Churchill against yellow fever for a proposed trip to Moscow through the Middle East (4). Because there would have been inadequate time for protection to develop, MacCallum counseled that there was little point in giving the vaccine. Later, MacCallum learned that someone who had received vaccine from the same batch as would have been used for Churchill had developed hepatitis 66 days after inoculation. It is a tantalizing question: what if Churchill had become incapacitated by hepatitis, with its

FIGURE 3 *Winston Churchill (left) with Franklin Delano Roosevelt (center) and Joseph Stalin at Yalta in 1945. Previously, F. O. MacCallum was asked to offer an opinion as to whether Winston Churchill should receive immunization against yellow fever on a trip through the Middle East to Moscow. MacCallum counseled against the immunization, based on inadequate time for immunity to develop. He may have spared the Prime Minister exposure to hepatitis virus as a contaminant of the vaccine with potential consequences for impairing his capacity to lead the war effort. Courtesy of the National Archives, Local Identifier: 111-SC-260486, National Archives Identifier: 531340.*

attendant depletion of energy stores and systemic symptoms? Would he have been able to stir Britain with his powerful rhetoric and play the pivotal role in wartime planning that he did with Roosevelt and Stalin (Fig. 3)?

Swept into the general understanding of serum jaundice in those years were reports of jaundice appearing months after transfusions of serum, blood, or plasma. In January 1943, medical officers of the Ministry of Health in the United Kingdom reported on 12 known cases of hepatitis which had developed after transfusions (8). In April of that year, Paul Beeson reported seven cases of jaundice occurring 1 to 7 months following transfusion (21). He pointed out the difficulties in identifying the connection between jaundice and administration of blood products because of the absence of a simultaneous experience, such as in a group of individuals immunized at the same time, and because of the long interval between the procedure and the appearance of disease. This complication would loom large in later years as transfusion-associated hepatitis would play havoc with the blood supply.

In the early 1940s, attention focused on the nature of the infections associated with jaundice, whether they were manifestations of different strains of the same agent or due to different agents (22). On the one hand, it was noted that the symptoms were clinically indistinguishable, but on the other hand, the incubation periods were clearly distinct: 20 to 40 days in infective (catarrhal) jaundice and 2 to 4 months in serum jaundice. Additionally, whereas infective jaundice was transmissible by contact, that did not appear to be the case with serum jaundice.

By mid-decade, J. R. Paul et al. were able to summarize transmission experiments, including experiments with human volunteers (23). Despite attempts by several groups of investigators, neither condition could be reliably transmitted to animals. The agent of each condition was found to be filterable and relatively resistant to inactivation by heat. Infectious hepatitis was demonstrated to be transmissible by the oral route. In 1947, MacCallum suggested that the designations hepatitis viruses A and B, now termed hepatitis A and B viruses (HAV and HBV), be used to identify the agents of infective and homologous serum hepatitis, respectively (24). While the conundrum of whether HAV and HBV were separate organisms or two strains of the same organism remained, the designations stuck. Two decades would pass before a particle would be found in sera associated with serum hepatitis, and that was a serendipitous observation with an immunological test (25).

BARUCH BLUMBERG, AUSTRALIA ANTIGEN, AND POSTTRANSFUSION HEPATITIS

The finding of the Australia antigen was the key to unlocking the puzzle of serum hepatitis and would result in the award of the Nobel Prize (26). Nobel Laureate Baruch Blumberg (Fig. 4) described the discovery as following a "Shandean" pattern of scientific research, after the novel *Tristram Shandy*, by Laurence Sterne. "Events ramble from one apparent irrelevancy to another, but a strange sense of order nevertheless emerges" (27). Blumberg, an American scientist, was seeking polymorphisms of proteins in human serum; he was not looking for the agent of serum hepatitis. Yet the particular strategy used to define serum polymorphisms, viewed in retrospect, was perfectly suited to the discovery of HBV. Using a double-diffusion-in-agar technique devised by O. Ouchterlony (28), Blumberg and his colleagues put individual sera from geographically diverse populations in wells in agar opposite wells containing individual sera from patients who had received multiple transfusions. They reasoned that multiply transfused individuals would have antibodies to genetic variations in serum components, allelic serum variations, and that the reactants would show up as antigen-antibody complex lines in agar (29). An unexpected finding emerged.

The antigen betraying the presence of a serum hepatitis virus was first described in a 1963 lecture by Blumberg as a "yet unidentified protein" (30). It was noted to be rare in Western populations but "not uncommon" in several other populations and found to be "quite common in patients with leukemia." No mention was made of the prevalence in hepatitis, and the Shandean ramble was on. In 1964,

FIGURE 4 *Baruch Blumberg. While seeking polymorphisms of proteins in human sera, he and his colleagues encountered a unique antigen which they termed the Australia antigen because of its origin from an Australian aborigine. It was later determined to be the key to unlocking the puzzle of a major type of serum hepatitis. (Image credit: NASA/Tom Trower.)*

Blumberg, H. J. Alter, and S. Visnich called the antigen the Australia antigen because of its origin in the serum of an Australian aborigine. They described its presence in about 10% of persons with leukemia (31) and suggested that this newly described antigen might have a role in the diagnosis of early, acute leukemia. By 1967 the investigators had expanded the scope of the prevalence of the Australia antigen to Down's syndrome, leukemia, and hepatitis (29). Then, in 1969 came the report which focused on the Australia antigen and acute viral hepatitis (32). Their studies demonstrated that the Australia antigen was not found in other diseases of the liver, nor was it simply a marker of liver damage. They reported a significantly higher frequency of Australia antigen, in 34.1% of posttransfusion hepatitis cases, in contrast to 13.1% in the remaining cases, which were termed infectious.

The presence of an antigen in the blood associated with hepatitis, in particular from blood donors, was reported from other labs. From the New York Blood

Center, A. M. Prince reported the finding of an antigen, which he called SH, during the incubation period of serum hepatitis (33). The SH antigen was found in seven of nine cases of serum hepatitis but in only 3 of 2,856 tested blood donors. Prince commented on the insensitivity of the immunodiffusion technique, noting that it "probably reveals only a small fraction of the total picture." He reported that the SH and Australia antigens were "closely related, and perhaps identical" (34). Soon after, Prince et al. found that over half of the patients with presumed infectious hepatitis, based on an absence of parenteral exposure, were SH antigen positive (35). This supported the findings of S. Krugman et al. that so-called infectious hepatitis was composed of two types: a short-incubation hepatitis and a long-incubation hepatitis. The latter resembled serum hepatitis but could be transmitted by means other than parenteral exposure (36). Reporting from a blood transfusion service in Japan, K. Okochi and S. Murakami estimated that the frequency of the Australia antigen in blood donors approached 1% (37).

Two years later, Okochi et al. reported that the majority of recipients of blood that had tested positive for Australia antigen had developed posttransfusion hepatitis (38). Hence, the Tokyo University Hospital excluded Australia antigen-positive blood for transfusion as of November 1968, and the Japan Red Cross Blood Centre followed suit in May 1969. Like Prince, the investigators commented on the need for a more sensitive assay to identify virus carriers. D. J. Gocke and N. B. Kavey also confirmed the transmission of hepatitis from hepatitis antigen-positive donors and noted the insensitivity of the technique for detection (39). To this point, although the possibility was strong that the antigen(s) represented a hepatitis virus or a component, it was not yet affirmed. This was due in no small part to the inability to grow a hepatitis virus in cell culture and the absence of a widely accepted animal model.

In April 1970, D. S. Dane et al. described "virus-like particles" by EM of about 42 nm in diameter in serum (40). The structure resembled known viruses, with an outer coat and an inner structure which appeared like a nucleocapsid with icosahedral symmetry. The smaller antigenic particles of 22 nm resembled the outer coat surface of the larger particle. Antibody caused aggregation of the small and large particles. Conclusions were drawn that the 42-nm particle was the virus of serum hepatitis and that the 22-nm-diameter forms, including the elongated types, were excess coat material. The larger forms became known as Dane particles, and their characteristics as viral particles were intensely studied in the ensuing few years (41). Just 3 years following the identification of the Dane particle of serum hepatitis, HBV, Stephen Feinstone and his colleagues at the Laboratory of Infectious Diseases of the National Institutes of Health (NIH) demonstrated the virus of infectious hepatitis, HAV, by immunoelectron microscopy (42).

There remained, however, an urgent need to develop sensitive assays for HBV that could be applied to a large number of samples. Virtually all investigators who had studied the Australia antigen or hepatitis-associated antigen, as it might be called, noted the insensitivity of the standard assay of double diffusion in gel. This meant that hepatitis-inducing donated blood could slip through if antigen was below the level of detection. Other means of antigen and antibody detection were developed, such as complement fixation (CF), which was more sensitive than agar gel diffusion (43), yet viral diagnosticians continued to seek ways to improve and speed up the detection of hepatitis-associated antigens and antibodies (44). Like the original finding of the Australia antigen, the great advance in detection of the antigen and its antibody came neither from the ranks of hepatologists nor from diagnostic virologists. It came from an endocrinology research lab.

ROSALYN YALOW AND SOLOMON BERSON: DEVELOPMENT OF RIA

Working with beef insulin, which they labeled with radioactive iodine, and antibodies against beef insulin, which they raised in guinea pigs, Rosalyn Yalow and Solomon Berson (Fig. 5) devised a highly sensitive assay to measure human plasma

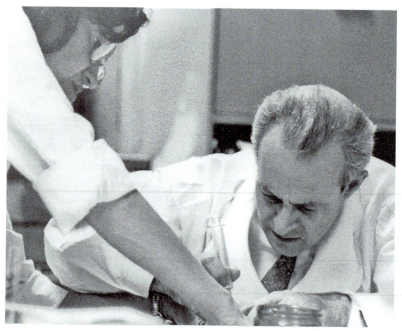

FIGURE 5 *Rosalyn Yalow and Solomon Berson. Working as an intensely collaborative team, Yalow and Berson devised the RIA with the capacity to measure remarkably small amounts of antigen. It revolutionized many fields in biomedicine, including the capacity to measure the Australia antigen and the antibody against it. Yalow received the Nobel Prize alone in 1977, Solomon Berson having passed away in 1972. (Courtesy of The Arthur H. Aufses, Jr. MD Archives, Icahn School of Medicine at Mount Sinai/Mount Sinai Health System, New York, N.Y.)*

insulin (45). That method, RIA, revolutionized immunology and the field of endocrinology and won for Yalow the Nobel Prize. Sadly, Berson had passed away before the time of the award (46, 47). RIA would provide the much-needed sensitive assay with which to screen donated blood for the hepatitis-inducing Australia antigen (48, 49). The story of RIA is one of two brilliant scientists who were exceptionally focused on their work together.

Rosalyn Yalow was born in New York City in 1921 and passed away in 2011. In her Nobel autobiography, she characterized her young self as a "stubborn, determined child" (46). Her early interests gravitated to math and chemistry, but in college she was drawn to physics, particularly nuclear physics. She had been inspired by the biography of Marie Curie and by a lecture by Enrico Fermi on nuclear fission, a process which made radioisotopes available. After graduation from Hunter College in New York City with a degree in physics and chemistry, she worked briefly as a secretary to a biochemist with the thought that it would provide a "back door" to graduate courses. However, she soon left for the University of Illinois, where she had been awarded a teaching assistantship in physics. There she met her husband, whom she married in 1943, and received her Ph.D. in 1945. She returned to New York, where she taught physics at Hunter College and served as a consultant to the Bronx Veterans Administration Hospital (VA). She ultimately took a full-time position to equip and develop the VA's radioisotope service. At that juncture in 1950, she realized that she needed a physician with whom to collaborate. She approached the Chief of the Medical Service, Bernard Straus, for a recommendation. Straus, who was the father of Eugene Straus, Yalow's biographer, recommended meeting Solomon Berson, whom he characterized as "the brightest physician I have ever trained" (50). While Berson had already accepted a position elsewhere, they had a remarkable meeting. It was reported among other things that they challenged each other with mathematical problems. Berson chose to work with Yalow.

Berson was born in New York City in 1918 and passed away in 1972 (51). Multitalented and exceptionally intelligent, he had aptitudes that included the violin and chess. He obtained his undergraduate degree from the City College of New York, took a master of science degree, and was awarded a teaching fellowship in anatomy at New York University. He did his medical school training there, after which he did his internship at the Boston City Hospital. Following a stint in the Army, he trained in medicine at the Bronx VA.

It quickly evolved that these two highly individualistic and talented people were remarkably compatible, had intense personal chemistry, and became prodigiously productive. Their first publication came in *Science*, just a year after their initial collaboration, on blood volume using radiolabeled red blood cells (52). Over the next 4 years they published numerous papers, most in the *Journal of Clinical*

Investigation, on a range of topics, including the physiological distributions of albumin and globin, thyroid function, and blood volume (51). Straus described their working relationship, in which they both generated ideas and relished the hands-on lab work: ". . . the speculations and ideas for new approaches or inventive methods would fly between them so that the air would become charged with energy until they would burst out to the laboratory to try something new" (53). The *raison d'être* for the decision to study the metabolism of insulin, the starting point for the development of RIA, is not entirely clear. Yalow's husband, Aaron, was dependent on insulin, but Yalow reportedly said that his condition did not influence the start of the studies (53). In A. Friedman's recounting, a fellow in the lab, Arthur Bauman, repeatedly approached Berson about studying the behavior of tagged insulin; Berson finally relented (50), and the critical studies followed.

From the theoretical perspective, the starting point was the hypothesis advanced by I. Arthur Mirsky that insulin insufficiency in diabetes results from increased destruction of insulin (53–55). It emerged that the fate of radioactively tagged insulin was dependent on whether an individual had been treated with insulin, either as a diabetic or as a schizophrenic patient to produce hypoglycemic shock. Rather than a more rapid disappearance from the bloodstream, as would have been predicted by the Mirsky hypothesis, they found that a low percentage of radioactive insulin persisted after 24 hours (50). In electrophoretic studies they found that the tagged insulin migrated with the globulin fraction of blood proteins in samples from patients who had been treated with insulin. The persistence of the administered tagged insulin was due to binding by a globulin, an "insulin transporting antibody" (56).

The findings were revolutionary in that they demonstrated that a peptide as small as insulin could induce an immunological response in the host. It was the takeoff point to develop a method to accurately measure very small quantities of antigen. Hitherto these low levels of antigen had not been measurable by relatively insensitive methods such as precipitin antigen-antibody complexes in agar gel. Yalow and Berson worked out the technique with mathematical and technical precision. This method, RIA, was dependent on the equilibrium constants of antigen-antibody reactions (47). The result was the ability to measure plasma insulin in humans (45, 57).

The development and proof of RIA ". . . took Yalow and Berson several years of work, including detailed studies and mathematical models of the quantitative aspects of the relation between insulin and antibody . . ." (53). Nonetheless, Yalow in her Nobel Prize speech noted that the RIA was "simple in principle" (47). In any antigen-antibody reaction there will be a ratio of antigen bound to antibody that can be compared to unbound or free antigen. One can determine that ratio by radiolabeling the antigen and measuring the radioactivity that is in the bound and free fractions. When increasing amounts of unlabeled antigen are added to the reaction in a competition assay, the ratio of bound to free labeled antigen will diminish in proportion to the amount of unlabeled antigen added. The inhibition produced by a clinical sample is compared to inhibition produced by standards of known amounts to then calculate the amount in the unknown, clinical sample. The method was found to have remarkable sensitivity, and its immunological basis provided a very high degree of specificity. Originally established for the measurement of human insulin, the assay was adapted in the Bronx VA lab to the measurement of numerous hormones, such as parathyroid hormone, growth hormone, adrenocorticotropin, and gastrin (51). RIA lent itself to the measurement of multiple samples at one time, thereby reducing cost and improving lab efficiency. This would be particularly important in blood banking, in which screening a large number of donors was needed to reduce and eliminate transmission of Australia antigen.

By 1969, investigations had been published demonstrating that a CF assay was more sensitive than agar gel diffusion for the detection of Australia antigen and antibody (43, 58). However, it was not of sufficient sensitivity to identify all Australia antigen-positive blood donors. Immunodiffusion in gel was refined through the use of an electric field, counterimmunoelectrophoresis (CIE), making it more sensitive than the original technique of agar gel diffusion (59). Yet it, too, was insufficiently sensitive.

An expert panel convened by the National Academy of Sciences-National Research Council of the United States on 31 October and 1 November 1969 reported ". . . the need for a laboratory test that would be suitable for general use and that would positively and reliably identify the prospective donor whose blood would infect susceptible recipients with viral hepatitis" (60). In retrospect, it would appear that the application of RIA to measurement of Australia antigen and its antibody at that time would have been the logical next step. RIA had been shown in the 1959 *Nature* paper by Yalow and Berson to be capable of detecting very small amounts of human serum insulin (45), and the assay had been finely honed and validated in succeeding years with other small peptide hormones.

That is, in fact, what happened. One of the investigators involved with a study of the CF assay for Australia antigen at the NIH, John Walsh, came to work as a research associate at the Bronx VA with Yalow and Berson. Walsh would go on to a distinguished research career in gastrointestinal physiology at the VA in Los Angeles and at the University of California, Los Angeles (61). While at the NIH, he had studied icteric and anicteric hepatitis following multiple blood transfusions during open heart surgery (62, 63). Walsh's work with Yalow and Berson to apply the RIA to the measurement of Australia antigen and antibody was rapidly successful, with two reports published in 1970 (48, 49). Using antigen and sera from colleagues at the NIH, Walsh, working in the labs of the Radioisotope Service of the Bronx VA, demonstrated that RIA was ". . . much more sensitive than the other methods presently used for the detection of Au and anti-Au" ("Au" was used as an abbreviation for Australia antigen) (48). Prophetically, the investigators noted that the very sensitive RIA could help determine whether ". . . a reservoir of hepatitis immunologically distinct from Au might remain."

The remarkable excess of stable antigen conferring antigenic specificity, variably called Australia antigen, hepatitis-associated antigen, and HBV surface antigen (HBsAg), offered unique diagnostic possibilities (64, 65). When combined with the remarkable sensitivity of isotope-labeled reagents and the exquisite specificity of immunological recognition, the way was opened to establish sufficiently sensitive assays to screen the blood supply. It is no surprise, then, that other assays using isotopes to label antigen or antibody, solid- as well as liquid-phase substrates, and direct rather than competitive methodology were rapidly developed. In addition, modifications of nonisotope assays emerged.

One of the initial efforts was to increase the sensitivity of the original agar gel diffusion assay for Australia antigen. CIE takes advantage of the opposite migration directions of Australia antigen and antibody in an electric field. The antigen and antibody wells are set up in an electrophoretic field positioned so that the migration toward each other is accelerated and precipitation is enhanced. The technique was applied to Australia antigen detection by D. J. Gocke and C. Howe (59). They demonstrated a 16-fold gain in sensitivity over the standard agar gel double-diffusion assay and a shortened time needed to complete the assay.

Another early assay took advantage of hemagglutination of red cells coupled to Australia antigen by chromic chloride. G. N. Vyas and N. R. Shulman demonstrated the presence of antigen in test samples through competitive inhibition of hemagglutination by standardized antigen (66). In yet another type of assay, the capacity of human red blood cells to bind the third component of complement was used to detect Australia antigen in an immune adherence hemagglutination assay by M. Mayumi et al. (67). They found it to be 40 times as sensitive as the CF assay and noted its simplicity and economy to screen blood donors. However, most of the efforts to identify Australia antigen-bearing individuals focused on variations of RIA.

In their application of RIA to Australia antigen detection, Walsh et al. adopted a number of features of their assays of peptide hormones (48, 49). A variety of modifications were also introduced by other investigators. One of the first of these was the use of a double-antibody system by R. D. Aach et al. (68). Test samples were mixed with radiolabeled Australia antigen and guinea pig antibody to

Australia antigen in a competitive binding step. Guinea pig antisera were more sensitive than were human antisera as had been used by Walsh et al. The application of a rabbit anti-guinea pig immunoglobulin G (IgG) allowed the labeled complex to be separated by centrifugation. F. B. Hollinger et al. tested a solid-phase assay system in which polystyrene tubes were coated with guinea pig antisera which were used to detect antigens in test samples compared to a set of standards (69).

The solid-phase antigen detection system was perfected by C. M. Ling and L. R. Overby in a noncompetitive two-step method for detection of HBV (70). Human serum samples for testing were incubated in polypropylene tubes coated with hyperimmune antisera raised against HBV in guinea pigs. After being washed, the tubes were incubated with ^{125}I-labeled antibody to HBV; excess, unbound antibody was washed away; and retained radioactivity was detected. Thus, retention of the labeled antibody was dependent on the capture of antigen in the serum by the unlabeled antibody adsorbed to the tube. The assay was of greater sensitivity than competitive RIA, agar gel diffusion, CF, and CIE. M. B. MacQuarrie et al. used immunodiffusion, CIE, and solid-phase RIA to detect HBsAg in a demonstration of transmission of HBV by a human bite (71). Solid-phase RIA was also applied to the identification of other viruses, such as for typing of herpes simplex virus types 1 and 2 (72).

J. J. Lander et al. applied a double antibody precipitation method to measure antibodies to hepatitis-associated antigen by RIA (73). Labeled antigen was incubated with the test serum, followed by incubation with antiserum directed against gamma globulin. Thus, the test antiserum combined with the labeled antigen, and the combination with the anti-gamma globulin allowed the complex to be precipitated by centrifugation. Not surprisingly, the test was termed radioimmunoprecipitation assay. Each of these radioimmuno-based assays, direct solid-phase assay and radioimmunoprecipitation assay, was considerably more sensitive than the assays with which they were compared and demonstrated high levels of antibody in several specific populations.

EIA AND ELISA

RIA produced such a marked increase in sensitivity compared to previous methods in the measurement of HBsAg that the various assays were classified as first, second, or third generation depending on their sensitivity (74). Using agar diffusion as the baseline or first generation, the second-generation tests included counterelectrophoresis, CF, and reverse passive latex agglutination and were 5 to 10 times more sensitive than first-generation assays. RIA, as the third generation, had an increased sensitivity, 50 to 100 times, over first-generation tests. Yet despite the marked increase in sensitivity, of particular importance in screening the blood supply, significant problems were presented by RIA for routine laboratory performance. Staffing, detailed protocols, and expensive equipment requirements were impediments to widespread implementation. Staff needed special training in the handling of isotopes; procedures needed to be followed for receipt and production of labeled components and for use and disposal of radioactive waste; and dedicated equipment, scintillation counters, needed to be purchased and maintained. Because of such complexity and expense, field applications were precluded and beyond reach for small or general labs.

The solution was to eliminate radioactivity in the assay, instead using enzymatic activity to produce a colored end product which could be subjectively assessed or measured by a spectrophotometer. The innovation was termed EIA by a commercially based lab (75) and ELISA by a university-based lab (76). The assay spawned a massive expansion in diagnostic testing, helped propel the shift from "homebrew" to commercially prepared reagents, and provided the basis for automated equipment systems capable of handling very large numbers of samples (77).

The pivotal development facilitating the later development of ELISA and EIA was the demonstration that an enzyme could be coupled to antibody and that a bifunctional conjugate would result. On the one hand, enzyme-labeled antibody needed to retain enzymatic activity to produce a colored reaction product in the presence of a substrate. On the other hand, and at the same time, immunological function, the recognition of antigenic sites, needed to be maintained. Concern that a large enzyme molecule would sterically hinder the antibody combining sites was a theoretical possibility. It is no small irony that the achievement of a bifunctional, conjugated antibody was occasioned by the attempt to address some of the perceived difficulties of fluorescent-antibody localization of antigens in tissues.

Improvement of Albert Hewlett Coons's technique of immunohistochemical localization (see chapter 7) was the goal. The difficulties of fluorescent-antibody detection of antigens in tissue sections were described by P. K. Nakane and G. B. Pierce, Jr., from the University of Michigan, in 1967 (78). The impermanent nature of fluorescence, which faded over time, militated against storage and reexamination of tissue section preparations. Certain tissues were high in autofluorescing components, complicating the interpretation of staining. The need for expensive fluorescence microscopes was a hindrance for small labs. Finally, application to electron microscopy "met with very limited success." Nakane and Pierce successfully coupled horseradish peroxidase to gamma globulin and used it in a procedure to localize epithelial basement membrane. Antibodies against the antigens were reacted with tissue sections, and enzyme-conjugated anti-species antibodies reacted with the first antibody. The peroxidase-labeled antibody technique appeared to give results as specific and sensitive as in the previously reported studies with immunofluorescence. A companion publication presented multiple light and electron microscopic images obtained using the technique (79), highlighting the amplifying effect of enzymatic activity for increased sensitivity.

In 1969, Stratis Avrameas, an immunochemist working at Villejuif, France, reported the use of glutaraldehyde as the coupling agent to conjugate any of several enzymes with proteins (80). Avrameas had previously reported the use of other coupling agents; however, glutaraldehyde proved to be the most effective in preserving significant enzymatic and immunological function. In comparison with fluorescent staining of spleen cells, enzymatic staining was found to be more sensitive, and examination of tissue sections was found to be "less tedious" than examination by fluorescence microscopy. Glutaraldehyde would become the coupling reagent of choice in studies by other groups in developing ELISA and EIA. Avrameas with B. Guilbert would report an "enzymo-immunologic method" (81). Avrameas would continue to make fundamental contributions to immunoassays (82).

Another foundation piece of what would become ELISA and EIA was the adsorption of antibodies to a solid phase which had originally been developed to simplify the RIA technique. Critical to the measurement of bound and free labeled antigen in the RIA competition assay was separation of the two fractions, often by complex, cumbersome methods. L. Wide and J. Porath described a simple technique based on coupling antibodies to Sephadex, an insoluble polymer (83). Following reaction with the specific antigen, the polymer-coupled antibody with labeled antigen was separated by centrifugation from unbound, labeled antigen. Incubation and centrifugation were carried out in the same tube, facilitating assay of large numbers of samples. Because there were fewer transfers and manipulations, it was especially useful in a clinical lab setting, in particular to measure urinary and serum luteinizing hormone, often a high-volume test. Further refinement and simplification were achieved by K. Catt and G. W. Tregear (84). They reported the coating of antibody to disposable polypropylene or polystyrene tubes, suitable for use in an automatic isotope counter. No special treatment was necessary to attach antibodies to the tubes, and the attachment was unaffected by the conditions of the assay. Also attractive was the suitability of the system for automation.

Eva Engvall, with Peter Perlmann (Fig. 6), the cocreator of the ELISA, later commented that she and Perlmann used the RIA format: labeled enzymes and coating of plastics with proteins as the basis for an immunoassay based on enzymatic readout (85). Thus, the pieces were in place for the reports on solid-phase enzyme-linked assays in 1971 from three labs: an academic lab in Stockholm, Sweden (76); a commercial lab in Oss, The Netherlands (75); and a research institute in Villejuif, France (81). The EIA project had developed as a result of Anton Schuurs's proposal to management at the company Organon to develop enzymes linked to antibodies or antigens for colorimetric readout (86). Organon patented the EIA process, whereas the University of Stockholm group did not patent the ELISA.

FIGURE 6 *Creators of the ELISA, Eva Engvall and Peter Perlmann, and the EIA, Anton Schuurs and Bauke van Weemen. Standing left to right: Engvall, Schuurs, Perlmann, and van Weemen with Johannes Buttner, President of the German Society of Clinical Chemistry. (Reprinted from Lequin RM. 2005. Clin Chem 51:2415–2418 by permission of Oxford University Press.)*

According to Bauke K. van Weemen, from The Netherlands, the licensing conditions were beneficial to laboratory medicine. Schuurs and van Weemen continued to make contributions to the field (87).

Peter Perlmann was born in 1919 in Sudetenland in the former Czechoslovakia and fled to Sweden as Hitler's army invaded (88). Here he trained in biology, chemistry, and immunology and would make wide-ranging contributions to developmental biology, immunology, and human medicine. He first studied fertilization in sea urchin eggs, explored membrane-associated complexes, and described what came to be called antibody-dependent cell-mediated cytotoxicity. In human medicine, he studied colon antigens in ulcerative colitis, examined the immunology of bladder cancer and tumor-associated antigens, investigated the immunology of malaria, and pursued the pathogenesis of cerebral malaria. In his 2006 obituary, Perlmann was described as broad-minded, "with intense dedication to science," and a man who ". . . realized that basic biomedical research would be an important way of improving the life quality for man" (88). It was the practical application of the immunological assay, the ELISA, that he developed with his graduate student, Eva Engvall, which was his most visible contribution to human well-being.

In a response to R. M. Lequin's note on the historical background of ELISA and EIA published in 2005 (77), Engvall recalled being accepted by Perlmann to do graduate work in his department (85). She wrote that ". . . on a cold winter day in early 1970, I had the first successful ELISA result." Three successive studies soon followed in which enzymes linked (conjugated) to antibodies were substituted for isotope-coupled antibodies. The first report, in 1971, noted that while enzymes coupled to antibodies had been used to identify antigens in tissue sections, they had not been used in the measurement of antigens and antibodies (76). Engvall and Perlmann used the format of the radioimmunosorbent technique in which rabbit IgG served as the antigen in a competitive assay. Sheep anti-rabbit IgG antibodies were coupled to microcrystalline cellulose as the solid phase. Substituting for an isotope, alkaline phosphatase was conjugated to the rabbit IgG with glutaraldehyde. In repeated experiments over a 6-month period, they found that the ELISA had sensitivities and precision similar to those of the radioimmunosorbent technique. Further, they reported that enzyme conjugates made with glutaraldehyde were stable for 6 months without loss of activity. The assay name and its acronym, ELISA, were introduced in the title of that paper (76). In their second 1971 paper (89), they credited Catt and Tregear's 1967 study using disposable plastic tubes as the solid phase in the RIA (84). In the study by Engvall et al., capture antibodies were adsorbed to disposable polystyrene tubes and the assay was carried out with the ELISA protocol. Performance of the assay in antibody-coated tubes greatly simplified the procedure. The third paper, published the following year, reported the use of ELISA to measure specific antibodies against human serum albumin or against the dinitrophenyl group (90). Good correlation was found in comparison with antibodies measured by quantitative precipitation. Overall, the investigators concluded that the ELISA was as sensitive as corresponding RIA techniques, with the advantages of reagent stability and less specialized equipment. The introduction of multiwell disposable plates as the solid substrate not only facilitated large-scale automation but also facilitated field studies where complex assays like RIA could not have been performed (91). The advantage of the ELISA was that results could be read visually as well as spectrophotometrically. Engvall credits Allister Voller with suggesting a field trial of malaria testing in East Africa under bush conditions (85), which proved successful. Another survey was conducted using multiwell microplates on samples from two regions of Colombia, in one of which mosquito eradication programs had been established (92). Malaria antibody levels detected by a field-based ELISA were much lower in the area in which the eradication program had been undertaken.

Multiwell microtiter plates for ELISA were applied to viral diagnosis, and the "sandwich" technique was implemented in place of the competitive assay format. In a study of rubella virus antibodies, Voller and D. E. Bidwell found a close correlation between micro-ELISA and hemagglutination inhibition values (93) and also determined that this format was well suited for the routine lab. It was simple and economical. The sandwich assay employs a primary antibody immobilized on a solid phase which "captures" specific antigen in specimens. A second, enzyme-conjugated antibody binds specifically to bound antigen captured by the primary antibody in sandwich fashion. Added substrate is converted by the bound enzyme to a visible signal which can then be evaluated spectrophotometrically, reflecting the amount of bound antigen. A variation allowing the second antibody to be unlabeled and detected by a third, labeled anti-species antibody reduces the number of labeled antibodies required in a lab.

The Organon Scientific Development Group turned their attention to development of a solid-phase EIA for the diagnosis of hepatitis B (94). In preliminary studies, they found that the EIA for HBsAg compared well with the RIA, sufficiently well, in fact, to organize a multicenter clinical trial showing that the EIA had higher sensitivity than a reverse hemagglutination assay and sensitivity similar to that of the RIA (95). However, they found that the EIA had a somewhat lower specificity than did either of the other two assays. It was the first marketed EIA and known as Hepanostika (77). A large virus reference laboratory in the United Kingdom found the test to be comparable to other available tests and did not have the disadvantages associated with the use of radioisotopes (96). The Organon group also developed an EIA sandwich assay for another antigen of HBV, e antigen (97). Over time, the diagnostic armamentarium for HBV would come to also include testing for antibody to core antigen and HBV DNA. The evolution of the various antigens, antibodies, nucleic acid, and liver enzymes would be used to define not only the presence of infection but also its natural history (98).

WESTERN BLOTS

In September 1979, Harry Towbin (Fig. 7), while working as a postdoctoral fellow in Julian Gordon's lab at the Friedrich Miescher Institute in Basel, Switzerland, published a paper on a technique that would become known as immunoblotting or Western blotting (99). With T. Staehelin, they described the electrophoretic transfer of proteins to nitrocellulose sheets from polyacrylamide gels. The technique would prove immensely useful to several fields in biology and chemistry. Hence, in the following decade it became the most highly cited paper published in the *Proceedings of the National Academy of Sciences of the United States of America* according to the *Science Citation Index* (100). The beauty of the technique was that it preserved the resolving power of electrophoretic separation of proteins in gels while combining it with the highly sensitive specificity of immunochemical identification.

FIGURE 7 *Creators of the Western blot, Harry Towbin and Julian Gordon. With T. Staehelin, Towbin and Gordon developed immunoblotting, or Western blotting. It combined the resolving power of electrophoretic separation of proteins in gels and the sensitivity of immunochemical reactions. Immensely useful in many scientific fields, the technique was applied to diagnostic virology, such as for diagnosis of HIV. In this photograph of Julian Gordon's group in 1979, Julian Gordon is center front and Harry Towbin is at the rear on the right. (Courtesy of Julian Gordon.)*

The problem which Towbin confronted was to identify antibodies to any of about 70 ribosomal proteins which had been separated in a two-dimensional gel (100). The most direct approach would have been to apply the antibodies to the gel on which the proteins had been separated. However, the density of the gel matrix was thought likely to interfere with the diffusion of the antibodies, and a means of producing a replica was considered. Precedent existed in studies of DNA fragments. Four years previously, E. M. Southern had described a technique of transferring DNA fragments that had been separated by gel electrophoresis to cellulose nitrate filters (101). Transfer to the filters was virtually complete, and the fragments were identified by hybridization with radiolabeled RNA. These transfers were termed Southern blots.

Attachment of proteins to membranous filters, through which sample solutions had been passed, had been shown by H. Kuno and H. K. Kihara to be virtually complete (102). The transfer of proteins from gels to activated cellulose by diffusion had been reported (103). Towbin realized that the process of transfer could be accelerated by transverse electrophoresis (100). He cobbled together an apparatus which brought "gel and nitrocellulose into a sandwich supported by a frame constructed from the ubiquitous plastic waste material generated in all cellular laboratories." The electrophoretic transfer operated efficiently and the proteins retained their immunogenicity. Various methods of visualizing the attached antibodies were tested, but they found enzyme amplification "particularly pleasing" due to the manner in which the colored bands developed. The marriage of the protein chemists' technique of protein separation by gel electrophoresis with the immunochemists' technique of EIA by immunoblotting would prove to be of considerable importance to virology.

Two years after the publication of the immunoblotting technique, W. N. Burnette applied it to the separation and identification of murine leukemia virus proteins from cellular lysates (104). He noted "with due respect to Southern . . . the established tradition of geographic naming of transfer techniques" and referred to the method as "Western" blotting. The appellation has stuck. In a review article published just 5 years after their original report, Towbin and Gordon were able to report "wide application in basic research and clinical applications" (105). These included several studies of human viral pathogens. In years to come, the clinical application of Western blotting would be of particular importance for the diagnosis of human immunodeficiency virus (HIV) infection (106, 107). HIV type 1 ELISA would be used as a screening test and Western blotting as a confirmatory test.

IMMUNOGLOBULIN CLASSES

The kinetics of antibody production, an increase in titer of virus-specific antibodies in acute- and convalescent-phase serum samples, has played a central role in diagnostic virology. Yet the interval over which that increase takes place, several days to a few weeks, is the crucial period of illness and recovery in most acute viral infections. Hence, the increase in antibody titers does not usually play a role in the clinical management of acute viral infections. The recognition that IgM appears early in infection before the IgG develops a diagnostic increase would come to play a role in the early diagnosis of viral infections. With IgM testing, a serological diagnosis could be achieved by a single serum sample taken during the acute phase of the illness.

The observations on the kinetics of the antibody response to viral infection were made in animal diseases and examined in animal models of human disease before being defined in humans. Working with strains of foot-and-mouth disease virus, F. Brown and J. H. Graves of the Research Institute for Animal Diseases at Pirbright, England, found differences in the characteristics of cattle sera taken at 7 days after infection compared to those taken at 14 and 21 days (108).

Additional studies on the antibodies in sera were undertaken in cattle and guinea pigs by Brown (109) (Fig. 8). Precipitating antibodies at 7 days after infection were found in the beta globulin region, whereas those taken at 14 days and

FIGURE 8 *Immunoglobulin class responses to viral infection. This photo was taken in 1998 at the 40th Anniversary of the World Reference Laboratory for FMD (foot-and-mouth disease) at the Institute for Animal Health, now The Pirbright Institute. It shows Fred Brown alongside the Institute's former Director, John B. Brooksby. With J. H. Graves, F. Brown determined that different types of antibodies developed in cattle at various times after infection with foot-and-mouth disease. Acute viral infections induce IgM, 19S antibodies, which are replaced later in the infection by IgG, 7S antibodies. (Courtesy of The Pirbright Institute.)*

later were found in the gamma globulin region. Density gradient ultracentrifugation was used by S.-E. Svehag and B. Mandel to separate 19S and 7S antibodies to poliovirus (110). At an antigen dose which resulted in an enduring antibody response, 19S antibody peaked at day 4 and started to decline by day 10, whereas 7S antibody was maximal at 3 weeks and remained so for at least 30 weeks. In a 1973 review, K. M. Cowan generalized the response to viruses as biphasic, with an early IgM (19S) response peaking at 5 to 8 days after challenge and the IgG (7S) response peaking at 15 to 20 days (111).

For humans, A. Schluederberg proposed using density gradient ultracentrifugation to measure the relative amounts of 19S and 7S antibody as a means of distinguishing recent viral infections and to also distinguish primary from secondary infections (112). Sedimentation analyses of sera from cases of measles virus infection showed that hemagglutination inhibition activity at 4 days after the rash was in two peaks corresponding to 19S and 7S fractions. By 19 days after the rash, much less 19S antibody was found; none was found more than 60 days after the rash. Treatment of the sera with 2-mercaptoethanol (2-ME), a sulfhydryl-reducing compound which inactivates 19S antibodies, reduced antibody titers in the 3 weeks following measles vaccination. Secondary antibody responses by measles-immune subjects to inactivated measles vaccine revealed an increase in 7S antibodies, but no 19S antibodies were detectable. Hence, these studies demonstrated both the usefulness of 19S as a marker for acute infection and the absence of a 19S response during secondary challenge.

The importance of documenting acute infection was brought to the fore with the 1964 rubella epidemic, which resulted in numerous cases of congenital rubella syndrome (113). Investigators in London pointed out that the clinical recognition of rubella was prone to error due to atypical or inapparent presentations. Furthermore, samples for serological investigation were often obtained when a rise in antibody titer could no longer be documented (114). One study demonstrated that about 85% of cases of acute rubella had a reduction of hemagglutinating antibody following treatment of sera with 2-ME, indicating the presence of IgM and hence acute infection. Investigators at the University of Michigan reported that studies with infants were complicated by transplacental transmission of maternal IgG (115). Using fluorescein-labeled anti-human immunoglobulins in an indirect assay, they demonstrated an IgM-specific response in a rubella virus-infected tissue culture system. The investigators advocated its use in serological laboratories that were already set up for fluorescence microscopy, noting the elimination of procedures to fractionate serum. Subsequently, Bagher Forghani and his colleagues at the California State Department of Health compared indirect immunofluorescence, sucrose density gradient separation, and 2-ME treatment to detect anti-rubella virus IgM antibodies (116). Treatment with 2-ME was found to be a relatively insensitive method. The indirect fluorescent-antibody technique was as sensitive as sucrose gradient separation when optimal conjugates were used, but high variability was found with anti-IgM conjugates from different commercial sources. Hence, the need remained for a simple and reliable assay for antiviral IgM antibodies.

In 1979, investigators from the Scientific Development Group at Organon reported the development of a new principle for the serological diagnosis of HAV (117). Microtiter plates coated with sheep anti-human IgM were exposed to test serum samples to capture IgM. HAV antigen was added to bind with the antigen-specific antibodies. In a variation designed to eliminate the false-positive complication of binding by rheumatoid factor, labeled antigen-specific Fab fragments were added. As performed by the investigators, labeling and measurement were by ELISA, but they noted that RIA or fluorescent-antibody testing could also be used. The authors observed that the new principle, use of a capture antibody on a solid phase, could be applied diagnostically to any of several antiviral IgMs. The IgM capture technique was applied to an ELISA for rubella virus by investigators from the Wellcome Research Laboratories in the United Kingdom, who used the hemagglutinin as the antigen (118). They too used an enzyme-labeled conjugate and found that the mean value for the rheumatoid factor-positive group was only slightly greater than for seronegative controls.

Subsequent to the classic work of D. H. Clarke and J. Casals, serological diagnosis of arboviral infections was based on hemagglutination inhibition (119). However, the requirement for precise assay conditions and fresh goose blood rendered the test suitable for only the most sophisticated laboratories. In addition, the requirement for acute- and convalescent-phase serum samples dictated, as in other viral infections, that acute infection could not be promptly diagnosed; however, the adaptation of IgM capture ELISA provided a solution.

B. L. Innis et al. found that antibody capture ELISA could replace hemagglutination inhibition as the diagnostic standard for dengue virus infection in regions where both dengue and Japanese encephalitis were present (120). Subsequently, T. Solomon et al. pointed out that IgM antibody capture ELISA, termed MAC ELISA, had become the diagnostic standard for Japanese encephalitis using serum and cerebrospinal fluid (121). They investigated an EIA for field diagnosis of Japanese encephalitis. The assay, termed MAC DOT, was reliable, required no specialized equipment, and was suitable for use in rural hospitals because of its simplicity and rapidity.

Serodiagnosis was no longer fixed as an after-the-fact diagnostic technique requiring two samples separated in time. IgM capture combined with ELISA methods moved serology into the clinical arena for diagnosis and management of acute viral infections.

MONOCLONAL ANTIBODIES

Until the early 1980s, antisera used for the various types of antibody-based viral diagnostic assays were polyclonal. There were several attendant disadvantages. Most problematic were the great variability from lot to lot and the limited quantities of a single lot of polyclonal antiserum. Production of polyclonal antisera was labor-intensive and expensive, involving repeated immunization of animals and significant efforts to standardize each batch of antiserum, and there were no guarantees of consistent immunological activity. This was particularly the case in producing antisera for fluorescent-antibody staining for rapid viral diagnosis, which also required absorption before use to remove the possibility of nonspecific staining. The limited supplies of immune serum affected other antibody-based techniques as well and impeded lab-to-lab comparisons for standardization of assays.

The expense, labor, and specialized knowledge required to produce polyclonal antisera prevented many general microbiology diagnostic labs from undertaking certain viral assays. This was to change when the technology was developed to understand the genetics and structure of antibody molecules, allowing monoclonal antibody production by the hybridoma technique. By the first half of the 1980s this technology already had a beneficial impact on basic and diagnostic virology (122–124).

The advance would come after a young postdoc, Georges Kohler (Fig. 9), joined the lab of a highly productive senior immunologist, Cesar Milstein (Fig. 9), at which the molecular mechanisms of antibody diversity were studied. Milstein was born and educated in Argentina, where his original doctoral work was in enzymology (125). He did postdoctoral work in the Biochemistry Department at Cambridge University in the United Kingdom, where he came under the influence of Nobelist Fred Sanger. Milstein credits Sanger with directing his research to the structure of antibody molecules. Milstein's work in Cambridge was so productive that he was awarded a second doctorate, and Sanger offered him a salary to remain longer in the department. As planned, Milstein returned to Buenos Aires; however, a military coup soon disrupted his scientific work, and he returned to Cambridge, where he would work for the remainder of his career.

Milstein's initial work upon his return was on the disulfide bridges in Bence-Jones proteins, but his research progressively focused on the mechanisms of antibody diversity. He worked on myeloma cells as models for antibody production. A postdoctoral investigator in his laboratory, Richard Cotton, succeeded in fusing two myeloma cell lines to study the mechanism of allelic exclusion, a mechanism whereby only one of two possible alleles is expressed (126). Fusion was achieved between a rat and a mouse line to produce a hybrid cell line, heralding the study of the cell line products. At this juncture, Milstein gave a seminar at the Basel

Institute of Immunology, where Georges Kohler completed his doctoral work on antibody molecules (127, 128). The seminar inspired Kohler to join Milstein's lab to work on mutations in antibody genes (128).

Kohler was born in Munich and educated in Kehl am Rhein. Little is known of Kohler's life aside from his scientific work and publications. He did his university studies at Freiburg before pursuing his doctoral work with Fritz Melchers at the Basel Institute. As explained by his biographer, Klaus Eichmann, "He was extremely private and notoriously reluctant to share personal feelings and private thoughts with others" and of "a contemplative type" (128). Apparently not driven, as would have been the traditional doctoral candidate, Kohler was said to have had a relaxed approach. Yet when they met, Milstein and Kohler were described as having "taken an immediate liking to one another, [and] their personalities were somehow complimentary in many ways." It would turn out that the Milstein lab,

FIGURE 9 *Georges Kohler (previous page) and Cesar Milstein (this page), creators of monoclonal antibodies. Despite what seemed to be insuperable theoretical barriers, Kohler and Milstein succeeded in fusing an antibody-producing cell with a myeloma cell line. Refinements led to the capacity to produce enormous quantities of highly specific antibodies. The advent of monoclonal antibodies was a great boon to diagnostic virology. (Courtesy of the Lasker Foundation.)*

Kohler's interests, their interactive creativity, and fate would produce an extraordinary advance in immunobiology.

Despite many potential theoretical impediments, Milstein and Kohler agreed to attempt fusion of a specific antibody-producing cell with myeloma cell lines. Selective media would allow the fused hybrid cells to grow, while the original component cells would die off. The antibody-secreting cells would be spleen cells from mice immunized with sheep red blood cells (SRBC) to be fused by inactivated Sendai virus with mouse myeloma cell lines. The presence of specific antibody-secreting cells would be determined in an SRBC hemolytic assay in soft agar (129). Remarkably, the experiment worked. Kohler later recalled that he had asked his wife to come along to the basement laboratory to read the initial results because he thought he would be doing the boring work of negative scoring (130). What he found instead were lytic halos indicating specific antibody production against SRBC.

Two publications followed (131, 132), and Kohler would return to the Basel Institute for Immunology. Modifications would be made in the technique, such as using nonsecreting myeloma cell lines as the immortality-conferring component and using polyethylene glycol for the fusion step. With Giovanni Galfrè, Milstein would later publish an explicit set of strategies and procedures for the preparation of monoclonal antibodies (133). This technique, which revolutionized many fields in biology and medicine, including virology, resulted in the award of the Nobel Prize to Kohler and Milstein with Niels Jerne in 1984 (1). In his presentation speech, Hans Wigzell of the Karolinska Institute described Jerne as "the great theoretician in modern immunology." Of Kohler and Milstein's contribution, Wigzell said that the "production of monoclonal antibodies have in less than a decade revolutionized the use of antibodies in healthcare and research." The antigenic specificity allowed by monoclonal antibodies was quickly put to use in virology. For influenza, in which antigenic drift is a cause of recurrent annual epidemics of respiratory disease, W. G. Laver et al. selected variants for peptide mapping and antigenic analysis with monoclonal antibodies (134). Within a few years, monoclonal antibodies were adapted for strain-specific identification of influenza virus isolates (135). Joyce McQuillin et al. adapted pools of monoclonal antibodies to the rapid identification of influenza A and B by an indirect immunofluorescence assay (136). They observed that "the shortage of suitable reagents of reproducible quality for rapid diagnostic techniques may be overcome by the preparation of pools of carefully selected monoclonal antibodies of appropriate specificity, even for the highly variable and unpredictable influenza viruses." This comment came from the laboratory which pioneered and perfected the use of immunofluorescence for rapid viral diagnosis.

The application of monoclonal antibodies was adapted to the improved performance of a number of other viral diagnostic assays. Among the first was the production by R. S. Tedder et al. of monoclonal IgG antibodies against rubella virus antigens (137). These IgG antibodies were radiolabeled for use in rubella virus-specific IgM capture assays and resulted in improved sensitivity. Assays for immunity to rubella were crucial for screening women and for determining whether immunization was needed (138). Another clinically important question to which monoclonal reagents were soon applied was the typing of herpes simplex virus isolates. Originally defined by clinical characteristics roughly corresponding to oral and genital isolates, herpes simplex virus isolates were distinguished in serological assays as types 1 and 2 based on polyclonal antisera; however, considerable cross-reactions were impediments to accurate identification. In a little over a year, in 1982–1983, several groups reported the successful serotyping of isolates using monoclonal reagents (139, 140) as well as the use of these reagents for rapid diagnosis (141).

The technology to produce virtually unlimited quantities of well-characterized monoclonal antibodies was readily adapted to commercial production. This allowed the development and standardization of viral diagnostic assays, improved quality control in viral diagnosis, and widespread implementation of testing in viral diagnostic laboratories. The ready availability of monoclonal antibodies was transformative for viral diagnosis, allowing rapid identification of virus isolates in cell culture, rapid culture methods to detect viral growth before cytopathic effects were evident, and detection of viruses directly in respiratory and skin lesion specimens. Thus, monoclonal antibody technology contributed to the widespread dissemination of rapid viral diagnosis and facilitated the development of specific antiviral reagents.

REFERENCES

1. **Wigzell H.**1984. The Nobel Prize in Physiology or Medicine 1984. Award ceremony speech. http://www.nobelprize.org/nobel_prizes/ medicine/laureates/1984/presentation-speech.html. Accessed 16 October 2010.
2. **Yolken RH.** 1980. Enzyme-linked immunosorbent assay (ELISA): a practical tool for rapid diagnosis of viruses and other infectious agents. *Yale J Biol Med* **53**:85–92.
3. **Virchow R.** 1865. Ueber des Vorkommen und den Nachweis des hepatogenen insbesondere des katarrhalischen Icterus. *Virchows Arch* **32**:117–125 http://dx.doi.org/10.1007/BF01929017.
4. **MacCallum FO.** 1972. 1971 International Symposium on Viral Hepatitis. Historical perspectives. *Can Med Assoc J* **106**:423–426.
5. **Cockayne EA.** 1912. Catarrhal jaundice, sporadic and epidemic, and its relation to acute yellow atrophy of the liver. *Q J Med* **6**:1–29.
6. **Mac Arthur W.** 1957. Historical notes on some epidemic diseases associated with jaundice. *Br Med Bull* **13**:146–149 http://dx.doi.org/10.1093/oxfordjournals.bmb.a069593.
7. **Lurman A.** 1885. Eine Icterusepidemic. *Berl Klin Wochenschr* **22**:20.
8. **Medical Officers of the Ministry of Health.** 1943. Homologous serum jaundice. *Lancet* **241**:83–88.
9. **Rapport R.** 2001. *Physician. The Life of Paul Beeson.* Barricade Books, Inc, Fort Lee, NJ.
10. **Beeson PB, Chesney G, McFarlan AM.** 1944. Hepatitis following injection of mumps convalescent plasma. *Lancet* **243**:814–815 http://dx.doi.org/10.1016/S0140-6736(00)75062-8.
11. **Theiler M, Smith HH.** 1937. The use of yellow fever virus modified by *in vitro* cultivation for human immunization. *J Exp Med* **65**:787–800 http://dx.doi.org/10.1084/jem.65.6.787.
12. **Findlay GM, MacCallum FO.** 1937. Note on acute hepatitis and yellow fever immunization. *Trans R Soc Trop Med Hyg* **31**:297–308 http://dx.doi.org/10.1016/S0035-9203(37)90055-5.
13. **Findlay GM, MacCallum FO, Murgatroyd F.** 1939. Observations bearing on the aetiology of infective hepatitis (so-called epidemic catarrhal jaundice). *Trans R Soc Trop Med Hyg* **32**:575–586 http://dx.doi.org/10.1016/S0035-9203(39)90018-0.

14. **Fox JP, Manso C, Penna HA, Madureira P.** 1942. Observations on the occurrence of icterus in Brazil following vaccination against yellow fever. *Am J Epidemiol* **36:**68–116 http://dx.doi.org/10.1093/oxfordjournals.aje.a118810.

15. **BMJ.** 1952. Obituary: GWM Findlay. *BMJ* **1:**658–660 http://dx.doi.org/10.1136/bmj.1.4759.658.

16. **Findlay GM.** 1938. B. Inclusion bodies and their relationship to viruses, p 292–368. *In* Doerr R, Hallauer C (ed), *Handbuch der Virusforschung, Erste Hälfte.* Julius Springer, Vienna, Austria. http://dx.doi.org/10.1007/978-3-662-25608-4_3.

17. **Grist NR.** 1995. Appreciations. Dr. Frederick Ogden MacCallum. *Bull R Coll Pathol* **67:**6–7.

18. **Howie JW.** 1965. The Public Health Laboratory Service. *Lancet* **1:**501–505 http://dx.doi.org/10.1016/S0140-6736(65)92013-1.

19. **Bedson SP, Downie AW, MacCallum FO, Stuart-Harris CH.** 1961. *Virus and Rickettsial Diseases of Man,* 3rd ed. Edward Arnold Publishers Ltd, London, United Kingdom.

20. **JAMA.** 1942. Jaundice following yellow fever vaccination. *JAMA* **119:**1110 http://dx.doi.org/10.1001/jama.1942.02830310044012. Editorial.

21. **Beeson PB.** 1943. Jaundice occurring one to four months after transfusion of blood or plasma. *JAMA* **121:**1332–1334 http://dx.doi.org/10.1001/jama.1943.02840170016005.

22. **Lancet.** 1943. Infective hepatitis and serum jaundice. *Lancet* **241:**683–684 http://dx.doi.org/10.1016/S0140-6736(00)42375-5. Editorial.

23. **Paul JR, Havens WP Jr, Sabin AB, Philip CB.** 1945. Transmission experiments in serum jaundice and infectious hepatitis. *JAMA* **128:**911–915 http://dx.doi.org/10.1001/jama.1945.02860300001001.

24. **Lancet.** 1947. Homologous serum hepatitis. *Lancet* **250:**691–692 http://dx.doi.org/10.1016/S0140-6736(47)90722-8. Editorial.

25. **Bayer ME, Blumberg BS, Werner B.** 1968. Particles associated with Australia antigen in the sera of patients with leukaemia, Down's syndrome and hepatitis. *Nature* **218:**1057–1059 http://dx.doi.org/10.1038/2181057a0.

26. **Blumberg BS.** 1976. Nobel Lecture: Australia antigen and the biology of hepatitis B. http://www.nobelprize.org/nobel_prizes/medicine/laureates/1976/blumberg-lecture.html?print=1. Accessed 30 August 2010.

27. **Blumberg BS.** 2002. *Hepatitis B. The Hunt for A Killer Virus.* Princeton University Press, Princeton, NJ. http://dx.doi.org/10.1515/9780691187235.

28. **Ouchterlony O.** 1948. In vitro method for testing the toxin-producing capacity of diphtheria bacteria. *Acta Pathol Microbiol Scand* **25:**186–191 http://dx.doi.org/10.1111/j.1699-0463.1948.tb00655.x.

29. **Blumberg BS, Gerstley BJ, Hungerford DA, London WT, Sutnick AI.** 1967. A serum antigen (Australia antigen) in Down's syndrome, leukemia, and hepatitis. *Ann Intern Med* **66:**924–931 http://dx.doi.org/10.7326/0003-4819-66-5-924.

30. **Blumberg BS.** 1964. Polymorphisms of the serum proteins and the development of iso-precipitins in transfused patients. *Bull N Y Acad Med* **40:**377–386.

31. **Blumberg BS, Alter HJ, Visnich S.** 1965. A "new" antigen in leukemic sera. *JAMA* **191:**101–106 http://dx.doi.org/10.1001/jama.1965.03080070025007.

32. **London WT, Sutnick AI, Blumberg BS.** 1969. Australia antigen and acute viral hepatitis. *Ann Intern Med* **70:**55–59 http://dx.doi.org/10.7326/0003-4819-70-1-55.

33. **Prince AM.** 1968. An antigen detected in the blood during the incubation period of serum hepatitis. *Proc Natl Acad Sci USA* **60:**814–821 http://dx.doi.org/10.1073/pnas.60.3.814.

34. **Prince AM.** 1968. Relation of Australia and SH antigens. *Lancet* **2:**462–463 http://dx.doi.org/10.1016/S0140-6736(68)90512-6.

35. **Prince AM, Hargrove RL, Szmuness W, Cherubin CE, Fontana VJ, Jeffries GH.** 1970. Immunologic distinction between infectious and serum hepatitis. *N Engl J Med* **282:**987–991 http://dx.doi.org/10.1056/NEJM197004302821801.

36. **Krugman S, Giles JP, Hammond J.** 1967. Infectious hepatitis. Evidence for two distinctive clinical, epidemiological, and immunological types of infection. *JAMA* **200:**365–373 http://dx.doi.org/10.1001/jama.1967.03120180053006.

37. **Okochi K, Murakami S.** 1968. Observations on Australia antigen in Japanese. *Vox Sang* **15:**374–385 http://dx.doi.org/10.1111/j.1423-0410.1968.tb04078.x.

38. **Okochi K, Murakami S, Ninomiya K, Kaneko M.** 1970. Australia antigen, transfusion, and hepatitis. *Vox Sang* **18:**289–300 http://dx.doi.org/10.1111/j.1423-0410.1970.tb01465.x.

39. **Gocke DJ, Kavey NB.** 1969. Hepatitis antigen. Correlation with disease and infectivity of blood-donors. *Lancet* **1:**1055–1059 http://dx.doi.org/10.1016/S0140-6736(69)91701-2.

40. **Dane DS, Cameron CH, Briggs M.** 1970. Virus-like particles in serum of patients with Australia-antigen-associated hepatitis. *Lancet* **1:**695–698 http://dx.doi.org/10.1016/S0140-6736(70)90926-8.

41. **Robinson WS, Lutwick LI.** 1976. The virus of hepatitis, type B (first of two parts). *N Engl J Med* **295:**1168–1175, 1232–1236 http://dx.doi.org/10.1056/NEJM197611182952105.

42. **Feinstone SM, Kapikian AZ, Purceli RH.** 1973. Hepatitis A: detection by immune electron microscopy of a viruslike antigen associated with acute illness. *Science* **182:**1026–1028 http://dx.doi.org/10.1126/science.182.4116.1026.

43. **Purcell RH, Holland PV, Walsh JH, Wong DC, Morrow AG, Chanock RM.** 1969. A complement-fixation test for measuring Australia antigen and antibody. *J Infect Dis* **120:**383–386 http://dx.doi.org/10.1093/infdis/120.3.383.

44. **Kelen AE, Hathaway AE, McLeod DA.** 1971. Rapid detection of Australia-SH antigen and antibody by a simple and sensitive technique of immunoelectronmicroscopy. *Can J Microbiol* **17:**993–1000 http://dx.doi.org/10.1139/m71-157.

45. **Yalow RS, Berson SA.** 1959. Assay of plasma insulin in human subjects by immunological methods. *Nature* **184**(Suppl 21)**:**1648–1649 http://dx.doi.org/10.1038/1841648b0.

46. **Yalow R.**1977. Rosalyn Yalow—autobiography. http://www.nobel prize.org/nobel_prizes/medicine/laureates/1977/yalow-autobio.html. Accessed 15 July 2010.

47. **Yalow RS.**1977. Nobel lecture. Radioimmunoassay: a probe for fine structure of biological systems. http://www.nobelprize.org/nobel_ prizes/medicine/laureates/1977/yalow-lecture.html?print=1. Accessed 15 July 2010.

48. **Walsh JH, Yalow R, Berson SA.** 1970. Detection of Australia antigen and antibody by means of radioimmunoassay techniques. *J Infect Dis* **121:**550–554 http://dx.doi.org/10.1093/infdis/121.5.550.

49. **Walsh JH, Yalow RS, Berson SA.** 1970. Radioimmunoassay of Australia antigen. *Vox Sang* **19:** 217–224 http://dx.doi.org/10.1111/j.1423-0410.1970.tb01515.x.

50. **Friedman A.** 2002. Remembrance: the Berson and Yalow saga. *J Clin Endocrinol Metab* **87:**1925–1928 http://dx.doi.org/10.1210/jcem.87.5.8602.

51. **Rall JE.** 1990. *Solomon A. Berson 1918–1972. A Biographical Memoir.* National Academy of Sciences, Washington, DC.

52. **Yalow RS, Berson SA.** 1951. The use of K42 tagged erythrocytes in blood volume determinations. *Science* **114:**14–15 http://dx.doi.org/10.1126/science.114.2949.14.

53. **Straus E.** 1998. *Rosalyn Yalow. Nobel Laureate. Her Life and Work in Medicine.* Perseus Books, Cambridge, MA.

54. **Mirsky IA.** 1952. The etiology of diabetes in man. *Recent Prog Horm Res* **7:**437–467.

55. **Mirsky IA, Perisutti G, Dixon FJ.** 1954. Destruction of I131 labeled insulin by liver slices. *Proc Soc Exp Biol Med* **86:**228–230 http://dx.doi.org/10.3181/00379727-86-21057.

56. **Berson SA, Yalow RS, Bauman A, Rothschild MA, Newerly K.** 1956. Insulin-I131 metabolism in human subjects: demonstration of insulin binding globulin in the circulation of insulin treated subjects. *J Clin Invest* **35:**170–190 http://dx.doi.org/10.1172/JCI103262.

57. **Yalow RS, Berson SA.** 1960. Immunoassay of endogenous plasma insulin in man. *J Clin Invest* **39:**1157–1175 http://dx.doi.org/10.1172/JCI104130.

58. **Shulman NR, Barker LF.** 1969. Virus-like antigen, antibody, and antigen-antibody complexes in hepatitis measured by complement fixation. *Science* **165:**304–306 http://dx.doi.org/10.1126/science.165.3890.304.

59. **Gocke DJ, Howe C.** 1970. Rapid detection of Australia antigen by counterimmunoelectrophoresis. *J Immunol* **104:**1031–1034.

60. **Transfusion J.** 1970. Statement on laboratory screening tests for identifying carriers of viral hepatitis in blood-banking and transfusion services, prepared by a panel of the Committee on Plasma and Plasma Substitutes of the Division of Medical Sciences, National Academy of Sciences--National Research Council. *Transfusion* **10:**1–2 http://dx.doi.org/10.1111/j.1537-2995.1970.tb00692.x.

61. **Record NIH.** 2000. Obituaries. Walsh, Director of NIDDK CURE Center, mourned. *NIH Rec.* **52:**1–2. http://nihrecord.od.nih.gov/news letters/10_31_2000/obits.htm. Accessed 4 September 2010.

62. **Walsh JH, Purcell RH, Morrow AG, Chanock RM, Schmidt PJ.** 1970. Posttransfusion hepatitis after open-heart operations. Incidence after the administration of blood from commercial and volunteer donor populations. *JAMA* **211:**261–265 http://dx.doi.org/10.1001/jama.1970.03170020025005.

63. **Walsh JH, Purcell RH, Morrow AG, Schmidt PJ.** 1968. Icteric and anicteric hepatitis following open heart surgery: a direct comparison of paid and voluntary blood donors. *Transfusion* **8:**318. Abstract.

64. **Dienstag JL.** 1980. Hepatitis viruses: characterization and diagnostic techniques. *Yale J Biol Med* **53:**61–69.

65. **Landry ML, Mayo DR, Hsiung GD.** 1982. The need for a rapid and accurate viral diagnosis. *Pharmacol Ther* **18:**107–132 http://dx.doi.org/10.1016/0163-7258(82)90028-6.

66. **Vyas GN, Shulman NR.** 1970. Hemagglutination assay for antigen and antibody associated with viral hepatitis. *Science* **170:**332–333 http://dx.doi.org/10.1126/science.170.3955.332.

67. **Mayumi M, Okochi K, Nishioka K.** 1971. Detection of Australia antigen by means of immune adherence haemagglutination test. *Vox Sang* **20:**178–181 http://dx.doi.org/10.1111/j.1423-0410.1971.tb00549.x.

68. **Aach RD, Grisham JW, Parker CW.** 1971. Detection of Australia antigen by radioimmunoassay. *Proc Natl Acad Sci USA* **68:**1056–1060 http://dx.doi.org/10.1073/pnas.68.5.1056.

69. **Hollinger FB, Vorndam V, Dreesman GR.** 1971. Assay of Australia antigen and antibody employing double-antibody and solid-phase radioimmunoassay techniques and comparison with the passive hemagglutination methods. *J Immunol* **107:**1099–1111.

70. **Ling CM, Overby LR.** 1972. Prevalence of hepatitis B virus antigen as revealed by direct radioimmune assay with 125 I-antibody. *J Immunol* **109:**834–841.

71. **MacQuarrie MB, Forghani B, Wolochow DA.** 1974. Hepatitis B transmitted by a human bite. *JAMA* **230:**723–724 http://dx.doi.org/10.1001/jama.1974.03240050051028.

72. **Forghani B, Schmidt NJ, Lennette EH.** 1974. Solid phase radioimmunoassay for identification of *Herpesvirus hominis* types 1 and 2 from clinical materials. *Appl Microbiol* **28:**661–667 http://dx.doi.org/10.1128/am.28.4.661-667.1974.

73. **Lander JJ, Alter HJ, Purcell RH.** 1971. Frequency of antibody to hepatitis-associated antigen as measured by a new radioimmunoassay technique. *J Immunol* **106:**1166–1171.

74. **Feinstone SM, Barker LF, Purcell RH.** 1979. Hepatitis A and B, p 879–925. *In* Lennette EH, Schmidt NJ (ed), *Diagnostic Procedures for Viral, Rickettsial and Chlamydial Infections*, 5th ed. American Public Health Association, Washington, DC.

75. **Van Weemen BK, Schuurs AHWM.** 1971. Immunoassay using antigen-enzyme conjugates. *FEBS Lett* **15:**232–236 http://dx.doi.org/10.1016/0014-5793(71)80319-8.

76. **Engvall E, Perlmann P.** 1971. Enzyme-linked immunosorbent assay (ELISA). Quantitative assay of immunoglobulin G. *Immunochemistry* **8:**871–874 http://dx.doi.org/10.1016/0019-2791(71)90454-X.

77. **Lequin RM.** 2005. Enzyme immunoassay (EIA)/enzyme-linked immunosorbent assay (ELISA). *Clin Chem* **51:**2415–2418 http://dx.doi.org/10.1373/clinchem.2005.051532.

78. **Nakane PK, Pierce GB Jr.** 1966. Enzyme-labeled antibodies: preparation and application for the localization of antigens. *J Histochem Cytochem* **14:**929–931 http://dx.doi.org/10.1177/14.12.929.

79. **Nakane PK, Pierce GB Jr.** 1967. Enzyme-labeled antibodies for the light and electron microscopic localization of tissue antigens. *J Cell Biol* **33:**307–318 http://dx.doi.org/10.1083/jcb.33.2.307.

80. **Avrameas S.** 1969. Coupling of enzymes to proteins with glutaraldehyde. Use of the conjugates for the detection of antigens and antibodies. *Immunochemistry* **6:**43–52 http://dx.doi.org/10.1016/0019-2791(69)90177-3.

81. **Avrameas S, Guilbert B.** 1971. Dosage enzymo-immunologique de proteines a l'aide d'immunoadsorbants et d'antigenes marques aux enzymes. *C R Acad Sci Paris Ser D* **273:**2705–2707.

82. **Avrameas S.** 1983. Enzyme immunoassays and related techniques: development and limitations. *Curr Top Microbiol Immunol* **104**:93–99 http://dx.doi.org/10.1007/978-3-642-68949-9_6.

83. **Wide L, Porath J.** 1966. Radioimmunoassay of proteins with the use of Sephadex-coupled antibodies. *Biochim Biophys Acta* **130**:257–260 http://dx.doi.org/10.1016/0304-4165(66)90032-8.

84. **Catt K, Tregear GW.** 1967. Solid-phase radioimmunoassay in antibody-coated tubes. *Science* **158**:1570–1572 http://dx.doi.org/10.1126/science.158.3808.1570.

85. **Engvall E.** 2005. Perspective on the historical note on EIA/ELISA by Dr. R.M. Lequin. *Clin Chem* **51**:2225 http://dx.doi.org/10.1373/clinchem.2005.059618.

86. **van Weemen BK.** 2005. The rise of EIA/ELISA. *Clin Chem* **51**:2226 http://dx.doi.org/10.1373/clinchem.2005.059626.

87. **Schuurs AHWM, van Weemen BK.** 1980. Enzyme-immunoassay: a powerful analytical tool. *J Immunoassay* **1**:229–249 http://dx.doi.org/10.1080/01971528008055786.

88. **Hammarström S, Berzins K, Biberfeld P, Engvall E, Hammarström ML, Holm G, Troye-Blomberg M, Wahlgren M.** 2006. Peter Perlmann 1919-2005. *Scand J Immunol* **63**:487–489 http://dx.doi.org/10.1111/j.1365-3083.2006.001769.x.

89. **Engvall E, Jonsson K, Perlmann P.** 1971. Enzyme-linked immunosorbent assay. II. Quantitative assay of protein antigen, immunoglobulin G, by means of enzyme-labelled antigen and antibody-coated tubes. *Biochim Biophys Acta* **251**:427–434 http://dx.doi.org/10.1016/0005-2795(71)90132-2.

90. **Engvall E, Perlmann P.** 1972. Enzyme-linked immunosorbent assay, Elisa. 3. Quantitation of specific antibodies by enzyme-labeled anti-immunoglobulin in antigen-coated tubes. *J Immunol* **109**:129–135.

91. **Voller A, Bidwell DE, Bartlett A.** 1976. Enzyme immunoassays in diagnostic medicine. Theory and practice. *Bull World Health Organ* **53**:55–65.

92. **Voller A, Bidwell D, Huldt G, Engvall E.** 1974. A microplate method of enzyme-linked immunosorbent assay and its application to malaria. *Bull World Health Organ* **51**:209–211.

93. **Voller A, Bidwell DE.** 1975. A simple method for detecting antibodies to rubella. *Br J Exp Pathol* **56**:338–339.

94. **Wolters G, Kuijpers L, Kacaki J, Schuurs A.** 1976. Solid-phase enzyme-immunoassay for detection of hepatitis B surface antigen. *J Clin Pathol* **29**:873–879 http://dx.doi.org/10.1136/jcp.29.10.873.

95. **Kacaki J, Wolters G, Kuijpers L, Stulemeyer S.** 1978. Results of a multicentre clinical trial of the solid-phase enzyme immunoassay for hepatitis B surface antigen. *Vox Sang* **35**:65–74 http://dx.doi.org/10.1111/j.1423-0410.1978.tb02902.x.

96. **Vandervelde EM, Cohen BJ, Cossart YE.** 1977. An enzyme-linked immunosorbent-assay test for hepatitis B surface antigen. *J Clin Pathol* **30**:714–716 http://dx.doi.org/10.1136/jcp.30.8.714.

97. **von der Waart M, Snelting A, Cichy J, Wolters G, Schuurs A.** 1978. Enzyme-immunoassay in diagnosis of hepatitis with emphasis on the detection of "e" antigen (HBeAg). *J Med Virol* **3**:43–49 http://dx.doi.org/10.1002/jmv.1890030111.

98. **Liaw Y-F, Chu C-M.** 2009. Hepatitis B virus infection. *Lancet* **373**:582–592 http://dx.doi.org/10.1016/S0140-6736(09)60207-5.

99. **Towbin H, Staehelin T, Gordon J.** 1979. Electrophoretic transfer of proteins from polyacrylamide gels to nitrocellulose sheets: procedure and some applications. *Proc Natl Acad Sci USA* **76**:4350–4354 http://dx.doi.org/10.1073/pnas.76.9.4350.

100. **Towbin H.** 1988. This week's citation classic. *Curr Contents* **31**:19.

101. **Southern EM.** 1975. Detection of specific sequences among DNA fragments separated by gel electrophoresis. *J Mol Biol* **98**:503–517 http://dx.doi.org/10.1016/S0022-2836(75)80083-0.

102. **Kuno H, Kihara HK.** 1967. Simple microassay of protein with membrane filter. *Nature* **215**:974–975 http://dx.doi.org/10.1038/215974a0.

103. **Renart J, Reiser J, Stark GR.** 1979. Transfer of proteins from gels to diazobenzyloxymethyl-paper and detection with antisera: a method for studying antibody specificity and antigen structure. *Proc Natl Acad Sci USA* **76**:3116–3120 http://dx.doi.org/10.1073/pnas.76.7.3116.

104. **Burnette WN.** 1981. "Western blotting": electrophoretic transfer of proteins from sodium dodecyl sulfate--polyacrylamide gels to unmodified nitrocellulose and radiographic detection

with antibody and radioiodinated protein A. *Anal Biochem* **112:**195–203 http://dx.doi.org/10.1016/0003-2697(81)90281-5.

105. **Towbin H, Gordon J.** 1984. Immunoblotting and dot immunobinding--current status and outlook. *J Immunol Methods* **72:**313–340 http://dx.doi.org/10.1016/0022-1759(84)90001-2.

106. **Lujan-Zilbermann J, Rodriguez CA, Emmanuel PJ.** 2006. Pediatric HIV infection: diagnostic laboratory methods. *Fetal Pediatr Pathol* **25:**249–260 http://dx.doi.org/10.1080/15513810601123367.

107. **Mylonakis E, Paliou M, Lally M, Flanigan TP, Rich JD.** 2000. Laboratory testing for infection with the human immunodeficiency virus: established and novel approaches. *Am J Med* **109:**568–576 http://dx.doi.org/10.1016/S0002-9343(00)00583-0.

108. **Brown F, Graves JH.** 1959. Changes in specificity and electrophoretic mobility of the precipitating antibodies present in the serum of cattle recovering from foot-and-mouth disease. *Nature* **183:**1688–1689 http://dx.doi.org/10.1038/1831688a0.

109. **Brown F.** 1960. A beta-globulin antibody in the sera of guinea pigs and cattle infected with foot-and-mouth disease. *J Immunol* **85:**298–303.

110. **Svehag S-E, Mandel B.** 1964. The formation and properties of poliovirus-neutralizing antibody. 1. 19S and 7S antibody formation: differences in kinetics and antigen dose requirement for induction. *J Exp Med* **119:**1–19 http://dx.doi.org/10.1084/jem.119.1.1.

111. **Cowan KM.** 1973. Antibody response to viral antigens. *Adv Immunol* **17:**195–253 http://dx.doi.org/10.1016/S0065-2776(08)60733-6.

112. **Schluederberg A.** 1965. Immune globulins in human viral infections. *Nature* **205:**1232–1233 http://dx.doi.org/10.1038/2051232a0.

113. **Bellanti JA, Artenstein MS, Olson LC, Buescher EL, Luhrs CE, Milstead KL.** 1965. Congenital rubella. Clinicopathologic, virologic, and immunologic studies. *Am J Dis Child* **110:**464–472 http://dx.doi.org/10.1001/archpedi.1965.02090030484020.

114. **Banatvala JE, Best JM, Kennedy EA, Smith EE, Spence ME.** 1967. A serological method for demonstrating recent infection by rubella virus. *BMJ* **3:**285–286 http://dx.doi.org/10.1136/bmj.3.5560.285.

115. **Baublis JV, Brown GC.** 1968. Specific response of the immunoglobulins to rubella infection. *Proc Soc Exp Biol Med* **128:**206–210 http://dx.doi.org/10.3181/00379727-128-32979.

116. **Forghani B, Schmidt NJ, Lennette EH.** 1973. Demonstration of rubella IgM antibody by indirect fluorescent antibody staining, sucrose density gradient centrifugation and mercaptoethanol reduction. *Intervirology* **1:**48–59 http://dx.doi.org/10.1159/000148832.

117. **Duermeyer W, Wielaard F, van der Veen J.** 1979. A new principle for the detection of specific IgM antibodies applied in an ELISA for hepatitis A. *J Med Virol* **4:**25–32 http://dx.doi.org/10.1002/jmv.1890040104.

118. **Diment JA, Chantler SM.** 1981. Enzyme immunoassay for detection of rubella specific IgM antibody. *Lancet* **1:**394–395 http://dx.doi.org/10.1016/S0140-6736(81)91723-2.

119. **Clarke DH, Casals J.** 1958. Techniques for hemagglutination and hemagglutination-inhibition with arthropod-borne viruses. *Am J Trop Med Hyg* **7:**561–573 http://dx.doi.org/10.4269/ajtmh.1958.7.561.

120. **Innis BL, Nisalak A, Nimmannitya S, Kusalerdchariya S, Chongswasdi V, Suntayakorn S, Puttisri P, Hoke CH.** 1989. An enzyme-linked immunosorbent assay to characterize dengue infections where dengue and Japanese encephalitis co-circulate. *Am J Trop Med Hyg* **40:**418–427 http://dx.doi.org/10.4269/ajtmh.1989.40.418.

121. **Solomon T, Thao LTT, Dung NM, Kneen R, Hung NT, Nisalak A, Vaughn DW, Farrar J, Hien TT, White NJ, Cardosa MJ.** 1998. Rapid diagnosis of Japanese encephalitis by using an immunoglobulin M dot enzyme immunoassay. *J Clin Microbiol* **36:**2030–2034 http://dx.doi.org/10.1128/JCM.36.7.2030-2034.1998.

122. **Oxford J.** 1982. The use of monoclonal antibodies in virology. *J Hyg (Lond)* **88:**361–368 http://dx.doi.org/10.1017/S0022172400070236.

123. **Schmidt NJ.** 1983. Rapid viral diagnosis. *Med Clin North Am* **67:**953–972 http://dx.doi.org/10.1016/S0025-7125(16)31161-0.

124. **Yolken RH.** 1983. Use of monoclonal antibodies for viral diagnosis. *Curr Top Microbiol Immunol* **104:**177–195 http://dx.doi.org/10.1007/978-3-642-68949-9_11.

125. **Neuberger MS, Askonas BA.** 2005. Cesar Milstein CH. 8 October 1927–24 March 2002. Elected FRS 1974. *Biogr Mem Fellows R Soc* **51:**268–289 http://dx.doi.org/10.1098/rsbm.2005.0017.

126. **Cotton RGH, Milstein C.** 1973. Letter: fusion of two immunoglobulin-producing myeloma cells. *Nature* **244:**42–43 http://dx.doi.org/10.1038/244042a0.

127. **Alkan SS.** 2004. Monoclonal antibodies: the story of a discovery that revolutionized science and medicine. *Nat Rev Immunol* **4:**153–156. (Response, **Karpas A.** 2004. Cesar Milstein and the discovery of monoclonal antibodies. Online correspondence. *Nat Rev Immunol* **4**. Accessed 18 October 2010. 10.1038/ nri1265-c1.)

128. **Eichmann K.** 2005. *Kohler's Invention*. Birkhauser Verlag, Basel, Switzerland.

129. **Köhler G, Milstein C.** 1975. Continuous cultures of fused cells secreting antibody of predefined specificity. *Nature* **256:**495–497 http://dx.doi.org/10.1038/256495a0.

130. **Wade N.** 1982. Hybridomas: the making of a revolution. *Science* **215:**1073–1075 http://dx.doi.org/10.1126/science.7038873.

131. **Köhler G, Howe SC, Milstein C.** 1976. Fusion between immunoglobulin-secreting and nonsecreting myeloma cell lines. *Eur J Immunol* **6:**292–295 http://dx.doi.org/10.1002/eji.1830060411.

132. **Köhler G, Milstein C.** 1976. Derivation of specific antibody-producing tissue culture and tumor lines by cell fusion. *Eur J Immunol* **6:**511–519 http://dx.doi.org/10.1002/eji.1830060713.

133. **Galfrè G, Milstein C.** 1981. Preparation of monoclonal antibodies: strategies and procedures. *Methods Enzymol* **73**(Pt B)**:**3–46 http://dx.doi.org/10.1016/0076-6879(81)73054-4.

134. **Laver WG, Gerhard W, Webster RG, Frankel ME, Air GM.** 1979. Antigenic drift in type A influenza virus: peptide mapping and antigenic analysis of A/PR/8/34 (HON1) variants selected with monoclonal antibodies. *Proc Natl Acad Sci USA* **76:**1425–1429 http://dx.doi.org/10.1073/pnas.76.3.1425.

135. **Schmidt NJ, Ota M, Gallo D, Fox VL.** 1982. Monoclonal antibodies for rapid, strain-specific identification of influenza virus isolates. *J Clin Microbiol* **16:**763–765 http://dx.doi.org/10.1128/jcm.16.4.763-765.1982.

136. **McQuillin J, Madeley CR, Kendal AP.** 1985. Monoclonal antibodies for the rapid diagnosis of influenza A and B virus infections by immunofluorescence. *Lancet* **2:**911–914 http://dx.doi.org/10.1016/S0140-6736(85)90849-9.

137. **Tedder RS, Yao JL, Anderson MJ.** 1982. The production of monoclonal antibodies to rubella haemagglutinin and their use in antibody-capture assays for rubella-specific IgM. *J Hyg (Lond)* **88:**335–350 http://dx.doi.org/10.1017/S0022172400070182.

138. **Pattison JR.** 1982. Rubella antibody screening. *J Hyg (Lond)* **88:**149–153 http://dx.doi.org/10.1017/S0022172400070029.

139. **Balachandran N, Frame B, Chernesky M, Kraiselburd E, Kouri Y, Garcia D, Lavery C, Rawls WE.** 1982. Identification and typing of herpes simplex viruses with monoclonal antibodies. *J Clin Microbiol* **16:**205–208 http://dx.doi.org/10.1128/jcm.16.1.205-208.1982.

140. **Pereira L, Dondero DV, Gallo D, Devlin V, Woodie JD.** 1982. Serological analysis of herpes simplex virus types 1 and 2 with monoclonal antibodies. *Infect Immun* **35:**363–367 http://dx.doi.org/10.1128/iai.35.1.363-367.1982.

141. **Goldstein LC, Corey L, McDougall JK, Tolentino E, Nowinski RC.** 1983. Monoclonal antibodies to herpes simplex viruses: use in antigenic typing and rapid diagnosis. *J Infect Dis* **147:**829–837 http://dx.doi.org/10.1093/infdis/147.5.829.

9 To the Barricades
THE MOLECULAR REVOLUTION

> *... at lunch Francis winged into the Eagle to tell everyone within hearing distance that we had found the secret of life.*
>
> James D. Watson, 1968 (1)

> *... it therefore seems likely that the precise sequence of bases is the code which carries the genetical information.*
>
> James D. Watson and Francis Crick, 1953 (2)

INTRODUCTION

The 1980s brought a catastrophic global viral epidemic and a scientific discovery which together transformed diagnostic virology. In 1981, two reports from the west and east coasts of the United States marked the beginning of the AIDS epidemic by the lethal, immunosuppressant retrovirus, human immunodeficiency virus (HIV) (3, 4). By the start of the 21st century, AIDS had killed over 21 million people worldwide. Deaths from this infection were lingering and painful, as opportunistic infections destroyed multiple organ systems. Just four years later, in 1985, Kary Mullis and his colleagues described the polymerase chain reaction (PCR) (5), which would contribute to an explosion of research on HIV and AIDS. PCR was the first and most important of the nucleic acid amplification technologies, which allowed the identification of minute quantities of genomic material

To Catch a Virus, Second Edition. Authored by John Booss and Marie Louise Landry.
© 2023 American Society for Microbiology. DOI: 10.1128/9781683673828.ch09

through exponential expansion of a target sequence. PCR would revolutionize not only diagnostic virology but also biology in general and related fields from embryology to anthropology. Fully automated commercial systems ultimately became available, which rapidly completed the diagnostic sequence from nucleic acid extraction to readout and reporting of the incriminated virus.

The principle of PCR came to Mullis, as he described it, on a moonlit mountain road into northern California's redwood country (6, 7). The development of the technique was in the laboratories of the Cetus Corporation, less romantic perhaps, but representative of the massive development of commercial biotechnology. Biological supply houses had long played a role in the evolution of diagnostic virology in providing reagents, cells, and media, but the growth of the biotech industry was of a different dimension. The massive development was fueled in part by the very large markets produced by the epidemics of HIV, hepatitis B virus (HBV), and hepatitis C virus (HCV) and by the opportunistic viral infections in transplant recipients and other immunosuppressed individuals. The markets were built on the increasing availability of antiviral medications with which to combat infections of individual patients. But other events, too, particularly patent decisions and legislation, had reassured capital investors in biotechnology. The decision in 1980 to award a patent for the process of creating and replicating recombinant DNA (genetic engineering), developed by Stanley N. Cohen and Herbert Boyer, has been described "as a turning point in the commercialization of molecular biology" (8). Political and legislative support for profit-making from scientific innovation in universities was inherent in the 1980 Bayh-Dole Act, which allowed universities to hold patents derived from research performed with federal funding.

Yet none of the developments in molecular biotechnology, including diagnostic molecular virology, could have occurred without the epochal developments in the molecular chemistry of the gene which began in the 1940s. In 1944, Oswald Avery (Fig. 1) and his colleagues reported that DNA was the substance which induced transformation in pneumococci and that the characteristic was transmissible (9). It was the first demonstration that heritable material was DNA. Several Nobel Prizes were to be awarded in molecular genetics, but none to Avery and his colleagues (10), which is regrettable. Subsequently, James D. Watson and Francis Crick, using data of Rosalind Franklin, successfully constructed the double helix of DNA (2); the genetic code of three nucleotides for each protein was deciphered (11); and the "central dogma" of the flow of genetic information from DNA to RNA to protein was articulated (12). Somewhat later, reverse transcription was described as a mechanism to allow replication of RNA tumor viruses through a DNA intermediate (13, 14). The demonstration of the enzyme, reverse transcriptase, accomplishing that reversal was to play a key role in the identification of HIV, the virus

FIGURE 1 *Oswald Avery. Working at the Rockefeller Institute in New York City, Avery demonstrated that transmission of a heritable characteristic was conveyed by DNA. Exacting studies published in 1944 with Colin MacLeod and Maclyn McCarty determined the nucleic acid basis of the transformation of pneumococcal colonies. It was a finding in advance of its time. (Courtesy of the Rockefeller Archive Center.)*

responsible for AIDS. Biotechnology was built on these processes and the enzymes that were discovered in the development of molecular biology.

The story begins with an overview of the molecular genetics revolution, which was fundamental to what ultimately became molecular diagnostic virology (15–19). The HIV/AIDS epidemic occurred at a time when the molecular foundation of knowledge was in place to address the vexing problem and served to drive rapid developments in clinical virology. Medical and public health urgencies and the political and legislative acumen also helped to propel this research. Technological advances, beginning with Kary Mullis's PCR, included rapid improvements in diagnostic assays, measurement of viral load (concentration of virus circulating in the blood), and characterization of viral mutants to assist clinical selection of antiretroviral regimens. While markedly reducing time to results and increasing throughput, automation in closed systems revolutionized the practice of diagnostic virology.

INHERITANCE, DNA, AND THE DOUBLE HELIX

Odd as it may seem now, up through the 1930s, the general scientific consensus was that inheritance information resided in proteins and that nucleic acids followed a boringly repetitive pattern structural in nature (20). Two crucial observations were made that reversed those flawed preconceptions. In the 1940s, Avery demonstrated that the basis of a heritable characteristic lay in DNA. The finding was not immediately incorporated into the scientific thinking of the time because of what Gunther Stent termed "prematurity and uniqueness in scientific discovery"; that is, the field was not prepared (21).

However, the finding was not lost on the brilliant biochemist Erwin Chargaff, who immediately reoriented the work of his laboratory to the chromatographic study of DNA (22, 23). Chargaff demonstrated that the molar relationships of purines to pyrimidines approximated unity, as did the ratios of adenine to thymine and guanine to cytosine (24, 25). Thus, Avery's work established that the crucial molecule of heredity was DNA, and Chargaff's ratios demonstrated the basis on which the genetic code was written.

The starting point for Avery's crucial report was the transformation in virulence of pneumococci, that is, reversion from the rough colonies of attenuated organisms to smooth colonies and virulence (17). The observation was first reported by Frederick Griffith in 1928 in a follow-up study of pneumococcal types (26). Although his report was published in 1944, Avery had worked on the pneumococcus since his arrival at the Rockefeller Institute in 1913 (27). In fact, Griffith's report included reference to Avery's pneumococcal work with Michael Heidelberger at the Rockefeller. These findings influenced Avery's lab to resume studies of the transforming factor which mediated the change in virulence. Interest in serological typing of pneumococcal isolates and the production of therapeutic antisera had dropped off markedly after the demonstration of the apparent effectiveness of antibiotic therapy (28). N. Russell attributed the resurgence of work in Avery's laboratory to the recognition that transformation could be a genetic effect (29).

The goal of Avery's studies, which appeared in the *Journal of Experimental Medicine* in 1944, was "to isolate the active principle from crude bacterial extracts and to identify if possible its chemical nature or at least to characterize it sufficiently to place it in a general group of known chemical substances" (9). Avery et al. used an arsenal of experimental techniques: "chemical, enzymatic, and serological analyses together with the results of preliminary studies by electrophoresis, ultracentrifugation, and ultraviolet spectroscopy." They found that the active transforming factor consisted "principally, if not solely, of a highly polymerized, viscous form of deoxyribonucleic acid." A coauthor, Maclyn McCarty, would write on a copy of the report that he sent to his mother "This is it, at long last" (30).

It was not immediately broadly accepted that genetic information was encoded in DNA, and the reasons why have been investigated (10). Oswald Avery had been described as "an effective bacteriologist . . . a quiet, self-effacing non-disputatious gentleman" (21). A. R. Dochez described Avery as "by nature somewhat reticent and [someone who] seldom permitted the ordinary distractions of life to divert him from those scientific problems in which he was so completely immersed" (27). Avery's colleague and biographer Rene Dubos noted that he very infrequently attended scientific meetings, even to accept prestigious awards (31). Those personality characteristics may explain in part the delayed appreciation of the importance of his work.

It took the maturation of the field a few years and confirmation in 1953 by A. D. Hershey and M. Chase in studies with bacteriophage for a certainty to emerge that genetic material consisted of DNA (32). In an ingeniously simple experiment, Hershey and Chase labeled phage DNA with radioactive phosphorus and phage protein with radioactive sulfur. They then looked at the radioactive contents of the progeny in infected bacteria. Progeny of radioactively labeled particles contained significant amounts of labeled phosphorus in nucleic acids but negligible amounts of labeled parental sulfur in proteins. Once nucleic acid was released from infecting particles, the bulk of the labeled protein had no role, showing that the genetic material was DNA, not protein.

One investigator who immediately understood the implications of the work was Erwin Chargaff (Fig. 2), a Vienna-educated biochemist based at Columbia University's College of Physicians and Surgeons. A scholar of daunting erudition, he was later to write of Avery's work, "This discovery, almost abruptly, appeared to foreshadow a chemistry of heredity and, moreover, made probable the nucleic acid character of the gene. . .For I saw before me in dark contours the beginning of a grammar of biology" (22, 23). To that point in his career, Chargaff had published extensively on several topics, including clotting mechanisms, radioactive organic tracer compounds, lipids, and lipoproteins. However, he abruptly "decided to

FIGURE 2 *Erwin Chargaff. A Vienna-educated biochemist working at the Columbia University College of Physicians and Surgeons, Chargaff had immediately understood the implications of Avery's work. He determined that the purine and pyrimidine bases composing nucleic acids had reproducible ratios. These findings became known as Chargaff's Rules, which defined base complementarity, a fundamental characteristic of heredity. (Courtesy of the Archives of Columbia University Medical Center and the National Library of Medicine.)*

relinquish all that we had been working on or to bring it to a quick conclusion." He realized that a new approach would be necessary to measure the very small amounts of the components of the nucleic acids, the purine and pyrimidine nucleotide bases. He adopted a newly described separative technique (paper chromatography), made quantitative measurements with an ultraviolet spectrometer, and organized a small team of investigators. The findings, key to the development of the Watson-Crick model of DNA, became known as "Chargaff's Rules." Four in number, they are as follows: first, the sums of the purines and pyrimidines are equal; second, the molar ratio of adenine to thymine is 1; third, the molar ratio of guanine to cytosine is 1; and fourth, as a consequence, the number of 6-amino groups (adenine and cytosine) is the same as the number of 6-keto groups (guanine and thymine) (23–25). The findings disproved the tetranucleotide thesis that DNA was a simple polymer of nucleotides with no particular specificity. Chargaff himself was humbled by what his group had uncovered, the basis for base complementarity (33). It was a pivotal accomplishment on the road to deciphering the structure of DNA.

Chargaff was a man of immense learning and gifts for language; he was said to have 15 languages at his command (34). He was often acerbic in his observations, as illustrated by the contrast of his responses to Avery on the one hand and Watson and Crick on the other (23). He characterized Avery's work as "beautiful experimentation" and said of him, "He was a quiet man; and it would have honored the world more, had it honored him more." In contrast, he said of a meeting with Watson and Crick, prior to the publication of their model, that ". . . they wanted, unencumbered by any knowledge of the chemistry involved, to fit DNA into a helix," and "It was clear to me that I was faced with a novelty: enormous ambition and aggressiveness, coupled with an almost complete ignorance of, and a contempt for, chemistry, that most real of exact sciences. . ."

It is an ironic contrast that Chargaff expressed deep respect for Avery, the quiet and careful experimenter, and disdain for Watson and Crick's lack of knowledge of chemistry. Yet Watson and Crick were able to integrate the biochemical findings of Erwin Chargaff and the X-ray crystallographic findings of Rosalind Franklin into a successful model of DNA which incorporated fundamental features of helical symmetry developed by Linus Pauling. It is in that synthetic quality that their genius lay, not in their work, or lack of it, as experimentalists.

Another pivotal component of the foundation of the double-helix story, X-ray crystallography, is the work of Rosalind Franklin (Fig. 3). There is general agreement that the excellence and clarity of Franklin's photographs of DNA were decisive, allowing Watson and Crick to construct an accurate predictive model of the structure of DNA. The photographs demonstrated that the phosphates had to be on the outside of the molecule to be encased in water, correcting a mistake in Watson and Crick's ill-fated first model. Watson and Crick also demonstrated the likelihood of a two-chain helical structure with chains running in opposite directions. This corrected Pauling and Robert Corey's original model, also ill fated, in which a three-chain molecule had been proposed. J. D. Dunitz, Pauling's memoirist for the National Academy of Sciences, wrote that if Linus Pauling had had access to

FIGURE 3 *Rosalind Franklin. A brilliant X-ray crystallographer, Franklin took photographs of DNA which provided the crucial pieces of data to decipher its structure. They demonstrated that DNA was a two-chain helical structure with the chains oriented in opposite directions. (From the National Portrait Gallery. Photo credit: Vittorio Luzzati.)*

Rosalind Franklin's X-ray photographs, he would have concluded that the molecule was composed of two chains running in opposite directions (35). Watson indicated that Franklin's data were decisive; however, Franklin was unaware that her data had been given to Watson and Crick. She was not included as an author in the decisive paper in *Nature*, nor were her data acknowledged as crucial to the construction of the model. Those omissions have prompted serious questions concerning the system of values in science at the time (36).

Rosalind Franklin, born in London in 1920, was raised in an upper-middle-class family and received education in elite English institutions, St. Paul's School and Newnham College, Cambridge (37, 38). She studied physical chemistry at Cambridge, and to support the war effort in World War II, she moved into studies on coal and its filtering capacities under various conditions. After the war, in 1945, she was awarded a Ph.D. in physical chemistry from Cambridge. She moved to Paris in 1947 for 3 years; here she learned X-ray diffraction methods. J. D. Bernal observed that it was her combination of chemical preparatory skills and X-ray analytic skills that allowed her to distinguish fundamental characteristics of various carbons (37). Turning to large biological molecules, she moved to King's College, London, where she did her work on the X-ray crystallography of DNA.

Franklin's work with R. G. Gosling, her student, was remarkably successful. It demonstrated a two-chain coaxial helical structure with the phosphate groups on the outside of the helix and sugar and base groups on the inside (39, 40). The crucial observation of the DNA B structure, a paracrystalline form of DNA, and the clarity of its X-ray crystallographic photographs allowed the construction of the Watson and Crick model. Because of conflict within the unit at King's College, London, she moved to Birbeck College, London, to work on tobacco mosaic virus. She again produced X-ray photographs of outstanding clarity which allowed conclusions about the structure of the virus. She passed away from cancer at the peak of her scientific powers in 1958, at the age of 37.

Considerable and sustained controversy followed regarding the indirect and originally unacknowledged way in which Franklin's data were obtained. Watson's characterization of Franklin in *The Double Helix* has been termed brutal by her sister, "possibly even libelous if she had still been alive" (41). Yet her achievements and the crucial role of her work in understanding the DNA molecule are her legacy. She held out for scientific proof, not going beyond what the data demonstrated. On being shown Watson and Crick's first model, she destroyed it conceptually, based on a fundamental chemical observation. On being shown their second and essentially correct model, her attitude, according to Gosling, was that it was very pretty, but how were they going to prove it (38)?

Linus Pauling (Fig. 4), considered by many the greatest chemist of the 20th century, exerted several important effects on determining the structure of DNA. Born poor in Condon, OR, in 1901, he had a prodigious memory and mathematical and geometrical intuition (35, 42). He came on the scene when two major advances were underway: X-ray analysis of compounds and the development of quantum mechanics. He applied these to an understanding of the chemical bond. In so doing, "he helped transform chemistry from a largely phenomenological subject to one based firmly on structural and quantum mechanical principles" (43). His 1939 book, *The Nature of the Chemical Bond*, has been described as "a magisterial statement on post-quantum theory evaluations of molecular and ionic structures which became the most influential chemistry text of the century" (43). He won a Nobel Prize in Chemistry in 1954 for his work on chemical bonds and is credited with the recognition that an illness, sickle cell anemia, resulted from a molecular change in hemoglobin. In addition, he originated the concept of a molecular clock, suggesting that a comparison of protein sequences of different species sheds light on the evolution of molecules and hence of organisms (42). In 1962, because of his work to eliminate atmospheric testing of nuclear devices, he won the Nobel Peace Prize (44). The last quarter of his life was devoted to his advocacy for the role of vitamin C in promoting health, to reduce the common cold, and to ward off cancer.

FIGURE 4 *Linus Pauling. Pauling's work on chemical bonds, including his text* The Nature of the Chemical Bond *(published in 1939), laid the basis for modern chemistry. For this work he won the Nobel Prize in 1954. His entry with Robert Corey in 1953 into the race to discover the structure of DNA, while flawed, accelerated Watson and Crick's work to fit the pieces of evidence together. (Courtesy of Ava Helen and Linus Pauling Papers, Oregon State University Libraries.)*

Pauling's work on chemical bonds and molecular structure laid the foundation for work by others on nucleic acids. In his discussion of the alpha helix in proteins, he argued that long polymers comprised of equivalent units had to have similar geometric positions, possible only in a helical form (42). In 1940, with Max Delbrück, he argued that the stability between identical or nearly identical molecules depended on their complementariness, that is, "two molecules with complementary structures in juxtaposition" (45). The forces that maintain stability in complementary structures include hydrogen bond formation. Pauling predicted that "it will be found that the significance of the hydrogen bond for physiology is greater than that of any other single structural feature" (42). After his successful work on the configuration of proteins, he was dramatically unsuccessful in predicting the structure of DNA. With Corey, he postulated a three-chain molecule with a reversed location of the phosphate groups (46). Yet the very fact that Pauling had taken up the structure of DNA greatly stimulated Watson and Crick's search for a DNA structure that would accommodate all the biochemical and X-ray crystallographic observations. In that search, they used a method exploited by Pauling, model building.

The drama of the double helix, both its accurate figuration after two flawed attempts and the personalities involved, continued to entrance well over half a century later (41). The combinations of James Watson and Francis Crick at the Cavendish Laboratory in Cambridge, Maurice Wilkins and Rosalind Franklin at King's College, London, and Linus Pauling and Robert Corey at Caltech in Pasadena, CA, were the players in the drama, each of whom contributed to the solution, and any pair or named individual might have hit on the solution first. It turned out that Watson and Crick won the race, but the decisive data came from Rosalind Franklin.

Maurice Wilkins, the self-described third man of the 1962 Nobel Prize with Watson and Crick, played a central role throughout (47). Watson was on a fellowship in Copenhagen, having received his Ph.D. on bacteriophage multiplication under Salvador E. Luria at Indiana University, when he heard Wilkins describe his early X-ray crystallographic studies of DNA. Deciding to learn X-ray diffraction techniques, Watson transferred his fellowship to the Cavendish Laboratory in September 1951. Max Perutz, molecular biologist and X-ray crystallographer, at that time working at the Cavendish Laboratory, recalled meeting Watson. "One day in 1950 a strange young head with a crew cut and bulging eyes popped through my door and asked, without so much as hello, 'can I come work here?'" (48). Watson ended up sharing an office with Francis Crick (20). The coupling of Watson and Crick (Fig. 5) turned out to be decisive. Each was convinced of the importance of solving the riddle of the structure of DNA, and each, again in the words of Max Perutz, ". . . shared the sublime arrogance of men who had rarely met their intellectual equals" (48). By the end of February 1953, the puzzle was solved; Crick announced in the Eagle Pub that he and Watson "had discovered the secret of life" (1), and they published it in *Nature* in April (2).

What were the pieces of the puzzle of the molecular structure of DNA, what characterized the flawed models, and how was the final model put aright? One component, the location of the phosphate-ribose backbone, was flawed in both a published model by Pauling and Corey (46) and the first and unpublished model by Watson

FIGURE 5 *Francis Crick (left) and James D. Watson. Working at the Cavendish Laboratories at Cambridge University, Watson and Crick determined the structure of DNA. Published in an elegant, brief paper in* Nature *in 1953, it provided the molecular structural basis to understand heredity. (Courtesy of the James D. Watson Collection, Cold Spring Harbor Laboratory Archives.)*

and Crick. Each had the phosphates on the inside of the molecule. Watson and Crick's model was quickly demolished by Rosalind Franklin's observation that interior phosphates were chemically impossible because of the presence of water (49, 50).

Watson and Crick were temporarily put off the DNA work but were allowed back on when it was seen that the Pauling-Corey model was also incorrect. Here Wilkins's role was again evident. Watson brought the Pauling-Corey manuscript to King's College, London, for Wilkins, and it was Wilkins who showed Watson Franklin's X-ray photograph. More complete information on Rosalind Franklin's X-ray crystallography work was contained in an MRC (Medical Research Council) report which was given to Crick and Watson by Max Perutz (50). The photograph indicated that the two coiled chains ran in opposite directions. As Aaron Klug wrote, "Of all the protagonists in the story, only Crick understood this" (50). Pauling's memoirist, Dunitz, suggested that Pauling too, had he seen the Franklin photograph, would have concluded that the molecule was composed of two chains running in opposite directions (35).

A crucial piece of the puzzle remained to be deciphered, the placement and relationships of the four bases, which would hold the chains together with hydrogen bonds. There were two purines, adenine and guanine, and two pyrimidines, thymidine and cytosine. Serendipity played a role in discovering the correct pairings.

During Watson's model making, a chemist, Jerry Donohue, who shared the office of Watson and Crick, observed that Watson was using the wrong tautomeric form of the bases. When that was corrected, Watson quickly found that the geometries of the pairs adenine-thymine and guanine-cytosine were virtually the same (50). That allowed the correct spacing of the bases on the antiparallel chains and was the "novel feature of the structure" by which the chains were held together (2). The specific pairings at once explained the equivalence of the purine and pyrimidine bases found by Chargaff and provided the essential complementarity for copying each DNA strand—hence, the secret of how life reproduced itself (Fig. 6).

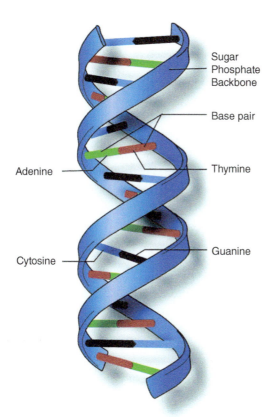

FIGURE 6 *DNA double helix. This drawing demonstrates the essential features of paired DNA strands: a backbone of sugars and phosphate groups, linked by bases pairing from opposite strands, and the strands coiled around each other oriented in opposite directions. Base pairing underlies the capacity to copy strands, that is, to reproduce genetic information. (Courtesy of the National Human Genome Research Institute.)*

The genetic implications of the structure were immediately apparent. "It has not escaped our notice that the specific pairing we have postulated immediately suggests a possible copying mechanism for the genetic material" (2). Watson and Crick did not immediately spell out the copying mechanism, fearing that the model might be incorrect. However, seeing the X-ray crystallographic data of Wilkins et al. (51) and the data of Franklin and her graduate student Gosling (39) in the same issue of *Nature*, they were reassured of the veracity of their model (50). The paper of Franklin and Gosling included an iconic image, photograph 51 of the B form of DNA. Hence, in a second paper in *Nature* in the next month, Watson and Crick spelled out some of the predictions of the model (52). They posited that "the precise sequence of the bases is the code which carries the genetical information" and that the specific order of the nucleotides on the complementary chains could be "how the deoxyribonucleic acid molecule might duplicate itself." They suggested that "prior to duplication the hydrogen bonds which held the chains together are broken and the two chains unwind and separate." Each chain would then serve "as a template for the formation on itself of a new companion chain." They speculated as to "whether a special enzyme is required to carry out the polymerization" or if a chain itself could serve as an enzyme. The processes, enzymes, intermediates, and products would consume molecular biologists and biochemists in the coming years (16, 53, 54). The work would result in an understanding of the biochemical code of life, lay the basis for the manipulation of the genome in genetic engineering, and in the process add a fundamentally new approach to the understanding, diagnosis, and treatment of viral infections. Watson, Crick, and Wilkins were awarded the 1962 Nobel Prize in Physiology or Medicine (55). Rosalind Franklin had passed away and was therefore ineligible for recognition.

Crick further elaborated on the specific order of the bases in what was called the sequence hypothesis (11). It turned out to be right. Namely, the code for the sequence of amino acids in a protein is determined by the sequence of bases in the nucleic acid. The first validation of that concept came in the work of Marshall W. Nirenberg and J. H. Matthaei. They demonstrated that polyuridylic acid coded for phenylalanine (56). "The coding problem" was to discover how four bases determined the amino acid sequence of the 20 known amino acids (11, 57). By 1966, all of the codons, the triplets of nucleotide bases which code for each amino acid, had been discovered (16). In 1968, the Nobel Prize was awarded to Nirenberg, Har Gobind Khorana, and Robert Holley (58). Khorana had devised methods for the synthesis of nucleic acids. Holley had been among those who discovered and described the structure of transfer RNA (tRNA), the molecule which conveys information for protein synthesis.

By all accounts, the final period of putting together the model of DNA by Watson and Crick was a time of great exhilaration with flashes of insight (1, 50). Also marked with flashes of insight and excitement was the successful demonstration of an unstable intermediate which carried the genetic information from DNA in the nucleus to ribosomes in the cytoplasm for the assembly of proteins. It had been shown in the 1950s that proteins were assembled in the cytoplasm on ribosomes, then called microsomes, but the mechanism was unknown. The story of the intermediate molecule, messenger RNA (mRNA), has been recounted by François Jacob in his autobiography, *The Statue Within* (59). With André Lwoff and Jacques Monod, Jacob received the Nobel Prize in 1965 for their studies of gene regulation.

At a small informal meeting in Sydney Brenner's apartment at King's College, Cambridge, in the spring of 1960, Jacob recounted recent experimental results which suggested the possibility of an unstable intermediary, probably RNA, which was crucial for protein assembly. Both Brenner and Crick leapt to their feet, each having made a connection to other experiments. As Jacob recalled of the two, they "Began to gesticulate. To argue at top speed in great agitation. A red-faced Francis. A Sydney with bristling eyebrows." Each had recognized a connection between an unstable intermediary, "X," and results in two different experimental systems (59).

After intense planning, what emerged was an agreement by Jacob and Brenner to work together briefly at Caltech in California, where each had been invited to visit. Working with Matthew Meselson, they suffered experimental failure after failure. Finally, on a Pacific beach, it came to Brenner that there had been insufficient magnesium to allow the ribosomes to cohere. Redone just before Brenner left Caltech, the experiment succeeded: ". . . as the last sample was counted, a double shout of joy shook the basement. Followed immediately by a wild double jig." Brenner completed further experiments on returning to Cambridge. These succeeded in demonstrating that ribosomal RNA (rRNA) did not code for protein synthesis but that RNA with a rapid turnover was synthesized and attached to preexisting ribosomes. It was predicted that the mRNA would be a copy of the gene (60).

Hence, the major components of molecular biology were in place: that information was transferred from DNA codons by mRNA to ribosomes for protein production. It would later be shown by Howard Temin and Satoshi Mizutani (14) and by David Baltimore (13) that RNA tumor viruses use an RNA template for the production of a DNA intermediate. The discovery of reverse transcriptase brings us back to the evolution of what Crick had called the "central dogma" of molecular biology (12). Information transfer was understood to go from DNA to RNA to protein. However, a unidirectional flow of information was in opposition to a reverse mechanism suggested by Howard Temin at the University of Wisconsin. He proposed that replication of the Rous sarcoma virus included a DNA-containing provirus step creating a template for progeny virus (61). The reversed flow of information from RNA to DNA would require an RNA-dependent DNA polymerase. This type of enzyme was found by Temin and Mizutani in Rous sarcoma virus (14) and by Baltimore in both Rauscher mouse leukemia virus and Rous sarcoma virus (13). It was, as an editorial commented, "an extraordinary personal vindication for Howard Temin" (62). Crick, commenting on the "central dogma" and the new findings, noted that there was no theoretical reason that information transfer could not sometimes go from RNA to DNA (12). Temin and Baltimore, with Renato Dulbecco, received the Nobel Prize in 1975 "for their discoveries concerning the interaction between tumor viruses and the genetic material of the cell" (63).

The enzymes necessary to carry out the processes of nucleic acid replication and transcription were the objects of intense study. The discovery of the enzymes and the capacity to manipulate their functions would be vital in the forthcoming genetic-engineering phase of the molecular revolution. Diagnostic virology would benefit from those advances. Clinical virology was soon to be challenged by the largest epidemic since the influenza pandemic of 1918–1919 with the emergence of HIV/AIDS. Molecular virology would assume great importance in the diagnosis, treatment, and clinical management of HIV.

HIV AND THE AIDS EPIDEMIC

The major viral epidemic of the late 20th century and continuing into the 21st has been caused by a chronic, immunosuppressive retrovirus, HIV (64) (Fig. 7). Techniques of molecular biology were fundamental to its discovery and the identification of its reverse transcriptase. Key to management of infected individuals has been determination of viral load, detected by nucleic acid amplification techniques such as the PCR.

The harbinger of the epidemic was *Pneumocystis carinii* (since renamed *Pneumocystis jirovecii*) pneumonia diagnosed in five young homosexual men (3, 65). The first report noted "profoundly depressed numbers of thymus-dependent lymphocyte cells" in the three patients who were tested. All of the patients were found to have had infection with cytomegalovirus (CMV), and the capacity of CMV to induce abnormalities of cellular immune function was noted. Close on the heels of that 5 June 1981 report came another report in the *MMWR Morbidity and Mortality Weekly Report* of 3 July 1981 of Kaposi's sarcoma (KS) with a fulminant course in 26 non-elderly homosexual men (4). This was unique in that KS usually occurred in elderly men, had a chronic course, and had a much lower

FIGURE 7 *Names Project AIDS Memorial Quilt on the National Mall in Washington, DC. The quilt is a massive memorial to individuals who have lost their lives to AIDS-related causes. (Courtesy of the National AIDS Memorial, http://www.aidsmemorial.com.)*

incidence. In all 12 patients for whom CMV testing was available, there was evidence of past or present infection. In addition, four homosexual men with severe, perianal herpes simplex virus (HSV) infections and cellular immunodeficiencies were reported in December of 1981 (66). An editorial in that December 1981 issue of the *New England Journal of Medicine* suggested that this constellation of findings described a new syndrome and that "some new factor may have distorted the host parasite relation." Doubt was expressed about the role of CMV in these multifaceted infections (67).

As studies evolved, the likelihood that AIDS was the result of a lethally mutated CMV or other known virus became less likely. Standard methods of viral isolation in tissue culture failed to reveal a novel agent. Certain laboratories started to investigate the possibility that the syndrome resulted from the emergence of a novel retrovirus. Robert Gallo recounted the sequence of events that led his laboratory at the National Institutes of Health (NIH) to investigate the possibility that AIDS might be caused by a novel retrovirus (68). His laboratory had discovered a T-cell growth factor, interleukin 2, which facilitated the isolation of the first human retrovirus, human T-cell lymphotropic virus type 1 (HTLV-1), from a patient with a

T-cell malignancy. Soon thereafter, another retrovirus, HTLV-2, was isolated from a case of hairy cell leukemia. Both viruses targeted T cells; hence, the search began for the cause of AIDS in this family of viruses.

Luc Montagnier (1932–2022) of the Cancer Oncology Unit and his colleagues at the Pasteur Institute in Paris, in cooperation with a group of clinicians and immunologists, also started an investigation of the possibility that a retrovirus was the cause of AIDS (69). At the start of January 1983, cells from a lymph node biopsy were cultured from a homosexual man with lymphadenopathy, recognized in those early years as a clinical precursor to the full-blown clinical syndrome. Fifteen days later, the cultures were found to contain reverse transcriptase, an indicator of the presence of a retrovirus. Simultaneously, published reports from the Pasteur Institute (70) and from the NIH (71) in 1983 described the isolation of a human retrovirus, respectively "Lymphadenopathy Associated Virus" (LAV) and HTLV-1. Questions later emerged about the source of the other NIH isolate, named HTLV-III and published in 1984 (138), when it was discovered that the French virus, LAV, and the NIH virus, HTLV-III, "appeared to be identical," prompting an investigation by the Office of the Inspector General of the U.S. Department of Health and Human Services (72). Years later, in 2008, the Nobel Prize was awarded to Françoise Barré-Sinoussi and Luc Montagnier (Fig. 8) of the Pasteur Institute "for their discovery of human immunodeficiency virus" (73).

Identifying the viral cause of AIDS opened the way to a diagnostic assay. This was important for defining risk groups and for designing protection strategies. A diagnostic assay was particularly urgent for screening the blood supply for virus. Transfusion-associated AIDS was reported by the CDC for 18 persons in whom no other risk factors could be identified, yet each had received blood components within 5 years of the onset of the illness (74). In follow-up studies of seven patients, it was found that a high-risk donor was the source of blood in each case. A brutally cruel situation emerged in hemophiliacs who were vitally dependent on blood products, particularly plasma factor concentrates. Not surprisingly and tragically, hemophiliacs were found to be at significant risk of developing AIDS (75). The first test for HIV was an enzyme-linked immunosorbent assay (33, 76). This, too, aroused dispute between the French and U.S. governments. In March 1985, the U.S. Secretary of Health and Human Services announced the licensing of an HIV blood test for antibody detection. It was later agreed that the virus used in the test had been provided to the NIH by the Pasteur Institute. A lengthy patent dispute evolved, and in a final resolution in 1994, the Pasteur Institute was to share the royalties with the NIH (77).

FIGURE 8 *Françoise Barré-Sinoussi (left) and Luc Montagnier (right). Working with clinicians who were attempting to find the cause of AIDS, Montagnier and Barré-Sinoussi at the Pasteur Institute in Paris found a retrovirus in a lymph node biopsy. Ultimately shown to be the cause of AIDS, the virus was to be called human immunodeficiency virus (HIV). Montagnier and Barré-Sinoussi were awarded the Nobel Prize in 2008 for their discovery. (Photos courtesy of Institut Pasteur. Photo credit: François Gardy.)*

While the risk groups prominently included intravenous drug abusers and male homosexuals in the United States and Western Europe, it was found that heterosexual transmission was significant in Africa. In an early report from Zaire, the male-to-female ratio was unity; homosexuality, blood transfusions, and intravenous drug abuse were not found to be risk factors in this population (78). The capacity to identify African patients with HIV infection by specific antibodies allowed the definition of an illness in Uganda called "slim disease" (79). It was manifested principally by gastrointestinal symptoms of diarrhea and weight loss associated with intermittent fevers and malaise. Retrospective viral studies of a plasma sample obtained in 1959 established Africa as the site with the earliest known HIV type 1 infection (80). By the 20-year anniversary in 2001 of the first reports of AIDS, sub-Saharan Africa had two-thirds of the world's people infected with HIV. The estimated global total of those living with HIV was 36 million people, with 23 million people worldwide already dead from the infection (81). In January 2000, the UN Security Council took up the issue of the sub-Saharan HIV/ AIDS epidemic as a possible threat to international stability (82).

Soon after the description of the syndrome and the isolation of the etiological virus, molecular methods played a crucial role in the design of effective antiretroviral therapy (ART). The timing of the evolution of molecular methods was fortuitous. Michael Morange, a biochemist and science historian, noted that "The new conceptual tools for analyzing biological phenomena were forged between 1940 and 1965. The consequent operational control was acquired between 1972 and 1980" (16). Hence, after the description of the syndrome in 1981, the identification of the etiological agent in 1983 facilitated the design of therapeutic agents. The enzymes required for infection, replication, and maturation could be targeted. Because HIV is a retrovirus, the first successful chemotherapeutic interventions targeted the reverse transcriptase.

Nucleoside and nonnucleoside reverse transcriptase inhibitors were evaluated. The first approved drug, zidovudine, was a nucleoside reverse transcriptase inhibitor (83). However, the capacity of HIV to mutate and develop resistance soon became evident (18). The advent of protease inhibitors and the use of combination ART produced a turning point in the epidemic in 1996. Palella et al. reported that mortality and the incidence of three opportunistic infections declined markedly between 1995 and 1997 (84). The three opportunistic infections which were studied and found to decline were *P. carinii* pneumonia, *Mycobacterium avium* complex disease, and CMV retinitis. No differences were found in the reductions with respect to risk factors, gender, or ethnicity. Despite problems which emerged as a consequence of therapy, such as the metabolic syndrome, it was a remarkable achievement. The advent of protease inhibitors in the context of combination ART changed HIV infection for most people with access to therapy from an assured premature death to a chronic, treatable disease.

Although HIV is a lentivirus, a "slow virus," D. D. Ho et al. demonstrated that HIV replication was anything but slow (85). Using a modified branched-DNA signal amplification technique in a study of a protease inhibitor, they found a very high rate of HIV production. CD4 lymphocytes were destroyed and replaced at a rate of 10^9 lymphocytes per day. Ultimately, without ART intervention, the race is lost by the host. In other studies, J. W. Mellors et al. found that the risk of progression to AIDS and death related directly to the viral load, which was a better predictor than was the CD4 T-cell count (86). In time, the measurement of viral load would become a key component for management of anti-HIV medications. As of mid-2022, elimination of virus has not yet become feasible as the HIV genome is integrated into the host cell genome.

EARLY DIAGNOSTIC APPLICATIONS OF MOLECULAR NUCLEIC ACID TECHNIQUES

Prior to the explosive commercial development of PCR and other nucleic acid amplification tests, certain molecular nucleic acid methods were explored for clinical diagnostic virology. While valuable for viral discovery and clinical research protocols, these techniques, including restriction endonuclease mapping and *in situ* nucleic acid hybridization, were generally too specialized, labor-intensive, and costly to be applied to routine diagnostic virology work (87).

The technique of restriction endonuclease mapping is based on the capacity of restriction enzymes to recognize and cleave DNA at specific locations. Separation of the cleaved products by gel electrophoresis distributes the fragments of different sizes. Hence, demonstration of different patterns of cleavage products allows the differentiation of individual viral isolates. The first DNA restriction enzyme was isolated from *Escherichia coli* and characterized in 1958 (88). Although these enzymes have a biological role in host defense against foreign DNA, they also play a fundamental role in molecular genetics and recombinant DNA technology. Thus, nucleic acid fragments from different sources can be joined to produce hybrid molecules (89). The discovery of these enzymes and their role in molecular technology resulted in the award of the Nobel Prize in 1978 to Werner Arber, Hamilton O. Smith, and David Nathans (90). In the Nobel Prize award ceremony speech, the technique was credited with revolutionizing our understanding of "the genetics of higher organisms and completely changed our ideas of the organization of their genes."

Clinically, restriction endonuclease mapping was used to evaluate chronic, perianal HSV infection in four homosexual males in 1981 (66). All four patients were found to have impaired cell-mediated immunity. Restriction endonuclease mapping was performed to determine if the isolates were epidemiologically related. It was found that they were all HSV type 2, yet unrelated. In an earlier (1978) study by T. G. Buchman et al., restriction endonuclease fingerprinting had been used as an epidemiological tool to investigate a temporal cluster of pediatric HSV infections (91). The technique was also used in an investigation of an apparent cluster of cases of herpes simplex encephalitis (92). Mapping showed the isolates to be distinct strains, excluding the possibility that a particularly virulent, neurotropic viral strain had emerged. Despite some success in research settings, the technique never really developed a foothold in clinical diagnostic virology labs.

In nucleic acid hybridization, labeled, single-stranded nucleic acid probes bind to complementary sequences in samples, thus allowing their detection, a useful technique for clinical specimens. Extensive research had been done seeking virus-specific DNA, such as Epstein-Barr virus (EBV) and human wart virus, in various human tumors (93, 94). The latter studies led to the demonstration by Harald zur Hausen (Fig. 9) of the association of human papillomavirus (HPV) and cervical cancer, work for which zur Hausen was awarded the Nobel Prize in 2008 (73).

Using cloned probes on cell spots and Southern blots, Warren Andiman and his colleagues examined clinical specimens for the presence of the EBV genome (95). Samples from 145 patients were examined, and nine were found to be positive for EBV. After considering the drawbacks of other techniques such as serology, virus isolation by immortalization of lymphocytes, or demonstration of EBNA, an EBV nuclear antigen, they commented that ". . . nucleic acid hybridization appears

FIGURE 9 *Harald zur Hausen. Suspected for years to be of viral origin, cervical cancer was demonstrated to be associated with HPV by zur Hausen. He had pioneered work with nucleic acid hybridization to detect viruses in clinical specimens of human tumors. He was awarded the Nobel Prize in 2008 for his discovery that HPVs cause human cervical cancer. (Courtesy of Harald zur Hausen, DKFZ, Heidelberg.)*

to have clear advantages as a method for describing the association of EBV with diverse clinical conditions." They suggested that hybridization techniques would facilitate understanding of the association of EBV and lymphoproliferative conditions among immunocompromised individuals. Thus, nucleic acid hybridization emerged as applicable to clinical research.

The same conclusion can be drawn from a study to detect CMV in urine by DNA hybridization. The report of S. Chou and T. C. Merigan, also published in 1983, was predicated on the need to rapidly identify virus candidates for antiviral agents in clinical trials (96). Urine specimens were immobilized on nitrocellulose filters and probed for CMV with a radiolabeled, cloned DNA fragment. Chou and Merigan achieved a rapid (24-hour) identification of CMV viruria in selected immunosuppressed patients, and no false positives were found when compared with culture-negative specimens. The authors acknowledged that the assay as described was appropriate for research settings but that application in clinical practice would require modifications.

One commercial kit was developed using hybridization to detect HPV in cervical specimens, since there were no culture, antigen, or serological alternatives for HPV. The test, which used spot hybridization with ^{32}P-labeled probes, was available in a few reference laboratories. However, hybridization was cumbersome, required expertise not often found in diagnostic labs, and was hazardous because of the need for radioisotopes. The last problem was circumvented by the development of biotin-labeled probes (97), but other problems persisted. Perhaps most importantly, nucleic acid hybridization was not more sensitive than nonmolecular methods for viral diagnosis in general. The emergence of highly sensitive molecular amplification methods for clinical purposes in the 1990s eclipsed other nucleic acid-based techniques.

PCR AND OTHER NUCLEIC ACID AMPLIFICATION TESTS

The origins of current-day diagnostic nucleic acid testing lay in the development of the PCR to amplify very small quantities of genomic sequences. Amplification was dependent on the isolation and characterization of DNA polymerase, which was achieved in 1955 in Arthur Kornberg's lab and published in 1958 (88, 98). Morange pointed out that PCR was a "post-mature" discovery as described by Harriet Zuckerman and Joshua Lederberg (99). That is, although Kary Mullis's description of PCR was in the 1980s, "all the necessary tools for its realization had existed in the 1960s" (16). When Mullis conceived the idea of the PCR in 1983, he thought it likely that someone else had already described it (6, 7).

Kary Mullis (Fig. 10) said of himself, "I've just been curious as hell all my life . . ." (100). As a youngster, he concocted a rocket fuel of sugar and potassium nitrite and made a 4-foot aluminum rocket which carried a frog safely back to Earth after a 2-mile journey into space. Apocryphal or not, the story demonstrates an interest in putting chemistry to practical use and a sense of adventure. He received his undergraduate degree in chemistry from the Georgia Institute of Technology and a doctorate in biochemistry from the University of California at Berkeley. While at Berkeley and independent of his graduate studies in biochemistry, he published a brief paper in *Nature*, "Cosmological Significance of Time Reversal" (101).

After postdoctoral training in Kansas and San Francisco and various positions on his return to Berkeley, Mullis was hired by Cetus Corporation in 1979 to synthesize oligonucleotide probes (6, 7). Soon thereafter, much of his oligonucleotide work was accomplished by the development of automation for probe production. As he put it, ". . . there was a fair amount of time to think and to putter." The problem he addressed was to devise a technique to identify nucleotides at specific locations in DNA. His thinking started from an established method of determining a DNA sequence in which dideoxynucleotide triphosphates were used to "cap" or halt elongation of a DNA sequence. Such strand elongation used a DNA polymerase on a template with a primer and nucleotide triphosphates which were then capped by the dideoxynucleotide triphosphate. He decided to modify the process by using two oligonucleotides, one on each strand, to get complementary sequencing information. He could identify specific sequences of DNA by bracketing them.

To counter an anticipated technical problem, he considered running the reaction twice when he realized that multiple repeats or cycles of the process would exponentially increase the number of copies produced. The cycles consisted of heating to separate the annealed strands and then cooling sufficiently to allow the polymerase to function, requiring an investigator to manually transfer reaction tubes back and forth between water baths to raise and lower temperatures. Hence, the later appearance of commercial thermal cyclers was a big advance. Originally the polymerase had to be added for each cycle. Then, a heat-tolerant polymerase was found in *Thermus aquaticus*, a hot springs living bacterium which tolerated the heating cycle (102). The end result was that a very small number of gene copies could be increased into millions very quickly, identifying the needle in the haystack or making a great deal of genomic product for other purposes.

Mullis recounted the story of the emergence of the idea of PCR as a revelation; he conceived of PCR while driving to a weekend retreat in the spring of 1983 (6, 7). He termed the process ". . . repeated reciprocal extension of two primers hybridized to the separate strands of a particular DNA sequence." Throughout the weekend

FIGURE 10 *Kary Mullis. The PCR takes advantage of the capacity of short lengths of nucleotides to find and hybridize with reciprocal pieces of nucleic acid. Once hybridized, multiple copies are made through repeating cycles of polymerase-assisted replication. PCR has transformed many fields in biology, including, notably, the capacity to discover minute amounts of viruses in clinical samples. Kary Mullis devised the process while driving one evening in 1983 and was awarded the Nobel Prize in 1993. (Courtesy of Dona Mapston, under license CC BY-SA 3.0.)*

he developed the idea, seeking reasons why it would not work. All the steps were known, and he thought it ought to have been tried previously, but a search of the literature the following week failed to reveal any prior reports. Nor did colleagues to whom he described the technique know of it having been tried. After months of sorting out the proper conditions, he got the assay to work in December 1983.

As a kind of postscript, Mullis described a poster presentation of his work the following year in which he asked Joshua Lederberg, the great molecular biologist, to look at the work. Lederberg then recounted a conversation of 20 years earlier with Arthur Kornberg, the discoverer of DNA polymerase, in which they had discussed using the polymerase to make large quantities of DNA but had not determined how to do it (6). Ironically, both *Nature* and *Science* rejected Mullis's paper on PCR (100). The first appearance of PCR in print was a report of the prenatal diagnosis of sickle cell anemia in 1985 (5). The technical report of the method appeared in 1987 (103). Mullis was awarded the Nobel Prize for Chemistry in 1993 in recognition of his work (104).

By the end of the 1980s, it became apparent that the PCR held real promise as a diagnostic virology tool. While noting the more spectacular use of PCR to identify DNA in extinct woolly mammoths and from a 7,000-year-old human brain, J. Marx pointed out its usefulness for viral diagnosis (89). In particular, the usefulness of PCR was emphasized for the detection of viruses which "go underground" and persist for years, such as HIV. An important step in that direction was achieved by the use of reverse transcription prior to PCR. In 1988, C.-Y. Ou et al. at the CDC demonstrated DNA amplification by PCR of HIV type 1 from DNA isolated directly from peripheral leukocytes of HIV-infected patients (105). Also in 1988, C. Hart et al. demonstrated HIV RNA in blood mononuclear cells (106). They argued that because of its speed and sensitivity, PCR could be used to monitor viral loads in patients receiving antiviral treatment. Even so, there was no immediate implementation of PCR in clinical diagnostic labs. On the technical side, the procedure was labor-intensive, there was a lack of expertise in clinical laboratories, and special equipment and separate PCR rooms were needed to prevent contamination. Exogenous nucleic acid contamination and the potential for false-positive results were seen as the greatest barriers to the application of PCR to the diagnosis of infectious diseases. As D. H. Persing put it, "the exquisite sensitivity of PCR proves to be its undoing" (107). Simplified methodologies were developed for nucleic acid extraction and purification from clinical samples. Other issues included establishing sensitivity cutoffs, rapid preparation of samples, and elimination of PCR inhibitors in blood and other specimens. Yet the future was sounded when Hoffmann-La Roche obtained a license from Cetus, which held the patent for PCR, for the development of a commercial system.

An early focus of PCR diagnosis was detection of HSV in cerebrospinal fluid as a more sensitive and less invasive alternative to brain biopsy and culture of brain tissue (108). Since each institution had few requests and no approved commercial assays were available, testing cerebrospinal fluid by PCR for HSV was confined to reference laboratories that developed and validated their assays for diagnostic use. Ultimately, the field of molecular diagnostic virology was fueled by the market opportunities presented by the epidemics of HIV, hepatitis (HBV and HCV), and opportunistic infections such as with CMV, EBV, and HSV. The need to manage medication adjustments in HIV-infected individuals demanded frequent assays on a large population of patients. From a business perspective, reimbursement payments by private insurers or government health care programs ensured significant returns on research investments for biomedical corporations.

Thus, a decade after PCR was developed at Cetus, several quantitative molecular assays were available for the measurement of HIV type 1 (109). These included a Roche system, Amplicor Monitor. It is based on reverse transcriptase PCR (RT-PCR), in which an initial reverse transcriptase step was required to transcribe the viral RNA into a complementary DNA (cDNA) copy. Another technology based on production of RNA transcripts by RNA polymerase, referred to as transcription-mediated amplification (TMA), and nucleic acid sequence-based amplification (NASBA) (110) were developed in the early 1990s and optimized for HIV (111). In contrast to RT-PCR, TMA and NASBA are isothermal and the amplified product is RNA, not DNA. In comparing the Amplicor Monitor and the NASBA systems, B. P. Griffith et al. found comparable concentrations of HIV RNA (112).

Another approach was to amplify the signal rather than amplifying the target gene. Chiron Corporation developed and refined the branched-DNA amplification system as a means of boosting the reporter signal (113). Amplifier probes hybridize to the target sequence in the sample and have multiple nucleotide branches for attachment of other probes (114). It is this feature which allows amplification of the signal, the "ornaments on a tree approach." As of 1994, Chiron had developed the branched-DNA quantitative assay for HBV, HIV, and CMV.

The ultimate usefulness of the nucleic acid amplification assays would depend on many factors, including accuracy, reproducibility, ease of use, cost, and clinical predictive value. Those issues would be addressed in the coming years as significant advances occurred in diagnostic virology (6, 66, 115). In addition, further refinements would result in more fully automated systems that reduced operator time and opportunities for error and contamination and incorporated automated reporting systems. As the 1990s moved along, the importance of measuring viral load to manage HIV antiretroviral treatment had become the standard of practice (116). Thus, the paradigm of diagnostic virology expanded considerably beyond the basic demonstration of infection to support crucial patient management decisions.

One of the significant problems in transitioning to an accurate, high-throughput process was post-PCR detection of the amplification product. This labor-intensive step could be gel electrophoresis or a hybridization procedure that increased the turnaround time to report results. Also of critical concern was the potential to release amplified product fragments into the laboratory when the reaction tube was opened to remove an aliquot for the detection step. The development of real-time PCR addressed these problems by combining the amplification and the detection steps in the same, unopened tube, allowing the steps to occur simultaneously in a completely enclosed system (117, 118). Strategies to measure the amplified products in "real time," as they are generated, included the use of SYBR green, a fluorescent dye that intercalates into double-stranded nucleic acid, and a variety of fluorogenic probes that can be detected as the reaction progresses through multiple cycles (119). The net result has been more rapid and less labor-intensive systems with reduced susceptibility for contamination. These features, combined with automated measurements and a greater dynamic range for quantitation, facilitate high-throughput and more user-friendly assays for clinical diagnostic purposes.

Development of multiplex PCR has simplified the diagnostic process for clinicians confronted with symptom complexes that can arise from any of several viruses. These highly multiplexed, molecular amplification tests detect more viruses than possible by culture methods and in a fraction of the time. An important example is respiratory disease: influenza, parainfluenza, respiratory syncytial, and other viruses can cause similar upper and lower respiratory tract diseases (120). The ability to rapidly detect multiple common pathogens in a single reaction greatly improves efficiency (121). Initial real-time PCR instruments were limited to differentiating no more than four different targets in a single reaction. To increase the number of targets detected, multiplex, conventional RT-PCRs were followed by a number of novel amplicon detection strategies, such as liquid-phase bead-based suspension arrays or automated microarray hybridization. These "highly multiplexed" tests permitted detection of more than 20 viruses in a single reaction; a number of these assays have been made commercially available (121, 122).

Viral infection of the central nervous system is another clinical area in which several different viruses, such as HSV type 1, HSV type 2, varicella-zoster virus, arboviruses, and enteroviruses, can present similar clinical pictures and in which a rapid diagnosis is required. Multiplex PCR screening has been applied to this problem as well (115). In yet another example, triplex nucleic acid testing has been used to identify HBV-, HIV-, and HCV-positive blood donors in the window period between exposure and seroconversion (123). Hence, multiplex testing such as in respiratory disease, central nervous system disease, and blood screening offers an approach for accurately identifying pathogens in specific clinical situations.

Molecular techniques have added yet more levels of sophistication to clinical management of persons with HIV, CMV, and HCV infections. For HIV, genotypic drug resistance testing is recommended both for those who are naïve with respect to antiretroviral drug treatment and for those in whom drug resistance is suspected or a change in treatment regimen is planned (124). Genotypic testing identifies known resistance mutations by sequence analysis of the reverse transcriptase and protease genes. Thus, more effective antiviral regimens can be selected. Similarly, sequencing to identify antiviral drug resistance mutations is standard practice for CMV, HCV, and HBV when initial treatment fails.

The decades following the discovery of PCR and the advent of the HIV epidemic have witnessed remarkable advances in viral diagnosis using molecular methods (125–127). Diagnostic virology has played a central role in patient management, in pharmacotherapy, and in researching the epidemiology of circulating viral infections. Likewise, management of infections with viruses such as HIV, CMV, HBV, and HCV has been significantly improved. The current trend is to devise user-friendly and walk-away molecular tests allowing more clinical laboratories to offer viral diagnostic services. For newer tests, the sample may simply be added to a pouch or cartridge and inserted into an instrument, and the operator walks away (128, 129). The duration of some assays is only 1 hour, and the instruments may be "random access" so that a cartridge accommodating additional samples may be added whenever there is capacity in the instrument or whenever a sample arrives in the lab. Such innovations allow 24-hour testing.

FUTURE DIRECTIONS FOR MOLECULAR DIAGNOSTICS AND VIRAL PATHOGEN DISCOVERY...THE NEXT CHAPTER

The challenges facing clinical virology include adequate distribution of diagnostic capacity and the prompt recognition and diagnosis of infections, including emerging viral diseases. Meeting these challenges is facilitated by the great leaps in technology that continue to revolutionize laboratory practices: reduced hands-on labor, decreased sample volume requirements, miniaturization of instrumentation, and enhancement of information transmission, storage, and retrieval. But the technological advances present very real challenges to laboratory directors. The acceleration of new and expensive commercial platforms and products makes the decision to purchase a complex one. Several questions arise: will viruses mutate from prototypes so that they will be missed by molecular technology; will new antiviral medications require a rapid adoption of a different diagnostic technology; and will new diagnostic algorithms, such as for HIV, require shifts in the spectrum of tests offered by diagnostic labs? The potentially most transforming question is whether diagnostic labs will continue as independent entities or be merged into molecular labs serving a variety of diagnostic goals. If institutional and educational impediments can be overcome, high-throughput sequencing may transform diagnostic microbiology (130). Thus, there is the possibility that clinical virology will be incorporated into molecular diagnostic facilities.

The greatest challenge to society will be to allocate sufficient resources to build and support the infrastructure necessary for labs of the future. Viral diagnosis is applied to both the diagnosis and management of individual patients, the province of medicine, and the recognition and prevention of the spread of viral infections, the province of public health. The responsibilities of medical care and public health can be best served if the capacity for viral diagnosis is broadly distributed.

A mechanism for broad distribution of diagnostic capacity is point-of-care (POC) testing. It has become increasingly useful in health care delivery. Such tests can be performed by personnel without specialized technical training, and they do not require cumbersome or sophisticated equipment (131). In general medical care, POC tests are often used to measure physiological elements, such as home-based blood sugar devices for the management of diabetes and office- and clinic-based tests to diagnose strep throat. If a "rapid strep test" is positive for streptococcus, antibiotic treatment can be started promptly. POC diagnostics have been applied with considerable benefit to viral infections. Initially, the majority of POC assays for viral infections detect viral antigens or antibodies by immunoassay (131). Then, in 2015, the first POC molecular test using nicking enzyme amplification was approved for influenza and RSV, followed by a real-time PCR test and then a large multiplex panel, all using small instruments that accept one sample at a time, with results in 15 to 60 minutes. As the technologies of miniaturization and fully contained diagnostic units become more advanced, the application of nucleic acid amplification devices, perhaps handheld, will greatly facilitate viral diagnosis at points away from traditional diagnostic labs.

POC devices have been developed for the early diagnosis of viral respiratory diseases. Such use demonstrates both the strengths and weaknesses of a POC approach to diagnosis. Like with any new technology, there are caveats that must be understood by health care practitioners so that they understand the applications of these devices. A consensus document on best practices for respiratory viral disease testing pointed out that clinicians need to be aware of the low sensitivity and marginal specificity, if disease prevalence is low, of rapid antigen direct tests (132). With time, new technologies may improve the utility and reliability of such testing. Although a negative result may not be clinically useful in a test of low sensitivity, a positive result in a test with acceptable specificity may help direct patient management and public health responses, as would subsequently be seen during the COVID-19 pandemic. Unnecessary hospitalizations, unnecessary antibiotics, and unnecessary testing can be avoided, with less inconvenience to the symptomatic individual and with much reduced costs of health care.

During the height of a respiratory disease season, POC testing in clinics and doctors' offices can reduce the testing burden on hospital-based, public health, and other clinical testing labs. Rapid results can improve the management of respiratory illnesses in settings in which there are cohorts of susceptible individuals, such as in long-term care facilities for the elderly or pediatric hospital services, or for immunocompromised and HIV-infected individuals. Result reporting to public health agencies can also provide an early warning system for viral disease outbreaks. What is lost with POC testing, nucleic acid applications, rapid antigen testing, and other types of viral detection approaches is the opportunity to recover a viral agent in culture for further characterization and growth and for antiviral susceptibility testing. With the ability to replicate nucleic acids for identification and testing, the traditional approach of culturing samples for viral isolation may soon be obsolete.

At the end of 2019, in contrast to HIV which emerged from a jungle setting perhaps no challenge so focuses the diagnostic process as the need to discover the cause of an epidemic infectious disease (Fig. 11). That lesson was learned at the beginning of this history, in the yellow fever epidemic in Philadelphia in 1793, in which miasma was said to have been the cause of the outbreak. Emerging and reemerging infectious diseases have been a feature of developed civilizations throughout the centuries (133). In recent years, ever-increasing numbers of diseases have emerged, many of them viral in nature (134, 135). One need only think of the severe acute respiratory syndrome (SARS), which jumped into humans from animals; West Nile virus infection, which emigrated to the New World by unknown means; Nipah virus encephalitis, which resulted from agricultural practices that put pigs in proximity to fruit-eating bats; the ever-present threat of new influenza pandemics; and, of course, the pandemic of HIV/AIDS, which made itself known in 1981. At the end of 2019, in contrast to HIV which emerged from a jungle setting, the next pandemic, COVID-19, would emerge in a massive urban environment. An advance in molecular technology, next generation sequencing (NGS) would identify and sequence the virus, SARS-CoV-2, within two weeks of medical acknowledgment of the syndrome. This astonishing feat would pave the way for designing molecular diagnostics, facilitate molecular epidemiology, and allow the creation of structure-based vaccines in record time. Yet, flawed governmental and global responses would result in a pandemic of monumental proportions. We tell that story in the next chapter.

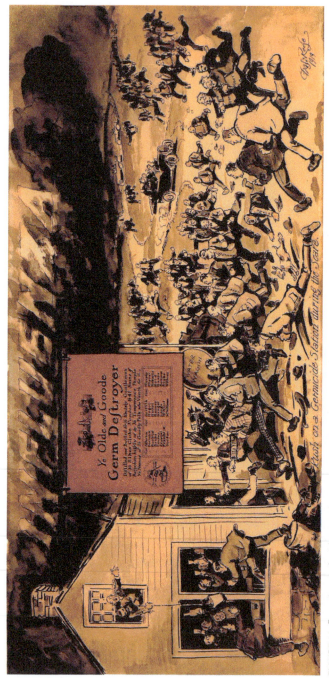

FIGURE 11 *Emerging viral epidemics. The appearance of a viral epidemic produces fear and panic as well as sickness and death. Next-generation sequencing (NGS) would provide the key to the rapid identification of SARS-CoV-2, as detailed in the next chapter. This political cartoon of 1919 by Charles Reese shows the panic of people rushing to get a "goode germ destroyer" in the context of the influenza pandemic. (Courtesy of the National Library of Medicine.)*

REFERENCES

1. **Watson JD.** 1968. *The Double Helix.* Atheneum, New York, NY.
2. **Watson JD, Crick FHC.** 1953. Molecular structure of nucleic acids; a structure for deoxyribose nucleic acid. *Nature* **171:**737–738 http://dx.doi.org/10.1038/171737a0.
3. **Centers for Disease Control (CDC).** 1981. *Pneumocystis* pneumonia--Los Angeles. *MMWR Morb Mortal Wkly Rep* **30:**250–252.
4. **Centers for Disease Control (CDC).** 1981. Kaposi's sarcoma and *Pneumocystis* pneumonia among homosexual men--New York City and California. *MMWR Morb Mortal Wkly Rep* **30:**305–308.
5. **Saiki RK, Scharf S, Faloona F, Mullis KB, Horn GT, Erlich HA, Arnheim N.** 1985. Enzymatic amplification of beta-globin genomic sequences and restriction site analysis for diagnosis of sickle cell anemia. *Science* **230:**1350–1354 http://dx.doi.org/10.1126/science.2999980.
6. **Mullis KB.** 1990. The unusual origin of the polymerase chain reaction. *Sci Am* **262:**56–61, 64–65 http://dx.doi.org/10.1038/scientificamerican0490-56.
7. **Mullis KB.** 1998. *Dancing Naked in the Mind Field.* Vintage Books, New York, NY.
8. **Hughes SS.** 2001. Making dollars out of DNA. The first major patent in biotechnology and the commercialization of molecular biology, 1974-1980. *Isis* **92:**541–575 http://dx.doi.org/10.1086/385281.
9. **Avery OT, MacLeod CM, McCarty M.** 1944. Studies on the chemical nature of the substance inducing transformation of pneumococcal types. Induction of transformation by a deoxyribonucleic acid fraction isolated from Pneumococcus type III. *J Exp Med* **79:**137–158 http://dx.doi.org/10.1084/jem.79.2.137.
10. **Wyatt HV.** 1972. When does information become knowledge? *Nature* **235:**86–89 http://dx.doi.org/10.1038/235086a0.
11. **Crick FHC, Barnett L, Brenner S, Watts-Tobin RJ.** 1961. General nature of the genetic code for proteins. *Nature* **192:**1227–1232 http://dx.doi.org/10.1038/1921227a0.
12. **Crick F.** 1970. Central dogma of molecular biology. *Nature* **227:**561–563 http://dx.doi.org/10.1038/227561a0.
13. **Baltimore D.** 1970. Viral RNA-dependent DNA polymerase. *Nature* **226:**1209–1211 http://dx.doi.org/10.1038/2261209a0.
14. **Temin HM, Mizutani S.** 1970. RNA-dependent DNA polymerase in virions of Rous sarcoma virus. *Nature* **226:**1211–1213 http://dx.doi.org/10.1038/2261211a0.
15. **Judson HF.** 1996. *The Eighth Day of Creation. Makers of the Revolution in Biology.* Cold Spring Harbor Laboratory Press, Cold Spring Harbor, NY.
16. **Morange M.** 1996. *A History of Molecular Biology.* (Cobb M, translator.) Harvard University Press, Cambridge, MA.
17. **Olby R.** 1994. *The Path to the Double Helix. The Discovery of DNA.* Dover Publications, Inc, New York, NY.
18. **Sepkowitz KA.** 2001. AIDS--the first 20 years. *N Engl J Med* **344:**1764–1772 http://dx.doi.org/10.1056/NEJM200106073442306.
19. **Stent GS.** 1968. That was the molecular biology that was. *Science* **160:**390–395 http://dx.doi.org/10.1126/science.160.3826.390.
20. **Judson HF.** 1993. Frederick Sanger, Erwin Chargaff, and the metamorphosis of specificity. *Gene* **135:**19–23 http://dx.doi.org/10.1016/0378-1119(93)90043-3.
21. **Stent GS.** 1972. Prematurity and uniqueness in scientific discovery. *Sci Am* **227:**84–93 http://dx.doi.org/10.1038/scientificamerican1272-84.
22. **Chargaff E.** 1971. Preface to a grammar of biology. A hundred years of nucleic acid research. *Science* **172:**637–642 http://dx.doi.org/10.1126/science.172.3984.637.
23. **Chargaff E.** 1978. *Heraclitean Fire. Sketches from a Life before Nature.* The Rockefeller University Press, New York, NY.
24. **Chargaff E.** 1950. Chemical specificity of nucleic acids and mechanism of their enzymatic degradation. *Experientia* **6:**201–209 http://dx.doi.org/10.1007/BF02173653.
25. **Zamenhof S, Brawerman G, Chargaff E.** 1952. On the desoxypentose nucleic acids from several microorganisms. *Biochim Biophys Acta* **9:**402–405 http://dx.doi.org/10.1016/0006-3002(52)90184-4.

26. **Griffith F.** 1928. The significance of pneumococcal types. *J Hyg (Lond)* **27**:113–159 http://dx.doi.org/10.1017/S0022172400031879.

27. **Dochez AR.** 1958. • • •, p 21–49. *In Oswald Theodore Avery. 1877–1955. National Academy of Sciences Biographical Memoir.* National Academies Press, Washington, DC.

28. **Austrian R.** 1999. Pneumococcus and the Brooklyn connection. *Am J Med* **107**(1A):2S–6S http://dx.doi.org/10.1016/S0002-9343(99)00108-4.

29. **Russell N.** 1988. Oswald Avery and the origin of molecular biology. *Br J Hist Sci* **21**:193–400 http://dx.doi.org/10.1017/S0007087400025310.

30. **Brenner S.** 1985. The rough and the smooth. *Nature* **317**:209–210 http://dx.doi.org/10.1038/317209a0.

31. **Dubos RJ.** 1976. *The Professor, the Institute, and DNA.* The Rockefeller University Press, New York, NY.

32. **Hershey AD, Chase M.** 1952. Independent functions of viral protein and nucleic acid in growth of bacteriophage. *J Gen Physiol* **36**:39–56 http://dx.doi.org/10.1085/jgp.36.1.39.

33. **Centers for Disease Control (CDC).** 1985. Update: Public Health Service workshop on human T-lymphotropic virus type III antibody testing--United States. *MMWR Morb Mortal Wkly Rep* **34**:477–478.

34. **Christy, N. P.** 2004. Faculty remembered. Erwin Chargaff 1905–2002. *P&S.* http://www.cumc.columbia.edu/psjournal/archive/winter-2004/ faculty.html.

35. **Dunitz JD.** 1997. •••, p 221–261. *In Linus Carl Pauling. February 28, 1901–August 19, 1994. National Academy of Sciences Biographical Memoir.* National Academies Press, Washington, DC.

36. **Abir-Am PG.** 1991. Nobelesse oblige: lives of molecular biologists. Essay review. *Isis* **82**:326–343 http://dx.doi.org/10.1086/355756.

37. **Bernal JD.** 1958. Dr. Rosalind E. Franklin. *Nature* **182**:154.

38. **Maddox B.** 2002. *Rosalind Franklin. The Dark Lady of DNA.* HarperCollins Publishers, New York, NY.

39. **Franklin RE, Gosling RG.** 1953. Molecular configuration in sodium thymonucleate. *Nature* **171**:740–741 http://dx.doi.org/10.1038/171740a0.

40. **Franklin RE, Gosling RG.** 1953. Evidence for 2-chain helix in crystalline structure of sodium deoxyribonucleate. *Nature* **172**:156–157 http://dx.doi.org/10.1038/172156a0.

41. **Glynn J.** 2008. Rosalind Franklin: 50 years on. *Notes Rec R Soc* **62**:253–255 http://dx.doi.org/10.1098/rsnr.2007.0052.

42. **Perutz MF.** 1994. Linus Pauling. 1901-1994. *Nat Struct Biol* **1**:667–671 http://dx.doi.org/10.1038/nsb1094-667.

43. **Davies M.** 22 August 1994. Obituary: Professor Linus Carl Pauling. *The Independent.* http://www.independent.co.uk/news/people/obituary-professor-linus-pauling-1377923.html. Accessed 11 January 2011.

44. **Severo R.** 21 August 1994. Obituary. Linus C. Pauling dies at 93; chemist and voice for peace. *New York Times.* http://www.nytimes.com/learning/general/onthisday/bday/0228.html. Accessed 11 January 2011.

45. **Pauling L, Delbrück M.** 1940. The nature of the intermolecular forces operative in biological processes. *Science* **92**:77–79 http://dx.doi.org/10.1126/science.92.2378.77.

46. **Pauling L, Corey RB.** 1953. A proposed structure for the nucleic acids. *Proc Natl Acad Sci USA* **39**:84–97 http://dx.doi.org/10.1073/pnas.39.2.84.

47. **Wilkins M.** 2003. *The Third Man of the Double Helix: Memoirs of a Life in Science.* Oxford University Press, Oxford, United Kingdom.

48. **Perutz MF.** 1989. Discoverers of the double helix, p 181–183. *In Is Science Necessary?* E P Dutton, New York, NY.

49. **Gann A, Witkowski J.** 2010. The lost correspondence of Francis Crick. *Nature* **467**:519–524 http://dx.doi.org/10.1038/467519a.

50. **Klug A.** 2004. The discovery of the DNA double helix. *J Mol Biol* **335**:3–26 http://dx.doi.org/10.1016/j.jmb.2003.11.015.

51. **Wilkins MH, Stokes AR, Wilson HR.** 1953. Molecular structure of deoxypentose nucleic acids. *Nature* **171**:738–740 http://dx.doi.org/10.1038/171738a0.

52. **Watson JD, Crick FHC.** 1953. Genetical implications of the structure of deoxyribonucleic acid. *Nature* **171:**964–967 http://dx.doi.org/10.1038/171964b0.
53. **Matthaei JH, Nirenberg MW.** 1961. Characteristics and stabilization of DNAase-sensitive protein synthesis in *E. coli* extracts. *Proc Natl Acad Sci USA* **47:**1580–1588 http://dx.doi.org/10.1073/pnas.47.10.1580.
54. **Meselson M, Yuan R.** 1968. DNA restriction enzyme from *E. coli*. *Nature* **217:**1110–1114 http://dx.doi.org/10.1038/2171110a0.
55. **Nobelprize.org.** Nobel Prize in Physiology or Medicine, 1962. http://www.nobelprize.org/. Accessed 7 November 2011.
56. **Nirenberg MW, Matthaei JH.** 1961. The dependence of cell-free protein synthesis in *E. coli* upon naturally occurring or synthetic polyribonucleotides. *Proc Natl Acad Sci USA* **47:**1588–1602 http://dx.doi.org/10.1073/pnas.47.10.1588.
57. **Crick FH, Griffith JS, Orgel LE.** 1957. Codes without commas. *Proc Natl Acad Sci USA* **43:**416–421 http://dx.doi.org/10.1073/pnas.43.5.416.
58. **Nobelprize.org.** Nobel Prize in Physiology or Medicine, 1968. http://www.nobelprize.org/. Accessed 15 January 2011.
59. **Jacob F.** 1988. *The Statue Within. An Autobiography.* Basic Books, Inc, Publisher, New York, NY.
60. **Brenner S, Jacob F, Meselson M.** 1961. An unstable intermediate carrying information from genes to ribosomes for protein synthesis. *Nature* **190:**576–581 http://dx.doi.org/10.1038/190576a0.
61. **Temin HM.** 1964. Malignant transformation in cell cultures. *Health Lab Sci* **1:**79–83.
62. **Nature.** 1970. Central dogma reversed. *Nature* **226:**1198–1199.
63. **Nobelprize.org.** Nobel Prize in Physiology or Medicine, 1975. http://www.nobelprize.org/. Accessed 27 January 2011.
64. **Gottlieb MS.** 2001. AIDS--past and future. *N Engl J Med* **344:**1788–1791 http://dx.doi.org/10.1056/NEJM200106073442312.
65. **Gottlieb MS, Schroff R, Schanker HM, Weisman JD, Fan PT, Wolf RA, Saxon A.** 1981. *Pneumocystis carinii* pneumonia and mucosal candidiasis in previously healthy homosexual men: evidence of a new acquired cellular immunodeficiency. *N Engl J Med* **305:**1425–1431 http://dx.doi.org/10.1056/NEJM198112103052401.
66. **Siegal FP, Lopez C, Hammer GS, Brown AE, Kornfeld SJ, Gold J, Hassett J, Hirschman SZ, Cunningham-Rundles C, Adelsberg BR, Parham DM, Siegal M, Cunningham-Rundles S, Armstrong D.** 1981. Severe acquired immunodeficiency in male homosexuals, manifested by chronic perianal ulcerative herpes simplex lesions. *N Engl J Med* **305:**1439–1444 http://dx.doi.org/10.1056/NEJM198112103052403.
67. **Durack DT.** 1981. Opportunistic infections and Kaposi's sarcoma in homosexual men. *N Engl J Med* **305:**1465–1467 http://dx.doi.org/10.1056/NEJM198112103052408.
68. **Gallo RC.** 2002. Historical essay. The early years of HIV/AIDS. *Science* **298:**1728–1730 http://dx.doi.org/10.1126/science.1078050.
69. **Montagnier L.** 2002. Historical essay. A history of HIV discovery. *Science* **298:**1727–1728 http://dx.doi.org/10.1126/science.1079027.
70. **Barré-Sinoussi F, Chermann JC, Rey F, Nugeyre MT, Chamaret S, Gruest J, Dauguet C, Axler-Blin C, Vézinet-Brun F, Rouzioux C, Rozenbaum W, Montagnier L.** 1983. Isolation of a T-lymphotropic retrovirus from a patient at risk for acquired immune deficiency syndrome (AIDS). *Science* **220:**868–871 http://dx.doi.org/10.1126/science.6189183.
71. **Gallo RC, Sarin PS, Gelmann EP, Robert-Guroff M, Richardson E, Kalyanaraman VS, Mann D, Sidhu GD, Stahl RE, Zolla-Pazner S, Leibowitch J, Popovic M.** 1983. Isolation of human T-cell leukemia virus in acquired immune deficiency syndrome (AIDS). *Science* **220:**865–867 http://dx.doi.org/10.1126/science.6601823.
72. **Cohen J.** 1994. A parting shot from a closed case. *Science* **265:**24 http://dx.doi.org/10.1126/science.265.5168.24.
73. **Nobelprize.org.** Nobel Prize in Physiology or Medicine, 2008. http://www.nobelprize.org/. Accessed 17 January 2011 and 5 November 2011.

74. **Curran JW, Lawrence DN, Jaffe H, Kaplan JE, Zyla LD, Chamberland M, Weinstein R, Lui K-J, Schonberger LB, Spira TJ, Alexander WJ, Swinger G, Ammann A, Solomon S, Auerbach D, Mildvan D, Stoneburner R, Jason JM, Haverkos HW, Evatt BL.** 1984. Acquired immunodeficiency syndrome (AIDS) associated with transfusions. *N Engl J Med* **310:**69–75 http://dx.doi.org/10.1056/NEJM198401123100201.

75. **Evatt BL, Ramsey RB, Lawrence DN, Zyla LD, Curran JW.** 1984. The acquired immunodeficiency syndrome in patients with hemophilia. *Ann Intern Med* **100:**499–504 http://dx.doi.org/10.7326/0003-4819-100-4-499.

76. **Pear R.** 3 March 1985. AIDS blood test to be available in 2 to 6 weeks. *The New York Times.* http://www.nytimes.com/1985/03/03/us/aids-blood-test-to-be-available-in-2-to-6-weeks.html. Accessed 17 December 2010.

77. **Cohen J, Marshall E.** 1994. AIDS blood-test royalties. NIH-Pasteur: a final rapprochement? *Science* **265:**313 http://dx.doi.org/10.1126/science.8023149.

78. **Piot P, Quinn TC, Taelman H, Feinsod FM, Minlangu KB, Wobin O, Mbendi N, Mazebo P, Ndangi K, Stevens W, Kalambayi K, Mitchell S, Bridts C, McCormick JB.** 1984. Acquired immunodeficiency syndrome in a heterosexual population in Zaire. *Lancet* **2:**65–69 http://dx.doi.org/10.1016/S0140-6736(84)90241-1.

79. **Serwadda D, Mugerwa RD, Sewankambo NK, Lwegaba A, Carswell JW, Kirya GB, Bayley AC, Downing RG, Tedder RS, Clayden SA, Weiss RA, Dalgleish AG.** 1985. Slim disease: a new disease in Uganda and its association with HTLV-III infection. *Lancet* **2:**849–852 http://dx.doi.org/10.1016/S0140-6736(85)90122-9.

80. **Zhu T, Korber BT, Nahmias AJ, Hooper E, Sharp PM, Ho DD.** 1998. An African HIV-1 sequence from 1959 and implications for the origin of the epidemic. *Nature* **391:**594–597 http://dx.doi.org/10.1038/35400.

81. **Weiss R.** 2001. AIDS: unbeatable 20 years on. *Lancet* **357:**2073–2074 http://dx.doi.org/10.1016/S0140-6736(00)05228-4.

82. **Birmingham K.** 2000. UN acknowledges HIV/AIDS as a threat to world peace. *Nat Med* **6:**117 http://dx.doi.org/10.1038/72177.

83. **Richman DD.** 2001. HIV chemotherapy. *Nature* **410:**995–1001 http://dx.doi.org/10.1038/35073673.

84. **Palella FJ Jr, Delaney KM, Moorman AC, Loveless MO, Fuhrer J, Satten GA, Aschman DJ, Holmberg SD, HIV Outpatient Study Investigators.** 1998. Declining morbidity and mortality among patients with advanced human immunodeficiency virus infection. *N Engl J Med* **338:**853–860 http://dx.doi.org/10.1056/NEJM199803263381301.

85. **Ho DD, Neumann AU, Perelson AS, Chen W, Leonard JM, Markowitz M.** 1995. Rapid turnover of plasma virions and CD4 lymphocytes in HIV-1 infection. *Nature* **373:**123–126 http://dx.doi.org/10.1038/373123a0.

86. **Mellors JW, Rinaldo CR Jr, Gupta P, White RM, Todd JA, Kingsley LA.** 1996. Prognosis in HIV-1 infection predicted by the quantity of virus in plasma. *Science* **272:**1167–1170 http://dx.doi.org/10.1126/science.272.5265.1167.

87. **Persing DH, Landry ML.** 1989. *In vitro* amplification techniques for the detection of nucleic acids: new tools for the diagnostic laboratory. *Yale J Biol Med* **62:**159–171.

88. **Lehman IR, Bessman MJ, Simms ES, Kornberg A.** 1958. Enzymatic synthesis of deoxyribonucleic acid. I. Preparation of substrates and partial purification of an enzyme from *Escherichia coli. J Biol Chem* **233:**163–170 http://dx.doi.org/10.1016/S0021-9258(19)68048-8.

89. **Marx JL.** 1988. Multiplying genes by leaps and bounds. *Science* **240:**1408–1410 http://dx.doi.org/10.1126/science.3375831.

90. **Nobelprize.org.** Nobel Prize in Physiology or Medicine, 1978. http://www.nobelprize.org/. Accessed 25 January 2011.

91. **Buchman TG, Roizman B, Adams G, Stover BH.** 1978. Restriction endonuclease fingerprinting of herpes simplex virus DNA: a novel epidemiological tool applied to a nosocomial outbreak. *J Infect Dis* **138:**488–498 http://dx.doi.org/10.1093/infdis/138.4.488.

92. **Landry ML, Berkovits N, Summers WP, Booss J, Hsiung GD, Summers WC.** 1983. Herpes simplex encephalitis: analysis of a cluster of cases by restriction endonuclease mapping of virus isolates. *Neurology* **33**:831–835 http://dx.doi.org/10.1212/WNL.33.7.831.

93. **Zur Hausen H, Schulte-Holthausen H.** 1970. Presence of EB virus nucleic acid homology in a "virus-free" line of Burkitt tumour cells. *Nature* **227**:245–248 http://dx.doi.org/10.1038/227245a0.

94. **zur Hausen H, Meinhof W, Scheiber W, Bornkamm GW.** 1974. Attempts to detect virus-secific DNA in human tumors. I. Nucleic acid hybridizations with complementary RNA of human wart virus. *Int J Cancer* **13**:650–656 http://dx.doi.org/10.1002/ijc.2910130509.

95. **Andiman W, Gradoville L, Heston L, Neydorff R, Savage ME, Kitchingman G, Shedd D, Miller G.** 1983. Use of cloned probes to detect Epstein-Barr viral DNA in tissues of patients with neoplastic and lymphoproliferative diseases. *J Infect Dis* **148**:967–977 http://dx.doi.org/10.1093/infdis/148.6.967.

96. **Chou S, Merigan TC.** 1983. Rapid detection and quantitation of human cytomegalovirus in urine through DNA hybridization. *N Engl J Med* **308**:921–925 http://dx.doi.org/10.1056/NEJM198304213081603.

97. **Brigati DJ, Myerson D, Leary JJ, Spalholz B, Travis SZ, Fong CKY, Hsiung GD, Ward DC.** 1983. Detection of viral genomes in cultured cells and paraffin-embedded tissue sections using biotin-labeled hybridization probes. *Virology* **126**:32–50 http://dx.doi.org/10.1016/0042-6822(83)90460-9.

98. **Bessman MJ, Lehman IR, Simms ES, Kornberg A.** 1958. Enzymatic synthesis of deoxyribonucleic acid. II. General properties of the reaction. *J Biol Chem* **233**:171–177.

99. **Zuckerman H, Lederberg J.** 1986. Forty years of genetic recombination in bacteria. Postmature scientific discovery? *Nature* **324**:629–631 http://dx.doi.org/10.1038/324629a0.

100. **Hargitti I.** 2002. Kary B. Mullis, p 182–195. *In Candid Science II: Conversations with Famous Biomedical Scientists.* Imperial College Press, London, United Kingdom.

101. **Mullis KB.** 1968. Cosmological significance of time reversal. *Nature* **218**:663–664 http://dx.doi.org/10.1038/218663b0.

102. **Saiki RK, Gelfand DH, Stoffel S, Scharf SJ, Higuchi R, Horn GT, Mullis KB, Erlich HA.** 1988. Primer-directed enzymatic amplification of DNA with a thermostable DNA polymerase. *Science* **239**:487–491 http://dx.doi.org/10.1126/science.2448875.

103. **Mullis KB, Faloona FA.** 1987. Specific synthesis of DNA in vitro via a polymerase-catalyzed chain reaction. *Methods Enzymol* **155**:335–350 http://dx.doi.org/10.1016/0076-6879(87)55023-6.

104. **Nobelprize.org.** Nobel Prize in Physiology or Medicine, 1993. http://www.nobelprize.org/. Accessed 6 November 2011.

105. **Ou C-Y, Kwok S, Mitchell SW, Mack DH, Sninsky JJ, Krebs JW, Feorino P, Warfield D, Schochetman G.** 1988. DNA amplification for direct detection of HIV-1 in DNA of peripheral blood mononuclear cells. *Science* **239**:295–297 http://dx.doi.org/10.1126/science.3336784.

106. **Hart C, Schochetman G, Spira T, Lifson A, Moore J, Galphin J, Sninsky J, Ou C-Y.** 1988. Direct detection of HIV RNA expression in seropositive subjects. *Lancet* **2**:596–599 http://dx.doi.org/10.1016/S0140-6736(88)90639-3.

107. **Persing DH.** 1991. Polymerase chain reaction: trenches to benches. *J Clin Microbiol* **29**:1281–1285 http://dx.doi.org/10.1128/jcm.29.7.1281-1285.1991.

108. **Rowley AH, Whitley RJ, Lakeman FD, Wolinsky SM.** 1990. Rapid detection of herpes-simplex-virus DNA in cerebrospinal fluid of patients with herpes simplex encephalitis. *Lancet* **335**:440–441 http://dx.doi.org/10.1016/0140-6736(90)90667-T.

109. **Hodinka RL.** 1998. The clinical utility of viral quantitation using molecular methods. *Clin Diagn Virol* **10**:25–47 http://dx.doi.org/10.1016/S0928-0197(98)00016-6.

110. **Compton J.** 1991. Nucleic acid sequence-based amplification. *Nature* **350**:91–92 http://dx.doi.org/10.1038/350091a0.

111. **Kievits T, van Gemen B, van Strijp D, Schukkink R, Dircks M, Adriaanse H, Malek L, Sooknanan R, Lens P.** 1991. NASBA isothermal enzymatic in vitro nucleic acid amplification optimized for the diagnosis of HIV-1 infection. *J Virol Methods* **35**:273–286 http://dx.doi.org/10.1016/0166-0934(91)90069-C.

112. **Griffith BP, Rigsby MO, Garner RB, Gordon MM, Chacko TM.** 1997. Comparison of the Amplicor HIV-1 monitor test and the nucleic acid sequence-based amplification assay for quantitation of human immunodeficiency virus RNA in plasma, serum, and plasma subjected to freeze-thaw cycles. *J Clin Microbiol* **35:**3288–3291 http://dx.doi.org/10.1128/jcm.35.12.3288-3291.1997.

113. **Urdea MS.** 1994. Branched DNA signal amplification. *Biotechnology (N Y)* **12:**926–928.

114. **Collins ML, Irvine B, Tyner D, Fine E, Zayati C, Chang C, Horn T, Ahle D, Detmer J, Shen L-P, Kolberg J, Bushnell S, Urdea MS, Ho DD.** 1997. A branched DNA signal amplification assay for quantification of nucleic acid targets below 100 molecules/ml. *Nucleic Acids Res* **25:**2979–2984 http://dx.doi.org/10.1093/nar/25.15.2979.

115. **Read SJ, Kurtz JB.** 1999. Laboratory diagnosis of common viral infections of the central nervous system by using a single multiplex PCR screening assay. *J Clin Microbiol* **37:**1352–1355 http://dx.doi.org/10.1128/JCM.37.5.1352-1355.1999.

116. **Nolte FS.** 1999. Impact of viral load testing on patient care. *Arch Pathol Lab Med* **123:**1011–1014 http://dx.doi.org/10.5858/1999-123-1011-IOVLTO.

117. **Gibson UE, Heid CA, Williams PM.** 1996. A novel method for real time quantitative RT-PCR. *Genome Res* **6:**995–1001 http://dx.doi.org/10.1101/gr.6.10.995.

118. **Heid CA, Stevens J, Livak KJ, Williams PM.** 1996. Real time quantitative PCR. *Genome Res* **6:**986–994 http://dx.doi.org/10.1101/gr.6.10.986.

119. **Cardullo RA, Agrawal S, Flores C, Zamecnik PC, Wolf DE.** 1988. Detection of nucleic acid hybridization by nonradiative fluorescence resonance energy transfer. *Proc Natl Acad Sci USA* **85:**8790–8794 http://dx.doi.org/10.1073/pnas.85.23.8790.

120. **Kehl SC, Henrickson KJ, Hua W, Fan J.** 2001. Evaluation of the Hexaplex assay for detection of respiratory viruses in children. *J Clin Microbiol* **39:**1696–1701 http://dx.doi.org/10.1128/JCM.39.5.1696-1701.2001.

121. **Mahony J, Chong S, Merante F, Yaghoubian S, Sinha T, Lisle C, Janeczko R.** 2007. Development of a respiratory virus panel test for detection of twenty human respiratory viruses by use of multiplex PCR and a fluid microbead-based assay. *J Clin Microbiol* **45:**2965–2970 http://dx.doi.org/10.1128/JCM.02436-06.

122. **Caliendo AM.** 2011. Multiplex PCR and emerging technologies for the detection of respiratory pathogens. *Clin Infect Dis* **52**(Suppl 4):S326–S330 http://dx.doi.org/10.1093/cid/cir047.

123. **Stramer SL, Wend U, Candotti D, Foster GA, Hollinger FB, Dodd RY, Allain J-P, Gerlich W.** 2011. Nucleic acid testing to detect HBV infection in blood donors. *N Engl J Med* **364:**236–247 http://dx.doi.org/10.1056/NEJMoa1007644.

124. **Panel on Antiretroviral Guidelines for Adults and Adolescents.** 10 January 2011. *Guidelines for the Use of Antiretroviral Agents in HIV-1 Infected Adults and Adolescents,* p 11–14. Department of Health and Human Services, Washington, DC. http://aidsinfo.nih.gov/content files/lvguidelines/adultandadolescentgl.pdf. Accessed 30 January 2011.

125. **Nolte FS.** 2009. Quantitative molecular techniques, p 169–184. *In* Specter S, Hodinka RL, Young SA, Wiedbrauk DL (ed), *Clinical Virology Manual,* 4th ed. ASM Press, Washington, DC.

126. **Storch GA.** 2007. Diagnostic virology, p 565–604. *In* Knipe DM, Howley PM, Griffin DE, Lamb RA, Straus SE, Martin MA, Roizman RA (ed), *Fields Virology,* 5th ed. Lippincott Williams & Wilkins, Philadelphia, PA.

127. **Wiedbrauk DL.** 2009. Nucleic acid amplification and detection methods, p 156–168. *In* Specter S, Hodinka RL, Young SA, Wiedebrauk DL (ed), *Clinical Virology Manual,* 4th ed. ASM Press, Washington, DC.

128. **de Crom SC, Obihara CC, van Loon AM, Argilagos-Alvarez AA, Peeters MF, van Furth AM, Rossen JWA.** 2012. Detection of enterovirus RNA in cerebrospinal fluid: comparison of two molecular assays. *J Virol Methods* **179:**104–107 http://dx.doi.org/10.1016/j.jviromet.2011.10.007.

129. **Pierce VM, Elkan M, Leet M, McGowan KL, Hodinka RL.** 2012. Comparison of the Idaho Technology FilmArray system to real-time PCR for detection of respiratory pathogens in children. *J Clin Microbiol* **50:**364–371 http://dx.doi.org/10.1128/JCM.05996-11.

130. **Pallen MJ, Loman NJ, Penn CW.** 2010. High-throughput sequencing and clinical microbiology: progress, opportunities and challenges. *Curr Opin Microbiol* **13:**625–631 http://dx.doi.org/10.1016/j.mib.2010.08.003.
131. **Blyth CC, Booy R, Dwyer DE.** 2011. Point of care testing: diagnosis outside the virology laboratory. *Methods Mol Biol* **665:**415–433 http://dx.doi.org/10.1007/978-1-60761-817-1_22.
132. **Ginocchio CC, McAdam AJ.** 2011. Current best practices for respiratory virus testing. *J Clin Microbiol* **49**(9_Supplement)**:**S44–S48 http://dx.doi.org/10.1128/JCM.00698-11.
133. **Morens DM, Folkers GK, Fauci AS.** 2008. Emerging infections: a perpetual challenge. *Lancet Infect Dis* **8:**710–719 http://dx.doi.org/10.1016/S1473-3099(08)70256-1.
134. **Daszak P, Lipkin WI.** 2011. The search for meaning in virus discovery. *Curr Opin Virol* **1:**620–623 http://dx.doi.org/10.1016/j.coviro.2011.10.010.
135. **Wolfe N.** 2011. *The Viral Storm. The Dawn of a New Pandemic Age.* Times Books, Henry Holt and Company, New York, NY.
136. **Kilbourne ED.** 1993. Afterword: a personal summary presented as a guide for discussion. *In* Morse SS (ed), *Emerging Viruses.* Oxford University Press, New York, NY.
137. **Lipkin WI.** 2010. Microbe hunting. *Microbiol Mol Biol Rev* **74:**363–377 http://dx.doi.org/10.1128/MMBR.00007-10.
138. **Gallo RC, Salahuddin SZ, Popovic M, Shearer GM, Kaplan M, Haynes BF, Palker TJ, Redfield R, Oleske J, Safai B, et al.** 1984. Frequent detection and isolation of cytopathic retroviruses (HTLV-III) from patients with AIDS and at risk for AIDS. *Science* **224:**500–503.

10 The World Changed

THE COVID-19 PANDEMIC

*"Economies nose-dived. Schools and workplaces closed.
Populations hid in their homes. Whole societies shut down."*

Clarke, Breathtaking, 2021 (1)

INTRODUCTION

After gestation in Wuhan, China, at the end of 2019, the disease that would come to be known as COVID-19 swept the globe. Strict measures of quarantine and lockdown initially protected some areas, but the pandemic ultimately crippled country after country as successive waves rippled around the world. There appeared to be a respite in early 2021 as vaccines, miraculously produced within a year, started to successfully hold the pandemic at bay. Then the Delta variant loomed with ferocious infectiousness and transmissibility. Countries or regions with lagging vaccination rates, poor vaccine access, or vulnerable populations suffered the most. At Thanksgiving time, at the end of November 2021, Omicron, a new variant with an alarming number of mutations and even greater transmissibility, was first reported in South Africa. It spread extremely rapidly around the world (2). Subvariants of Omicron would emerge and spread even more rapidly.

Although this chapter continues the book's fundamental thread to tell the story of the ways in which viruses are captured and identified, it differs in some respects from earlier chapters due to the pandemic nature of COVID-19. It builds on the molecular revolution described in the preceding chapter, with the key advance

To Catch a Virus, Second Edition. Authored by John Booss and Marie Louise Landry.
© 2023 American Society for Microbiology. DOI: 10.1128/9781683673828.ch10

being the application of next-generation high-throughput sequencing. "Next-generation" sequencing has superseded previous (Sanger) sequencing methods and produces much more data, much more rapidly, which facilitated the rapid determination of the genomic sequence of SARS-CoV-2 (severe acute respiratory syndrome coronavirus 2). In turn, that allowed the rapid detection of the spread of the virus, the development of diagnostic assays, the recognition of new variants, and the structural design of COVID-19 vaccines. The other key advance has been the rapid implementation of new vaccine platforms.

We begin with a description of key events, highlighting how we learned about the behavior of this massive scourge. We felt that it was necessary for responding to future pandemics to track the factors impacting the spread of the virus. We have devoted a considerable amount of attention to the cultural aspects of the pandemic. The pandemic has told us a great deal about how societies function, well or poorly, in the face of a significant threat. On the downside, it has revealed chasms of inequity. On the upside, it has revealed the transformative power of bioscience. This chapter places greater reliance than previous ones on biographies of the key individuals who have fought the pandemic. They reveal courage, insight, and perseverance in facing the pandemic and in solving scientific challenges.

HISTORY OF THE PANDEMIC

Wuhan, China—. . . . where it all began

The first international notice that something was amiss came in the last two days of December 2019 when ProMed posted news of an unexplained pneumonia in Wuhan, China (3). ProMed, short for the Program for Monitoring Emerging Diseases, was established in 1994 as an internet service to identify infectious disease events. The first alert of 30 December 2019 reported 4 cases. The following day, ProMed reported an emergency symposium of Wuhan hospitals in which 27 cases of viral pneumonia of unknown etiology were found; some were associated with South China (Huanan) Seafood City, a local "wet" market. Wuhan is the capital of Hubei Province, lying in the heart of China at the confluence of the Yangtze and Han Rivers. A city of over 11 million people, Wuhan is a manufacturing and transportation hub and has the potential to serve as a large incubator and distribution center for infectious diseases.

Li Wenliang was a 33-year-old husband, father, son, and ophthalmologist who became a physician-hero of the pandemic and an enduring symbol for open communication. Before the epidemic nature of cases had been recognized, he sent former medical school colleagues a message warning them of what he had seen clinically—an illness resembling the SARS outbreak of 2003—and asking them to be careful (4). His message somehow became public, which ran him afoul of the authorities. He was forced to sign a confession that he had made false statements that disturbed the public order. He returned to work at the Wuhan Central Hospital, where he soon contracted COVID-19 from an asymptomatic patient. In an interview with Elsie Chen for *The New York Times* shortly before his death, he said that he felt that he had been wronged, that it would have been better if the officials had disclosed information about the epidemic earlier. He said that "there should be more openness and transparency." Before he passed, he told his story on a microblogging website, sparking public outrage (5). He passed away 7 February 2020 to immense public sorrow. He left a young son and a pregnant wife. His parents, who had also become infected, survived (4). After his death, the authorities formally exonerated him and apologized to his family. Four months after his passing, his wife delivered his second son. His memory endures internationally, and he has become a symbol of clinical courage and the importance of public openness and speaking truth to power.

A team led by Zhang Yong-Zhen (see brief biography below) very quickly isolated and sequenced the virus and, prompted by Edward Holmes, posted its genome to an open access site on 10 January 2020 (6). A report on the clinical, laboratory, and radiological features of 41 patients seen up to 2 January 2020 was published online in *The Lancet* on 24 January 2020 (7). The virus would be named SARS-CoV-2 and its disease COVID-19. The wealth of virological and clinical information should have led to prompt and massive public health interventions. However, something went tragically wrong; strong measures to contain the outbreak were not immediately undertaken.

According to a U.S. intelligence document reported by *The New York Times* on 20 August 2020, "Officials in Beijing were kept in the dark for weeks about potential devastation of the virus by officials in Central China" (8). The consensus U.S. intelligence report noted too that Beijing officials, while trying to get more information themselves, withheld information from the World Health Organization (WHO). Following the emergence of a cluster of cases in Shenzhen and a visit of experts to Wuhan, Beijing officials ordered a lockdown of Wuhan on 23 January 2020. Soon thereafter, mass quarantine was extended across 13 additional cities, encompassing 36 million people (9).

Investigative reporting by Chris Buckley, David D. Kirkpatrick, Amy Qin, and Javier C. Hernandez published by *The New York Times* on the last day of 2020 documented China's bureaucratic failure to effectively recognize and respond to the growing infectious scourge within Wuhan (10). This resulted in the 25-day delay before the city was shut down, during which time the virus escaped from Wuhan. It was the power structure, first at the local level and then nationally, that denied clinical and scientific findings and then interfered with their recognition and promulgation. At one point, for example, laboratories that had isolated and sequenced the virus were instructed to hand over their samples to the authorities or to destroy them. Despite growing evidence for person-to-person spread of the virus, it was denied. By the time it was officially acknowledged and Wuhan was closed down, it was too late. The virus had spread to other cities in China and started its international travels to 10 other countries (9).

It was a decisive 8 days. On 3 January 2020, with a mysterious pneumonia raging ominously in Wuhan, Dr. Zhang Yongzhen's virology laboratory in Shanghai received a clinical sample. On 5 January, less than 2 days later, his laboratory was able to sequence a virus related to SARS. That same day he notified the authorities in Shanghai and uploaded the sequence to a repository and waited for a reply. At that point, however, the virus's blueprint, its genome, was not available to the international scientific community. Zhang visited Wuhan to learn of the situation firsthand. On 10 January, the decisive moment arrived. While seated on an airplane about to leave for Beijing, he received a telephone call from his collaborator Edward Holmes. Holmes, an evolutionary virologist from the University of Sydney, asked Zhang to make the data public. In this pivotal moment, while the flight attendant was asking Zhang to turn off his phone, he authorized Holmes to release the data. Holmes promptly uploaded it to the website virological.org (11–13). The world now had a crucial weapon at its disposal (Fig. 1).

The impact of the release of the sequence data was immediate and dramatic. Thailand used the genome data to determine that the virus had crossed into its territory (11). Other laboratories used the genome sequence to quickly develop PCR diagnostic assays. The greatest benefit of the genome data was that several organizations used it to develop vaccines with lightning speed. The messenger RNA (mRNA) vaccines, for example, were developed, clinically tested, and approved for use within 11 months. Distribution started within the same year, 2020. Zhang, whose expertise and courage provided a turning point, remained humble. He asked an interviewer if they could mention that "if I have made a contribution to society, it is because of the support of my wife" (13). His wife had passed in October 2019, before the world knew of COVID-19, let alone of her husband's monumental contribution.

After the lockdown of Wuhan on 23 January, the virus reached France, Germany, and the United Kingdom by the end of the month (9). The first deaths attributed to COVID-19 in Iran were reported on 19 February in the city of Qom (14). It was thought that the virus may have been brought to Qom by a merchant who had traveled to Wuhan. Italy, as Horton put it, "endured the first humanitarian catastrophe outside of China" (see the section on Lombardy below). The country went into lockdown on 9 March and the Lombardy region was put into quarantine. On 11 March, WHO declared SARS-CoV-2 a pandemic. Spain was put into lockdown on 14 March. More lockdowns were to come.

FIGURE 1 *Professor Zhang Yongzhen (center). With the assistance of colleague Edward Holmes, he uploaded the genome sequence of what became known as SARS-CoV-2 for the global community of scientists. (Courtesy of GigaScience, under license CC BY.)*

Truths from a cruise ship

Ironically, a number of truths about the pandemic emerged from the voyage of a cruise ship. Leaving Yokohama, Japan, on 20 January, before China's lockdown of Wuhan, the cruise ship, the *Diamond Princess*, had a total of 3,711 persons on board, counting passengers and crew. It was to make stops at Viet Nam, Taiwan, and Japan (15). On 23 January, a passenger developed a cough; the passenger disembarked 2 days later and was diagnosed with COVID-19 on 1 February. The ship returned to Yoko-hama on 3 February, where authorities refused to allow disembarkation. After test-ing showed 10 passengers were infected, those aboard were placed in cabin-based quarantine. Ultimately, 712 persons were found to be infected. As many as fourteen died; all were elderly.

Of the infected passengers, 410 were asymptomatic when originally tested (16). From the asymptomatic patients 96 were studied more thoroughly. Eleven became symptomatic; hence they had been presymptomatic. The significant incidence of infection in asymptomatic persons would be a characteristic of the COVID-19 pandemic and a key to understanding its rapid and stealthy spread. A second char-acteristic of the pandemic was ultimately discovered by modeling studies, those in which epidemiologists set up a set of circumstances and then determine the poten-tial outcomes. It was found that "aerosol inhalation was likely the dominant con-tributor to COVID-19 transmission among passengers" (17). A third observation, also true of the pandemic in general, was the vulnerability to SARS-CoV-2 infec-tion of persons in poorly ventilated, closed spaces. This would be particularly true of those in nursing homes and prisons. Starting from one infected passenger, many were infected on the *Diamond Princess* and several hallmarks of the ensuing pan-demic were recognized.

Lombardy, Italy, shattered the European sense of safety

The outbreak of COVID-19 in Lombardy beginning in February 2020 "shattered the sense of safety and distance" that Europe had felt. In this highly populated, indus-trialized, and wealthy region in northern Italy, towns were locked down, schools were closed, sporting events were called off, and the Venice Carnival was cancelled 2 days early (18). The open borders policy in Europe would be challenged. Despite its wealth, Lombardy was poorly prepared. Its low number of intensive care unit (ICU) beds, well below the European average, quickly filled after the first "home-grown" case of 21 February 2021 (Fig. 2). Many patients were managed at home and many died. Other factors also contributed to the collapse of health care in Lombardy, including a dearth of clinical guidelines, a paucity of personal protective equipment (PPE), limited diagnostic testing, and a large population of those over 65 (19). Out-comes in nursing homes, of which Lombardy had the greatest number in Italy, were so extreme that they have been characterized as a "massacre." On the one hand, nurs-ing homes received patients recovering from COVID-19 from hospitals, and on the other, the homes were discouraged from transferring patients over 75 to the hospital. Both policies were designed to reduce pressure on hospitals but instead enhanced transmission of the virus. In a nursing home in Nembro, Italy, for example, 37 of 87 residents passed away in February and March (19).

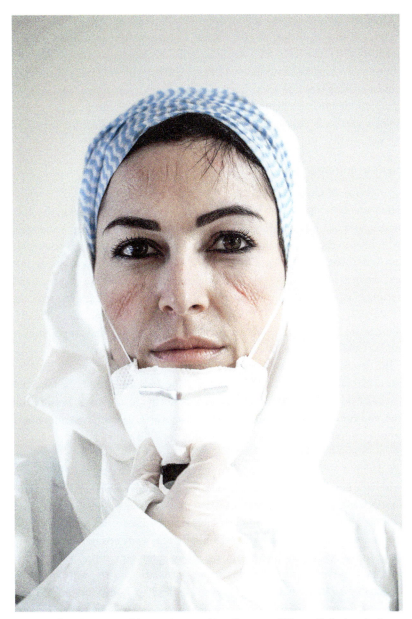

FIGURE 2 *Italian ICU nurse with bruises from masking. (Courtesy of Alberto Giuliani, under license CC BY-SA 4.0.)*

The first wave of infection ended in May. Particularly vulnerable were older adults and those with chronic conditions, "a phenomenon known as the harvesting effect" (20). Events in Italy were watched with great apprehension in the rest of Europe. An English clinician would write, "Perhaps the most chilling news of all from Lombardy was that some hospitals are having to resort to a form of rationing more suited to battlefields than modern health services, denying some patients intensive care beds not because they are beyond saving but because the supply of beds has run dry" (1).

American epicenter: New York City

The outbreak of COVID-19 in New York City in the spring of 2020 riveted the world's attention. Imagine an empty Times Square, lights turned off in the Theater District, no sports fans cheering at Madison Square Garden, no visitors taking in the city from atop the Empire State Building, and no children staring in fascination at animals in habitats at the Bronx Zoo. As had 9/11, the pandemic was an assault on the country as well as on the city. Viral genome sequencing for the origins of viruses infecting NYC inhabitants was undertaken. It demonstrated multiple independent introductions of SARS-CoV-2 into NYC from Europe and from elsewhere in the United States (21). With the emergence of community-based infections, most were found to be related to viruses circulating in Europe.

The city had become the epicenter of the COVID-19 outbreak in the United States (22). The pandemic in NYC would shine a bright light on the inequalities of rich and poor, racial disparities, and the impact of age, underlying health conditions, and residence in nursing homes. As has been true from time immemorial, the well-to-do were able to flee to safer locales. "The most fortunate residents were among the first to abandon the city." In contrast, "armies of service workers were deemed 'essential' but often went unseen and unprotected" (23). Those living in high-poverty neighborhoods suffered higher incidences of disease, hospitalization, and death (22). So too for Black and Hispanic populations, the elderly, and those with underlying conditions such as heart disease, diabetes, obesity, and chronic kidney disease. The authors noted that these trends reflected the failure of "policies, practices, and cultural messages" to eliminate cultural inequities. Further, they observed that Black and Hispanic populations were often in low-paying jobs, unable to implement social distancing, and without work-related health insurance. Overall, the city would experience close to 203,800 recognized cases, of which about 54,200 required hospitalization, and suffer almost 18,700 deaths in the period of March to May. Of the deaths, 22.5% were known to have occurred in nursing homes.

India: a tale of two pandemics

The world was shocked and saddened at the cataclysmic eruption of a second COVID-19 outbreak in India, from March to July 2021, peaking sharply in June. India had apparently sailed through its first outbreak from May 2020 to January 2021, with a flattened curve that peaked in September and a low impact on its health care system. Prime Minister Narendra Modi had prematurely said that his nation had triumphed over coronavirus (24). The effects of the second outbreak were horrific. The health system was overcome; patients died while waiting for admission to the hospital and without the supplemental oxygen necessary to sustain life. Deaths soared, far surpassing official estimates, overloading the capacity of funeral pyre arrangements.

What happened? Why did a cataclysmic outbreak follow a well-controlled outbreak? While the lockdown for the first outbreak was flawed, many commentators deemed it a success. Prime Minister Modi proclaimed a national lockdown for 21 days on 24 March 2020, with only 4 hours' advance notice (25). It "brought daily life to an immediate grinding halt—and also seemed extremely effective in limiting the spread of the virus." It was effective because it was enforced by police. But "the restrictions were economically devastating, putting tens of millions out of work" (24).

Cases had fallen significantly by the start of the new year. In February 2021, in the opinion of one analyst, the government, "cheered on by a pliant triumphalist media, prematurely declared victory against the pandemic" (26). Unable to reimpose a national lockdown, the government also failed to limit mass gatherings such as campaign rallies and a Hindu religious festival at the Ganges River, the Kumbh Mela (24). The festival, named for a vessel containing divine nectar, brought three million people to one of the events. Since the festival drew people from all over the country, it allowed distribution of viral variants prevalent in different regions (27).

The highly infectious Delta variant that had emerged in India in December 2020 would lay siege to the world. Added to the difficulties was a low percentage of vaccine coverage, despite India being a major center of vaccine production internationally. As one commentator reported, the surge could have been due to "an unfortunate convergence of factors, including the emergence of particularly infectious variants, a rise in unrestricted social interactions, and low vaccine coverage" (28). The outbreaks were worse than appreciated at the time, according to one analysis that relied on examination of excess deaths (29). Estimates ranged from 3.4 million to 4.9 million deaths, in contrast to the official count of 400,000. Further, the first wave was found to have caused more deaths than had been believed.

South Africa: start of the whirlwind Omicron variant

A whirlwind blew into the world through South Africa at the end of November 2021. South African scientists and officials, much to their credit, promptly alerted the world to a new viral variant. Soon to be named Omicron (2), it quickly eclipsed the Delta variant in its spread globally. It was found to be very rapidly transmissible, likely due to its high number of mutations, particularly in the spike gene (30). In South Africa, the Omicron wave followed those of the ancestral variant, Beta, and Delta waves (31). Clinically, infections were found to be associated with reduced odds of hospitalization and severe disease (32). When hospitalized there was a lower need of oxygen therapy, mechanical ventilation, and ICU admission, and there was a significantly lower death rate (31). It was speculated that previous immunity could play a role in the reduced severity. Experimental studies would suggest that another reason for less severe disease might be due to a reduced ability of the virus to replicate in the lungs (33). As dramatically as the whirlwind blew in, it quickly subsided. Cases peaked at very high levels in South Africa in December 2021 and had fallen precipitously by the start of January 2022 (34). The pattern would repeat itself elsewhere, followed by successive waves of Omicron subvariants.

THE VIRUS AND ITS VARIANTS

After SARS emerged in Guangdong, China, in late 2002 and Middle East respiratory syndrome (MERS) in Saudi Arabia in 2012, the scientific and public health communities were acutely aware that bat coronaviruses could cross over to humans via an intermediate host. Thus, basic research on SARS, MERS, and other bat coronaviruses commenced, including work on therapy and vaccination strategies. However, as SARS was eliminated and MERS cases remained limited despite endemic infection of camels in several countries, funding waned. While the knowledge gained contributed to a more rapid deployment of tests, therapies, and vaccines for SARS-CoV-2, past experience with these two viruses was also misleading. SARS and MERS were found primarily in the lower respiratory tract, not in the nose. Transmission of SARS and MERS was generally inefficient, requiring close contact with symptomatic cases. Despite more than 8,000 SARS cases worldwide in 2003, only 27 cases were diagnosed in the United States, with no deaths. While 2,500 cases of MERS have been confirmed in 27 countries since 2012, only 2 mild cases in travelers were diagnosed in the United States. Testing for SARS and MERS was largely confined to the Centers for Disease Control and Prevention (CDC). In contrast, SARS-CoV-2 is found in very high titers in the nose and can be readily transmitted via the airborne route, including by asymptomatic individuals, and widespread testing is key to its control. These key differences were not initially appreciated and contributed to the delays in test development and to the seemingly contradictory recommendations on the need for masking of individuals without symptoms.

The virus

Coronaviruses are RNA viruses named for their resemblance to the solar corona on electron microscopy (EM) (see chapter 5, Fig. 10). The surface spike proteins, which are so striking on EM images, are key to the ability of coronaviruses to attach to and then infect host cells. Many coronaviruses are found in animals, including bats and birds. Four seasonal or endemic coronaviruses (designated 229E, OC43, NL63, and HKU1) commonly circulate in human populations and cause mild respiratory illness, with a peak in the mid-winter months in temperate zones. SARS, SARS-CoV-2, and MERS, which can cause severe respiratory disease in humans, belong to the coronavirus genus *Betacoronavirus*, with SARS and SARS-CoV-2 within the subgenus *Sarbecovirus*. SARS-CoV-2 has 79% genetic sequence similarity to SARS-CoV, but only 50% similarity to MERS-CoV. As with SARS-CoV, bats are the likely reservoir of SARS-CoV-2 in nature. At the onset of the pandemic, the coronavirus RaTG13, which infects the horseshoe bat *Rhinolophus affinis*, was found to be closest to SARS-CoV-2 at 96.1% similarity, and in late 2021, additional horseshoe bat coronaviruses were found in Laos with high sequence similarity, especially in the critical receptor binding domain (35, 36). While an intermediate host is hypothesized, one had not yet been identified as of mid-2022.

At this writing, the origins of SARS-CoV-2 remain a mystery. While years often pass before the source of a new virus in nature is identified, the lack of transparency in the initial investigation, and the fact that the Wuhan Institute of Virology is the center for bat coronavirus research in China, contributed to speculation that an inadvertent lab leak unleashed the virus into the population (37, 38). Nonetheless, most scientists believe a natural phenomenon is most likely, given the number of animals susceptible to SARS-CoV-2 and the potential link of the first case to the Huanan Market (39, 40). Two extensive studies by Michael Worobey and his colleagues strongly suggested that the virus spilled over into humans from live mammals at the Huanan Seafood Market. A third largely Chinese study suggested the virus was imported there (41). Nonetheless, until more information comes to light, or the source of the virus in nature is identified, this will remain an area of controversy.

The SARS-CoV-2 genome is about 30,000 nucleotide bases in length. Two-thirds of the genome consists of the open reading frames (ORFs) 1a and 1b, which contain the genetic code for so-called "nonstructural" proteins that are not part of the structure of the virus particle (Fig. 3). These include viral proteases and the RNA-dependent RNA polymerase (RdRp) or RNA replicase, enzymes essential

ORF1a   ORF1b   Spike   E M   N

FIGURE 3 *Diagram of the SARS-CoV-2 genome.*

to creating new virus particles, and thus targets of antiviral therapy. The remaining third of the viral genome includes genes that direct the production of proteins that make up the structure of the virus particle, including the nucleoprotein (N), matrix (M), envelope (E), and spike (S) proteins, as well as nonstructural accessory proteins that help the virus evade the immune response or interfere with normal processes within the cell.

A region of the SARS-CoV-2 spike protein called the receptor binding domain, or RBD, consists of a core structure and a receptor binding motif (RBM). The RBD attaches to the angiotensin-converting enzyme 2 (ACE-2) receptor on the host cell surface, the first step in entering the cell (Fig. 4). Thus, an antibody that binds to the spike protein will prevent the virus from infecting the cell. Once the virus is inside the cell, the host cell machinery is hijacked to produce new virus particles (42, 43).

Variants

The propensity of SARS-CoV-2 to mutate was a concern from the beginning of the pandemic. RNA viruses are prone to "spelling errors" when the genetic code is copied during the production of new virus particles and also have imperfect "proofreading" functions to correct the errors. Although SARS-CoV-2 has an exonuclease enzyme that corrects many errors, mutations still accumulate over time.

Mutations in the genetic sequence can be of three types: substitutions, deletions, or insertions in the genetic code. While many mutations are "silent," some may be deleterious to the virus, and others can lead to minor or significant changes in the transmissibility of the virus and severity of disease, or in the effectiveness of

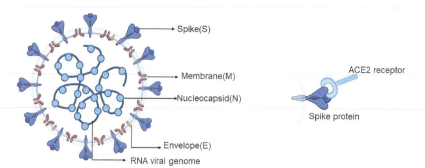

Spike(S)

Membrane(M)

Nucleocapsid(N)

ACE2 receptor

Spike protein

Envelope(E)

RNA viral genome

FIGURE 4 *Diagram of SARS-CoV-2 virus particle (left) and spike protein receptor binding domain (RBD) binding with ACE-2 receptor on host cell (right). (Modified from Min L, Sun Q. 2021.* Front Mol Biosci *8:671633. © 2021 Min and Sun.)*

diagnostics, therapies, and vaccinations. Additionally, coronaviruses can recombine in nature, giving rise to new variants.

Spike protein genetic material is the sole SARS-CoV-2 ingredient of mRNA vaccines (Moderna and Pfizer) and adenovirus vector vaccines (AstraZeneca and Johnson & Johnson/Janssen), is a component of whole-virus inactivated vaccines, and is the target of monoclonal antibody therapies given early after infection. Thus, mutations that alter the spike protein, and potentially the ability of antibodies to bind, are of greatest concern.

Despite concerns about viral mutation, there was no coordinated national effort in the United States in the early months of the pandemic to build SARS-CoV-2 sequencing capacity. This was in contrast to the United Kingdom, due in no small measure to Sharon Peacock, a Cambridge University professor and microbiologist who has based her career on sequencing pathogens to track outbreaks of infectious diseases (Fig. 5). In early March 2020, shortly after COVID-19 arrived in the United Kingdom, Professor Peacock contacted colleagues to gauge their interest in a bold concept: to use widespread sequencing of the genetic code as a means to track the new virus. Getting an affirmative response, within a week she had convened a larger group of experts to explore how this could be accomplished. Within days, they had submitted a grant application to the government's Chief Scientific Officer, and by 1 April 2020, COVID-19 Genomics UK Consortium (COG-UK) came into being.

Professor Peacock, who is widely recognized as key to the creation and organization of the COG-UK consortium, was almost excluded from a career

FIGURE 5 *Professor Sharon Peacock. (Photo courtesy of Sharon Peacock.)*

in science (44). Born in 1959 into a working-class family in Sussex, England, she knew no one who had attended university. At 11 years of age, she took the national exam that placed children on three educational paths: academic, technical, or functional. She failed and was directed to a nonacademic functional school, and thus was not able to pursue her interest in biology. At 16, her schooling finished, she got a job in a local store. However, at 17 she found work as a dental assistant, then transitioned to training as a medical nurse despite lacking the right science qualifications, and, while working as a nurse in end-of-life care, was inspired to become a physician. She took the necessary coursework part-time, but then failed twice to get accepted into medical school. Undeterred, she called Southampton University, arranged a meeting with the admissions officer, and convinced him to give her a chance. After graduating in medicine and completing postgraduate medical training, she trained in clinical microbiology and obtained a Ph.D. From there, she embarked on what was to become a very distinguished career in research and diagnostic and public health microbiology, focused on pathogen genomics and sequencing to track infectious disease outbreaks and antimicrobial resistance.

The COG-UK Consortium is led by Peacock and administered by the University of Cambridge. It is a collaborative network that involves four public health agencies, the Wellcome Sanger Institute, numerous academic institutions, multiple National Health Service laboratories, as well as diagnostic laboratory partners.

When asked why the United Kingdom has been such a world leader in SARS-CoV-2 genomics, Peacock replied that she had puzzled over this question herself and suggested several contributing factors: pathogen sequencing for public health has been widely adopted as routine practice in the United Kingdom; many of the world's experts in pathogen genomics reside there; there is a strong network of laboratories with sequencing capabilities; and the information generated is freely shared (45).

At the beginning of the pandemic, the focus on organized large-scale monitoring of emerging variants as part of the pandemic response was novel. As Peacock recounted, "When we first started, our ambition was to determine whether sequencing data could actually help inform public health actions and policy decisions. We all felt it was a bold step to take at the time. But actually, we've demonstrated that in spades" (45).

Variant naming

The naming of the variants can be confusing, even among the scientific and medical communities. When a mutation in a gene sequence leads to a change in the protein it encodes, it is identified by the original "wild-type" amino acid, the location of the mutation on the protein, then by the new amino acid (e.g., D614G). A group of closely related viruses from a common ancestor is called a lineage. A variant may contain one or more mutations, and the same mutation may be found in multiple variants. There are three current variant naming systems. Phylogenetic Assignment of Named Global Outbreak Lineages (PANGOLIN), named after the pangolin (an animal thought to be the intermediate host for SARS CoV-2 early in the pandemic) and now known as PANGO, includes a letter and then up to three numbers separated by periods to include sublineages. The PANGO system led to initial designations of the B.1.1.7 (U.K.) and B.1.627.2 (India) variants. Nextstrain is an open-source collaborative project intended to support public health by facilitating the surveillance of pathogen evolution and spread with a different naming strategy. Lastly, GISAID (Global Initiative on Sharing Avian Influenza Data) has its own nomenclature. In May 2021, for simplification, and to avoid stigma regarding place of origin, WHO introduced Greek letters to label variants. Variants are further classified by WHO and the CDC as variants under monitoring (VUM), variants of interest (VOIs), variants of concern (VOCs), and variants of high consequence (VOHCs), based on increased transmissibility, disease severity, effect on diagnostics, or reduced neutralization by antibodies. As VOIs or VOCs recede in importance, they transition to variants being monitored (VBMs). At the time of this writing, July 2022, there are no VOHCs, but five VOCs have been recognized globally: Alpha or lineage B.1.1.7 (U.K.), Beta or lineage B.1.351 (South Africa), Gamma or lineage P.1 (Japan/Brazil), Delta or lineage B.1.617.2 (India), and Omicron or lineages B.1.1.529, BA.1, BA.2, BA.3, BA.4, BA.5 (South Africa) (46) (Table 1). The Technical Advisory Group on SARS CoV-2 Virus Evolution (TAG-VE) was formed to advise WHO on virus evolution, mutations, and variants.

The original reference virus isolated in China is called Wuhan-Hu-1. Early in the pandemic, a mutation in the spike protein emerged, designated as D614G, that enhanced replication in the upper respiratory tract and transmissibility, and thus improved the "fitness" of the virus. While D614G was rare before March 2020, it was common by May and quickly spread worldwide. From December 2019 until the early fall of 2020, the virus acquired about two mutations per month (47). Then, multiple variants emerged, most notably in the United Kingdom, South Africa, and Brazil, and all contained the D614G mutation as well as others. The attention of the world on the importance of variants was captured with the recognition of variant B.1.1.7, also called the U.K. variant, and now referred to as Alpha. With 23 differences in its genetic code compared to the original virus, it was an outlier on the phylogenetic or "family" tree that scientists create to compare genomes with each other and follow their evolution, and it was more transmissible than the parent strain. It soon swept the United Kingdom.

TABLE 1 Examples of variants of concern (VOCs) of global significance[a] (46)

WHO label	PANGO lineage[b]	GISAID clade	Next-strain clade	Earliest documented samples	Date of designation
Alpha	B.1.1.7	GRY	20I (V1)	United Kingdom, September 2020	18-Dec-2020
Beta	B.1.351	GH/501Y.V2	20H (V2)	South Africa, May 2020	18-Dec-2020
Gamma	P.1	GR/501Y.V3	20J (V3)	Brazil, November 2020	11-Jan-2021
Delta	B.1.617.2	G/478K.V1	21A, 21I, 21J	India, October 2020	VOI-4-Apr-2021 VOC-11-May-2021
Omicron[c]	B.1.1.529	GR/484A	21K, 21L, 21M, 22A, 22B, 22C	South Africa, November 2021	VUM-24-Nov-2021 VOC-26-Nov-2021

[a] Modified from WHO reference 46, updated 18 July 2022
[b] Includes all descendent lineages.
[c] Includes BA.1, BA.2, BA.3, BA.4, BA.5 and descendent lineages.

The news of the U.K. variant reported by COG-UK created a heightened sense of urgency, and by the fall of 2020, the CDC in the United States had initiated partnerships with academic medical centers, universities, public health laboratories, and industry to expand SARS-CoV-2 genetic sequencing capacity, share information, and better track the impact of emerging strains. However, federal guidance and funding for the sequencing effort lagged. By early 2021, COG-UK had contributed almost half of all SARS-CoV-2 sequence data submitted from around the globe to GISAID, the international sequence database originally established in 2008 to share influenza data. The United States, despite having the most identified COVID cases, was 43rd in proportion of viruses sequenced (0.3%). In contrast, the United Kingdom had sequenced 7.4% of its cases (48).

While the Alpha variant was more transmissible and spread quickly in the United States once introduced in the early months of 2021, it had minimal impact on the efficacy of vaccinations or monoclonal antibody therapies. When a vaccinated person did have a "breakthrough" infection, the infection was usually asymptomatic and the amount of virus in the respiratory tract was minimal and not transmissible. The Alpha variant was associated with failure to detect one target (the spike gene) in a common PCR test; fortunately, the other two gene targets included in the test were unaffected. This same spike gene target failure (SGTF) was later identified in the Omicron variant.

As more of the population became vaccinated and rates of new infection fell, states across the United States loosened restrictions on masking, especially for the vaccinated, and social gatherings, indoor dining, and travel began to resume in May and June of 2021. Then variant B.1.617.2, known as Delta, took hold and quickly became dominant across the United States, and it became clear that the virus had changed. With Delta's transmissibility being twice as efficient as that of previous strains, hospitalizations increased primarily among the unvaccinated, but vaccine breakthroughs also became more frequent, and vaccinated individuals who became infected usually had symptoms and very high levels of virus in the nose, similar to unvaccinated persons. The public felt whiplashed as masking and other restrictions were reinstated. However, despite some reduction in protection, vaccination remained highly protective, especially against hospitalizations and severe disease.

Nevertheless, there is concern going forward that as more of the world's population develops immunity, the pressure on the virus to evade protective antibodies will increase. In addition, immunocompromised individuals, who shed virus for long periods and receive monoclonal antibody or other therapies, may inadvertently generate more heavily mutated variants (49, 50). If two SARS-CoV-2 variants infect an individual simultaneously, recombination can also occur and lead to a more potent variant (51).

Indeed, the Omicron variant identified in late 2021 in South Africa had an alarming number of mutations, with 30 in the spike gene alone, and its recognition triggered an international alert. This highly mutated virus was postulated to have arisen either from an immunosuppressed individual with a chronic infection, such as a human immunodeficiency virus (HIV)-infected patient, from transmission to an animal host then back to humans (reverse zoonosis), or from an infected patient treated with the mutagenic drug molnupiravir. It also shared sequences with seasonal coronaviruses, raising the issue of recombination. Omicron was quickly confirmed to be more transmissible than Delta. Within weeks, laboratory studies using the blood of vaccinated patients predicted that current vaccines based on the Wuhan strain would be less effective at preventing infection with Omicron, but a third dose of an mRNA vaccine was shown to restore protection, especially against severe disease (52). Subvariants of Omicron would emerge, such as BA.2.12.1, BA.4, and particularly BA.5, that would prove both evasive of established immunity and highly infectious. They were called "escape artists." While a fourth vaccine dose provided protection against BA.2.12.1, it was less effective for BA.4 and BA.5.

As variant sequencing capacity increases, the need to rapidly correlate the genetic sequence, or "genotype," of a newly discovered variant with its behavioral traits, or "phenotype," will be key. To explain which phenotypic traits in a new variant are important to identify quickly, Peacock employs a mnemonic, TILT, where T is for transmissibility, I is for immunity, L is for lethality, and T is for testing, as these are the factors that determine the variant's ultimate impact. Is it more transmissible, does it escape natural or vaccine-induced immunity or immune therapies, is it more lethal, and is it missed by current tests (45)? This is the next challenge for which a systematic approach is needed, as well as innovation in diagnostic methods, since quickly determining the behavior of a variant allows early interventions to mitigate its impact, including targeted control measures, selection of the most effective therapies, and updating of vaccines (53).

VACCINES: THE NEW "ARMS RACE"

To stop the pandemic, vaccines were urgently needed. However, there had been no prior vaccines for coronaviruses approved for humans and only one vaccine for viral respiratory disease, namely for influenza. Similar to influenza, the primary goal for SARS-CoV-2 was to prevent hospitalization and death. Fortunately, research on vaccines for SARS and MERS provided key insights and allowed rapid progress once the sequence of the viral genome was posted. Vaccine development typically takes years to complete, and time was of the essence. For several decades, much effort had been spent developing new vaccine technologies to provide more effective and longer-lasting protection and to allow for rapid response to a new pandemic, which due to international travel was expected to spread quickly.

The simplest traditional approach for viral vaccine preparation, which is also supported by existing production and distribution infrastructure, utilizes whole virus grown in culture, followed by inactivation by formalin. A concern for SARS-CoV-2 was the potential for vaccine-induced hypersensitivity in the lungs following subsequent exposure to the wild-type virus, as this had been observed with inactivated SARS and MERS vaccines in animal models. Strategies to overcome this side effect had been devised, and two inactivated vaccines that did not show this side effect were quickly developed in China by Sinopharm and Sinovac. After they were approved at the end of December 2020, these vaccines were made widely available to other countries and were effective in reducing severe disease. However, while whole-virus inactivated vaccines can induce strong neutralizing antibodies to multiple virus components, are stable, and are inexpensive, immunity may decline faster than with more-targeted vaccines.

The alternative to using the whole virus is to focus on components of the virus that are the main targets of the immune system, such as the spike protein. Different vaccines have used different approaches, from the whole native S protein, to the RBD only, to an S protein modified by addition of two proline residues for stabilization and optimization of its stimulation of the immune system.

New vaccine platforms

The development of novel vaccine platforms reflects the work of many dedicated scientists. That story is best introduced through brief biographies of the four given

FIGURE 6 *Barney Graham (left) and Jason McLellan (right). (Left panel courtesy of NIAID. Photo credit for right panel: Vivian Abagiu/University of Texas at Austin.)*

below, two each working collaboratively. We start with Drs. Barney Graham and Jason McLellan, who developed the structure on which the vaccine immunogen was based (Fig. 6). We move next to Drs. Katalin Karikó and Drew Weissman, who solved the dilemma of how mRNA could be used therapeutically and in vaccines (Fig. 7). Their work resulted in the extremely rapid and extremely effective production of mRNA vaccines against COVID-19.

Barney Graham, now retired from the Vaccine Research Center (VRC) at the National Institutes of Health (NIH), was the driving and organizing force in developing the Moderna SARS-CoV-2 mRNA vaccine. He, along with Jason McLellan, created a genetic construct for the Moderna mRNA vaccine, which would go on to be used in the Pfizer-BioNTech mRNA vaccine as well. The key feature of the construct was the structural configuration of the immunogen, the viral spike protein, the molecular configuration of which was defined and set thanks to protein imaging technology.

Following graduation from medical school at the University of Kansas in 1979, Graham decamped for Vanderbilt University Medical Center, where he would stay for 20 years. After residency, he did an infectious diseases fellowship and began studies on respiratory syncytial virus (RSV), which is a primary cause of hospitalization for children under 5 years of age and can be lethal in the elderly. Graham also directed Vanderbilt's unit in the government's network to research HIV vaccines before leaving for the NIH in 2000 (54).

Fast forward to 2008 at the NIH VRC, where "Graham had a stroke of luck" (55). Jason McLellan, a postdoctoral fellow in structural biology in Peter Kwong's lab at the VRC, moved into Graham's lab. Kwong had focused on structure-based vaccine design for potential HIV vaccines (56). In Graham's lab, McLellan would apply these structure-based concepts to RSV and would move on to coronaviruses, which would help prepare them for their vital work on SARS-CoV-2 many years later.

Jason McLellan had grown up in St. Clair Shores, a small city in Michigan. His original goal had been to become a pediatrician, but that changed during his undergraduate studies at Wayne State University in Detroit from 1999 to 2003, where his interests shifted to research. He graduated with a B.S. degree in chemistry and moved on to a Ph.D. in structural biology under Dr. Daniel Leahy at the Johns Hopkins University from 2003 to 2008. There he trained in X-ray crystallography (57, 58). He moved to the National Institute of Allergy and Infectious Diseases (NIAID) in 2008 as a postdoctoral fellow.

Work on coronaviruses by collaborators of McLellan and Graham, and by others, laid specific groundwork for the future design of a SARS-CoV-2 vaccine. In 2016, Kirchdoerfer et al. zeroed in on the spike protein of a human coronavirus that causes mild respiratory disease, HKU-1 (59). McLellan, who by then had moved to Dartmouth, and Andrew Ward's lab at the Scripps Research Institute were able to stabilize the prefusion (prior to fusion of the virus with the cell) spike antigen of MERS-CoV by adding two proline amino acids at critical locations (60). Highly detailed cryoelectron microscopy (cryo-EM) studies of the stabilized SARS-CoV-1 spike followed (61). The 2017 Nobel Prize in chemistry had been awarded to Jacques Dubochet, Joachim Frank, and Richard Henderson for cryo-electron microscopy (62).

The stage was set. Once the genetic sequence of the virus causing the Wuhan outbreak was posted online on 10 January 2020, Graham and McLellan compared it with the genomes of SARS-CoV-1 and MERS-CoV. They found the crucial location to modify the code for the spike protein (63). The sequence was then sent off to Moderna on 14 January 2020 (64). The cryo-EM structure of the virus's spike was published from McLellan's lab at the University of Texas in Austin on 19 February 2020 (65).

The first patient in the phase 1 Moderna trial was injected on 16 March 2020 (66). In parallel, BioNTech with Pfizer developed their vaccine also using the Graham-McLellan spike protein construct. Both companies used an mRNA platform, and both were granted Emergency Use Authorization (EUA) by the U.S. Food and Drug Administration (FDA) before the year was out. It was a stunning achievement.

The potential of mRNA technology to prevent and cure human disease is enormous. So too was the frustration when the initial efforts to develop it failed repeatedly. The puzzle was that synthetic mRNA triggered a destructive inflammatory response, vitiating any attempts at biological intervention. Katalin Karikó and Drew Weissman would solve this puzzle and open the doors to the rapid development of the mRNA vaccines against COVID-19. The paths that brought them together at the University of Pennsylvania in Philadelphia were serendipitous.

FIGURE 7 *Drew Weissman (left) and Katalin Karikó (right). (Courtesy of Penn Medicine.)*

Katalin Karikó was born the daughter of a butcher in communist Hungary in 1955. Fascinated by science, she was awarded an undergraduate degree in biology at the University of Szeged in 1978 as a scholarship student. She took a Ph.D. in biochemistry at the same institution in 1982 and stayed on at Szeged as a postdoc. In 1985, she and her husband, Bela Francia, and their 2-year-old daughter, Zsuzs (Susan), immigrated to Philadelphia. Before leaving Hungary, they sold their car on the black market and Katalin sewed the money into Susan's teddy bear to bring with them to America. Of note, daughter Susan was to become a two-time Olympic gold medalist for the United States in rowing (67–71).

In Philadelphia, Karikó worked as a postdoctoral fellow in biochemistry at Temple University from 1985 to 1988. In 1989, she moved over to the University of Pennsylvania armed with excellent biochemical skills and a passion for mRNA. She had many years of productive work but was unsuccessful in obtaining funding. Consequently, she was moved off the full professor academic ladder. Yet she decided to stay at Penn. It was a good thing, for she met Drew Weissman, a respected and funded immunologist, who had come to Penn in 1997.

Drew Weissman obtained an M.D. and Ph.D. from Boston University in 1987. After training in internal medicine, he did a fellowship on HIV with Anthony Fauci at the NIH (72). Karikó and Weissman met by chance at a copy machine and, after discussing their mutual interests, agreed to work together on mRNA (70). The major problem was that mRNA made experimental animals sick. "Their [mice] fur got ruffled, they hunched up, they stopped eating, they stopped running" (68).

The investigators made a key observation that led to a crucial change in the experimental protocol. They found that it was the nucleotide composition of the experimental mRNA that triggered the immune response. In an experimental control they observed that pseudouridine in transfer RNA did not provoke an overreaction (68). With a key paper published in 2005, they demonstrated that modified (methylated) nucleosides or pseudouridine reduced the deleterious activity of dendritic cells (73). They then demonstrated that incorporation of pseudouridine into mRNA decreased immunogenicity *in vivo* and increased translational capacity (74).

While one would have thought that these findings constituted the holy grail of mRNA research, they did not create much of a stir. However, they would prove foundational to two biotech companies that would go on to develop the mRNA vaccines against SARS-CoV-2, Moderna and BioNTech (75). Moderna and BioNTech both licensed the mRNA technology developed by Karikó and Weissman. In fact, BioNTech hired Karikó as senior vice president. When the genome sequence of what would be named SARS-CoV-2 was determined by Zhang Yongzhen and put online on 10 January 2020, BioNTech and Moderna were off to the races. BioNTech arranged to partner with Pfizer for manufacturing and testing of their vaccine. Both companies would develop, test, and obtain EUA for their COVID-19 mRNA vaccines before the year was out. The vaccines proved to have very high efficacy, a testament to the experimental insight and intense work of Karikó and Weissman.

On 18 November 2020, just 10 months after the genome of SARS-CoV-2 was posted by Chinese scientists, Pfizer-BioNTech announced that their mRNA vaccine was 95% effective against COVID-19 without serious side effects (76). The mRNA vaccine from Pfizer-BioNTech was the first SARS-CoV-2 vaccine approved in the West, first in the United Kingdom and a few days later in the United States, followed by the Moderna/NIH mRNA vaccine in the United States (77). Although there had been many animal studies and several early clinical human trials for other pathogens, the COVID vaccines were the first mRNA vaccines approved for human use. Both used a two-dose regimen, and protection against symptomatic infection was reported at a remarkable 94 to 95%, higher than for other early vaccines. A new era in the battle against COVID-19 had begun. The first vaccines were administered on Monday, 14 December 2020, to considerable hope. Yet it was also a time of dread, as the U.S. death toll had risen to 300,000, the greatest number of deaths in any country. *The New York Times* headlined on its first page, "'Healing Is Coming': US Vaccinations Begin. Dread Persists as Death Toll Tops 300,000" (78).

Of the new technologies, mRNA and adenovirus vector vaccines were the earliest approved and by some measures were the most effective vaccines against

SARS-CoV-2. Importantly, these new vaccine platforms allow rapid updates to the vaccine if needed. A disadvantage of mRNA vaccines, however, is the need for extreme cold storage to prevent degradation of the mRNA, thus limiting utility in low-resource areas. Virus vector vaccines are generally more stable to temperature and use a "harmless" virus that is different from the virus being targeted to deliver key information to cells. The virus vector is combined with a key segment of target virus genetic code, usually the gene segment needed to stimulate immunity. Virus vector vaccines have been previously used clinically in humans to prevent Ebolavirus infection. The first such vaccine approved (rVSV-ZEBOV) uses vesicular stomatitis virus (VSV) as the vector virus combined with the attachment protein gene from Zaire ebolavirus. rVSV-ZEBOV was developed at the Canadian National Microbiology Laboratory, then by Merck Pharmaceuticals, and approved in the United States and Europe in late 2019 (79). The vaccine has saved many lives in the recurring and devastating Ebolavirus epidemics in Africa since 2015. A second Ebolavirus vaccine developed by Johnson & Johnson at Janssen Pharmaceutical includes an adenovirus type 26 (Ad26) vector vaccine as the first dose of a two-dose regimen and was approved by the European Commission in July 2020 (80).

Three adenovirus vector vaccines were among the earliest SARS-CoV-2 vaccines available. All include the SARS-CoV-2 spike protein gene. The Oxford-AstraZeneca vaccine from the United Kingdom uses a chimpanzee adenovirus vector in a two-dose regimen (ChAdOx1) (81). The Johnson & Johnson/Janssen vaccine from the United States uses the Ad26 vector as a single dose, which was viewed as an advantage for some populations (82). A Russian vaccine, Sputnik V, uses Ad26 for the vector as the first dose and the Ad5 vector as the second dose. While Sputnik V was the earliest vaccine to be approved by its government, in August 2020, the clinical trial data in Russia were very minimal, leading to some skepticism from the scientific community and from its own citizens (83, 84). Following the invasion of Ukraine, Russian facilities could not be inspected and WHO postponed approval in March 2022 indefinitely (85).

Operation Warp Speed. In the United States, Operation Warp Speed had been established to facilitate rapid development of vaccines and to bring them to market. At a government level it was a partnership of the Department of Health and Human Services and the Defense Department (86). Peter Marks of the FDA emphasized simultaneous pharmaceutical development and manufacturing (87). The idea was to manufacture "at risk," that is, to manufacture vaccines before full proof of safety and effectiveness was demonstrated. If that was not achieved, the vaccine was to be discarded (88). The name Warp Speed was chosen, said Marks, "to shake us out of our everyday mode of doing things." The former head of the FDA, Scott

Gottlieb, appointed by President Donald Trump, has called it "one of the greatest public health achievements in modern times" (89). While the vaccines brought hope and relief, trouble lay ahead. First, within the United States and elsewhere, an old nemesis reared its ugly head, vaccine hesitancy. Second, coupled with vaccine hesitancy, were the threats posed by the Delta variant and subsequently Omicron and its subvariants. Finally, on a global scale, there was insufficient funding and distribution capacity, as well as widespread vaccine hesitancy. Providing vaccine coverage to all nations was crucial; for as long as populations remained unvaccinated, global spread of new virulent variants would remain a threat.

Vaccine hesitancy and resistance

From the beginning, Edward Jenner's instillation of cowpox to prevent smallpox, there has been intense opposition to vaccines. See, for example, the cartoon from the time (Fig. 3 in chapter 3) concerning the introduction into France of smallpox vaccination. In recent times, prior to COVID-19, outbreaks of measles have occurred among unvaccinated persons, many of whom cite philosophical or religious objections to vaccination. The authors noted "[v]accine hesitancy is pervasive, affecting a quarter to a third of US parents" (90).

Well over a year after the emergency use authorizations for the mRNA vaccines, a low but significant proportion of eligible people in the U.S. remained unvaccinated. A useful means of understanding the causes is to examine vaccine hesitancy and resistance (91). Hesitancy refers to individuals who are still on the fence. They may distrust the government, the American medical establishment for past actions such as the Tuskegee syphilis experiment in Black men, the speed with which the vaccines were developed, and the media. Social media have been the source of a torrent of misinformation. Resistance includes those who are resolutely opposed to getting vaccinated. Numerous characteristics such as age, race, and education status have served to distinguish the groups resisting vaccination in the U.S. But according to a report from the Kaiser Family Foundation, political partisanship has been found to be a stronger predictor of whether someone is vaccinated. Additionally, among the vaccinated, a majority (73%) of Republicans describe getting vaccinated as a personal choice, whereas a strong majority (81%) of vaccinated Democrats see it as everyone's responsibility to protect the health of others" (92).

COVID-19 came to the United States at a time of deep cultural division. It became an object of that division and a factor in amplifying it. For example, D. A. Graham argued that it was not COVID-19 vaccine hesitancy, but rather, COVID-19 denialism that underlay reluctance for some to be vaccinated (93). Anthony Fauci, Director of the National Institute of Allergy and Infectious Diseases (see bio below), said that he had never seen such distrust and anger in the country.

"Political divisiveness doesn't lend itself to having a coordinated, cooperative, collaborative response against a common enemy. There is also this pushback in society against anything authoritative. And scientists are perceived as being authority, so that's the reason I believe we have an anti-science trend, which leads to an anti-vaccine trend" (55).

Intense anti-vaccine attitudes are a well-documented international concern. A December 2021 essay in *The Atlantic* noted that "anti-vaccine sentiment has become a global phenomenon at precisely the wrong time" (94). A systematic review reported low rates of vaccine acceptance, including several European countries, Africa, the Middle East, and Russia (95). In early 2022, in the world's poorest countries vaccine access remained an issue and complicating that was emerging resistance to vaccination (96).

Resistance to vaccines was so pervasive in the U.S. that President Biden imposed vaccine mandates on large employers and on health care workers. However, the Supreme Court struck down the large employers mandate while retaining the health care worker mandate (97).

Global control. With apologies to the poet John Donne, no nation or continent is an island, that is, survives as an island of immunized peoples in a world of the susceptible. Without effective vaccination globally, there is increased risk of continual reseeding of vaccinated nations and, more distressingly, the generation of virulent viral variants. To that end, both altruistic and self-preservation reasons underlay the formation of the COVID-19 Vaccines Global Access (COVAX) initiative, whose goal was to prevent vaccine nationalism and to distribute vaccines equitably among rich and poor nations. It was supported by Gavi (the Global Alliance for Vaccines and Immunizations), the Coalition for Preparedness Innovations (CEPI), and WHO, but unfortunately, that goal has not come to pass. As of early 2022, funding had fallen far short (98). Distribution of vaccine to all corners of the world calls to mind the successful global eradication of smallpox, certified in 1979 (see chapter 3).

Duration of immunity and impact of variants

After initial euphoria due to the high efficacy and earlier-than-expected rollout of vaccines, the focus increasingly shifted to questions of duration of protection, the impact of variants on protection, the need for boosters and for whom, and when to change the antigens in vaccines. By early 2022, the value of boosters to protect against Omicron became clear and the higher toll of the virus in the U.S. compared to Europe was at least partly attributed to the low uptake of booster shots in the U.S. But as new subvariants of Omicron emerged, escape from immunological control emerged as a particularly difficult problem. The U.S. FDA recommended adding Omicron subvariant targets BA.4 and BA.5 to the original formulation of the vaccines (99). As time passes and new variants continue to appear, determining the answers to the questions above will challenge the scientific, medical, and public health communities, and the governments under which they operate.

Additional vaccines

Additional vaccines were developed around the world and approved by their national governments for use. Selected vaccines were recognized by WHO based on quality of clinical trial data and ongoing monitoring. The WHO website tracks COVID vaccines in development, and in early 2022 listed more than 300 additional vaccines, employing 11 different strategies, with 146 in clinical development. Protein subunit vaccines were the most common platform, at 33% of those in clinical trials, and may be more readily accepted by some (100). In July 2022, Novavax's adjuvanted protein vaccine was approved in the U.S.

DIAGNOSTIC TESTS

Molecular tests for SARS-CoV-2

From the onset of the pandemic, laboratory testing was recognized as critical to identifying cases, implementing effective prevention and treatment measures, and saving lives. Usually operating in the background and taken for granted, the laboratory became the early focus of attention as its essential role in the pandemic was keenly felt.

Early in the pandemic, the U.S. FDA exerted regulatory oversight over all diagnostic testing, including laboratory-developed tests in high-complexity laboratories. The requirements for FDA EUA, in addition to the inherent challenges of developing a test for a new pathogen without access to either the virus or to positive clinical samples, proved an insurmountable barrier. Instead, samples that met very narrow eligibility criteria for testing had to be routed via state health departments to the CDC. The initial CDC assay involved 4 separate PCRs for each sample, 3 viral gene targets and 1 human cell control for sample quality to ensure accurate detection. Furthermore, the CDC recommended testing 2 to 4 different samples for each patient: sputum, bronchoalveolar lavage, nasopharyngeal, and oropharyngeal swabs to determine the best sample to test for this new virus. Casting such a wide net necessitated setting up 8 to 16 PCRs per patient, greatly reducing testing capacity (101).

When the SARS-CoV-2 test kits distributed by the CDC to the state and public health laboratories on 4 February 2020 were found to be flawed, the United

States was essentially without testing outside of the CDC until 28 February, when revised test kits, requiring 3 instead of 4 PCRs per sample, became operational in state labs. This 24-day delay in testing capacity contributed to a false sense of security until the last week in February, when it was recognized that SARS-CoV-2 had been circulating undetected in U.S. communities. Thus, on 29 February the FDA revised the EUA criteria and provided an expedited template for laboratories to validate laboratory-developed tests. However, actually obtaining the SARS-CoV-2 virus, as well as essential reagents and instruments needed to validate an assay, remained onerous, and testing required highly trained staff with experience in molecular techniques. Early molecular assays were largely based on either the CDC- or WHO-recommended gene targets.

Commercial manufacturers were also provided an expedited path to EUA for their tests. By the end of March/early April 2020, simpler, more automated, multiplexed molecular tests received FDA EUA. However, inability to obtain instruments as well as supply chain issues for all testing components, from reagents to buffers to plasticware, continued to limit test availability. If the instrument was not already on site in the laboratory, it might not be delivered for 8 to 12 months. All reagents were "on allocation," meaning only a limited number of test kits would be shipped, and that number varied unpredictably from week to week. Laboratories were compelled to contract with multiple companies to stay operational. Validating each test and training staff on multiple platforms during multiple shifts was challenging. Committees of clinicians and laboratorians were established to set priorities for various patient groups based on test availability and time needed for the result. Supplies were stacked in hallways, offices, and conference rooms. Third shifts were opened to provide 24/7 molecular testing. Staffing was expanded by pulling staff from other laboratories or by hiring persons without laboratory certification and training them. Testing, first limited to hospitalized patients, was able to be expanded to symptomatic individuals in the community. However, demand continued to exceed capacity, and time to results was days to weeks.

Testing asymptomatic individuals

Early in the pandemic, it was shown that, in contrast to SARS and MERS, the SARS-CoV-2 virus could be transmitted by persons before symptoms developed or by persons who never developed symptoms (102). Thus, to reduce transmission and prevent outbreaks, testing individuals with no COVID symptoms became a priority—for at-risk nursing home and congregate living residents, including prisons and colleges; persons exposed to infected individuals; all patients admitted to the hospital; women in labor and delivery; individuals scheduled for surgical or other medical procedures; and persons traveling internationally on airplanes. To be able to expand testing despite scarce reagents and limited equipment and staffing, some laboratories pooled 4 to 10 samples in a single tube as an initial screen, and then tested the individual samples only if the initial pool of samples tested positive.

Tests and sensitivity

Tests that first amplify the virus genetic material in a sample a millionfold or more as part of the testing process are referred to as nucleic acid amplification tests (NAATs) or molecular tests and are the most sensitive option (103). Polymerase chain reaction (PCR), with an initial reverse transcription (RT) step to create a DNA copy from the SARS CoV-2 RNA, is the most common but not the sole NAAT method (see chapter 9).

The preferred PCR method is most often real-time RT-PCR (rRT-PCR) due to its high sensitivity and the generation of a "cycle threshold," or Ct value, that provides a rough estimate of the amount of virus in the sample. PCR tests typically utilize 40 cycles of amplification, and in each cycle, the viral target sequence is doubled. In real-time PCR, the cycle number when the test signal crosses a threshold to register as a positive result is recorded and is inversely related to the amount of virus in the original sample. If the signal crosses to positive at an early cycle (low Ct value), more virus is present in the sample. In contrast, if the signal crosses at a very high cycle number (high Ct value), it indicates a very low amount of viral RNA is present. Although absolute Ct values vary among different tests and sample types, when serial specimens are compared using the same sample type and test, trends in Ct values can indicate whether the amount of virus is declining with time or in response to treatment.

To devise molecular diagnostic tests that remain accurate, despite the continual evolution in circulating viruses, two or three regions of the SARS-CoV-2 genome are usually targeted in molecular test kits, rather than just one region, as is standard practice for most other pathogens. The genome regions that are conserved and less prone to variation between strains are selected. Nevertheless, early in the pandemic, despite the use of highly sensitive assays, some cases of COVID-19 were missed, even in very sick hospitalized patients (104). Concerns were reported in the popular press in April of 2020, namely that PCR was only 70% sensitive for detecting new infections (105). Reasons for falsely negative PCR results included variations in tests, collecting samples too early or too late in illness when virus was not present in the sampled nasal passages, poor sample collection technique, or testing suboptimal sample types. The optimal sample with the highest yield is a thin swab inserted deep through the nasal passage to the back of the throat (referred to as a nasopharyngeal swab), which is uncomfortable for patients and necessitates a skilled collector.

Despite initial concerns that PCR was insufficiently sensitive, when screening of asymptomatic low-risk groups became common in the summer of 2020, it was then argued that PCR was too sensitive and that the cutoff for positive results should be scaled back (106). In addition, sending samples to reference laboratories led to delays of days to even weeks, which greatly reduced their utility in controlling transmission. This led to increased interest in simpler rapid antigen tests that could be done at the "point of care" outside of laboratories, with results in 10 to 30 minutes. Rapid antigen tests that directly detect viral proteins without amplification had long been used for influenza and respiratory syncytial viruses, but, due to the lack of an amplification step, required significantly more virus in the sample to register a positive result. In recent years, antigen tests had been largely replaced by more-sensitive molecular methods. Now the same methodologies were applied to COVID-19. Antigen tests could provide rapid results when detecting only strong positives was sufficient, when testing was repeated in asymptomatic low-risk persons at frequent intervals such as at colleges, or as a preliminary step to rapidly identify the most infectious individuals within minutes, with follow-up testing of negative results when needed using a more sensitive NAAT (107).

Ultimately, it became acceptable that sample types and test sensitivities could vary depending on the risk for serious disease, the medical setting, whether single or repeated testing would be performed, and the consequences of missing a low-positive result.

Infectious virus

Prolonged detection of SARS-CoV-2 RNA after infection, for weeks and sometimes months, led to early questions about the correlation of PCR results with ability to transmit virus to others. Both molecular and antigen detection methods detect components of the virus but not whether the virus is infectious. For infectivity, culture is needed, which for SARS-CoV-2 requires higher-level laboratory biosafety practices and facilities than is standard for clinical laboratories. In addition, most clinical laboratories had discontinued viral culture methods prior to the COVID pandemic in favor of more rapid and efficient molecular tests. For both reasons, SARS-CoV-2 culture was not available for clinical management.

Sample collection

Due to concerns early in the pandemic about the safety of sample collection in closed and inadequately ventilated spaces, it was recommended that only negative pressure rooms be used, and that staff wear enhanced personal protective equipment (PPE) including N95 respirators, gowns, gloves, and eye protection. The usual practice of discarding PPE after every patient encounter quickly resulted in PPE shortages in hospitals, especially N95 masks, which had to be saved and reused.

Physicians' offices were unable to collect samples from their patients or have infected patients in their waiting rooms. Thus, as commercial tests became avail-

FIGURE 8 *Drive-through sample collection tent for COVID-19 testing. (Image courtesy of Michelmond, Shutterstock ID: 1705973104).*

able and test capacity increased locally and at reference labs, drive-through sites, as pioneered by South Korea, were set up outdoors to collect samples where infectious aerosols would be diluted and dispersed (Fig. 8). Patients could remain in their cars, and staff did not continually have to change PPE. Securing test sites, setting up facilities, training staff, and optimizing logistics required substantial effort. In addition, scheduling appointments and reporting results were challenges. Patients also needed a car, which excluded a portion of the population. Subsequently, mobile vans were deployed to underserved communities, as well as transient "pop-up" test sites in places where people gathered. Eventually, drive-up pharmacy windows were able to provide nasal swabs with instructions for self-collection and sample drop-off boxes.

While nasopharyngeal swabs were considered the best sample, the collection kits became unavailable. Laboratory staff were deployed to prepare collection kits using a variety of different swabs and media, whatever was available, as well as to prove their accuracy (108). 3-D printing of swabs was entertained to increase capacity, and in some instances was successfully implemented (109).

Since sampling saliva does not require a swab, is more comfortable for the patient, and the sample can be self-collected, it gained attention especially for settings where repetitive testing was indicated such as college campuses and schools (110). However, since no commercial kits were approved for testing saliva,

saliva requires special processing in the lab, and reported results were conflicting, only some laboratories were willing or able to offer this testing.

Self-collection of samples that are then shipped to a certified laboratory for molecular testing was approved for some tests. Antigen tests for home use began to be sold in stores or online to diagnose COVID. Although simple and convenient, antigen tests are known to be less accurate, particularly when used in persons without symptoms. While the usual concern is false-negative results, one kit was recalled for false-positives (111). Then in late 2021 and early 2022 with the Delta and Omicron surges and the recognition that antigen test positivity correlated with peak transmissibility, home use of antigen tests became widely accepted as a valuable tool for rapid and convenient diagnosis of symptomatic persons and for quickly identifying highly infectious persons prior to social gatherings. As molecular testing demand again outstripped capacity, the U.S. federal government began to provide free antigen tests on request and to require health plans to reimburse for tests if purchased.

More sensitive do-it-yourself at-home molecular test kits were predicted to be available in the fall of 2020 to use in persons with or without symptoms but proved more difficult than anticipated. However, in 2021, several at-home molecular tests received EUA in the United States. One used loop-mediated isothermal amplification (LAMP), a simpler, lower-cost NAAT alternative to PCR that is more sensitive than unamplified antigen tests and has been used at the point of care in low-resource settings.

Diagnostic innovation

To speed the development, validation, and commercialization of innovative point-of-care and home-based tests, encourage new approaches, as well as improve and increase capacity for laboratory tests for direct virus detection, the NIH initiated Rapid Acceleration of Diagnostics (RADx) programs. Among the new methods showing promise to provide simpler, faster, more sensitive and accurate point-of-care testing are CRISPR/Cas-based nucleic acid methods (112).

Other innovations accelerated by the COVID pandemic that could prove useful in future pandemics include wearable sensing technologies that can detect SARS CoV-2 in the air, monitoring for aberrant physiological parameters that detect presymptomatic cases of COVID-19, remotely tracking progress in recovering patients, or analyzing volatile substances in breath to rapidly identify COVID-19 infections (113). Dogs have also been trained to scent-identify volatile organic compounds generated during SARS CoV-2 infection (114). Artificial intelligence and new digital technologies, including smartphone apps, also played an important role in supporting COVID medical and public health efforts, tracking transmission, providing exposure alerts to individuals, and assisting the remote monitoring of positive patients (115).

SARS-CoV-2 antibody testing

Early in the pandemic, the popular press touted the imminent availability of SARS-CoV-2 antibody testing as the passport to a return to normal activities. If a person had a positive antibody test, they would be presumed to be immune to reinfection and not at risk of transmitting infection to others, similar to how viral antibody tests are used to assess immunity to hepatitis B virus, measles, mumps, or rubella. However, scientists quickly pointed out that the level of antibody that correlated with protection from reinfection and lack of infectiousness to others was not yet known for SARS-CoV-2. Furthermore, duration of immunity to SARS-CoV-2 and protection against newly emerging variants were uncertain and could be quite short-lived in persons with mild natural infection. In addition, numerous antibody tests of uncertain accuracy flooded the market, and testing scams proliferated. Some early tests used the internal nucleoprotein (N), because antibodies to N are a sensitive indicator of past infection with SARS-CoV-2. However, many vaccines only induce antibodies to the SARS-CoV-2 spike (S) protein, which mediates virus attachment to host cells. Thus, additional tests were developed to detect antibody to the spike protein since it is an indicator of both vaccine response and natural infection. Most antibody test results are reported simply as positive or negative. For some tests, referred to as semiquantitative, a value is given but the correlation of a specific value with protection is unknown, particularly since it will vary depending on the variant and the test used (116).

Changing test demands

After months of efforts to increase testing capacity, with the availability of vaccination and the effectiveness of masking and social distancing, test volumes plummeted in the spring of 2021. Stockpiled test kits began to expire, underutilized COVID testing staff were reassigned to other laboratories, some excess supplies were donated or discarded, and equipment sat idle. Many collection tents closed due to low demand and their staff were reassigned or let go. Then in the summer of 2021, the Delta variant took hold, infections surged, and demand for testing rebounded. By this time, manufacturers had been able to increase production of instruments and test kits, and time to result in most laboratories was within 24 to 48 hours, which was fortunate since in January 2021, the U.S. government Centers for Medicare & Medicaid Services instituted a policy that would reduce payment for tests that took more than 48 hours for results (117). With Omicron on the heels of Delta in early 2022, demand surged yet again. Fortunately, home tests were available to ease some of the burden on laboratories.

Staffing crisis

As the pandemic continued to take its toll, burnout led to laboratory staff resignations, retirements, or transfers to less stressful, higher-paying or remote work options (118). Furthermore, many health systems instituted vaccination require-ments, leading to termination of unvaccinated health care workers, contributing fur-ther to staffing shortages, including in the laboratory.

Viral genome sequencing

Until the COVID pandemic, viral whole-genome sequencing (WGS) was mostly limited to research settings. Suddenly, investment in building sequencing capacity and expertise became an urgent focus. First and foremost, genetic sequencing was needed for public health surveillance, where the testing could be done outside of clinically regulated laboratories; however, results could be reported only to public health officials and not to the patient's physician or medical record. Since variants could be resistant to some monoclonal antibody therapies, physicians were inter-ested to know which variant caused disease in their patients. This put pressure on clinical laboratories to develop viral sequencing capacity, an onerous task with uncer-tain reimbursement outside of government contracts (119).

There are three options for identification of SARS-CoV-2 variants. The least expensive and fastest option is to design PCR assays that target specific mutations of interest. However, targeted PCR will detect only selected known mutations and will not identify the full set of mutations that define a variant. A second option is to sequence only the region of interest, such as the spike gene, using the tradi-tional first-generation Sanger or chain termination sequencing method developed in 1977 by two-time Nobel laureate Fred Sanger and colleagues, for which equip-ment and expertise may be more widely available. Here too the data generated will be more limited and time to result will be days to a week. The third option is next-generation WGS to determine the full genetic sequence of the virus. WGS requires multiple overlapping PCR reactions to cover the entire viral genome, then assemblage of all the data generated using computer informatics, followed by comparison of the sequence generated to databases with reference strains to identify relevant mutations and lineages. A substantial investment in equipment, bioinformatics and secure data storage, and personnel with technical expertise is required. Time to results is currently several days to several weeks. If, as the pan-demic matures, the identification of the infecting variants impacts therapy, more-rapid results will be needed and the clinical need may serve as an incentive for technological advances to simplify, automate, and accelerate sequencing methods. This happened as the effectiveness of various monoclonal therapies was lost with the emergence of certain Omicron subvariants.

Surveillance for SARS CoV-2 in wastewater

In the early months of the pandemic, it was recognized that SARS COV-2 is com-monly shed in stool, and that wastewater-based epidemiological methods developed by environmental scientists could play a key role in mass surveillance of COVID-19 (120). Wastewater surveillance for SARS CoV-2 RNA can detect both symptomatic and asymptomatic infections, serve as an early warning system, be used to trend levels of SARS CoV-2 at the community level and allocate testing resources, and monitor the effectiveness of public health interventions. Initial wastewater surveillance efforts in the U.S. were variable and included some academic researchers, commercial laboratories, and utilities, with limited public health involvement. In September 2020, the CDC launched the first National Wastewater Surveillance System (NWSS) to coordinate and build capacity for wastewater surveillance in support of local and national public health efforts (121). While NWSS participation continues to grow in response to COVID, in the future it can be extended to other pathogens to identify emerging threats and to help prepare for future pandemics.

CLINICAL DILEMMAS

Adult COVID-19

The world has been transfixed by images of severe disease and death from different locales and times in the COVID-19 pandemic: overwhelmed hospitals with ICUs filled to capacity, hospitalized patients on ventilators, staff pressed beyond their limits, refrigerated trucks used as overflow morgues lined up on city streets, and plots of densely packed burning funeral pyres. Yet the case fatality rate in the United States as of the start of February 2022 was 1.18% (122). How does one explain the apparent disparities? The images of overwhelmed health care facilities and fatalities are from surges of infection. As time has gone on, the populations afflicted have changed, with many patients now younger, care has improved and diagnostic testing too has made a significant difference, with more ready access and more rapid turnaround time.

COVID-19 is primarily a respiratory disease with highly variable outcomes, from asymptomatic infection to death. In asymptomatic and mild cases, it is primarily confined to the upper respiratory tract, with loss of taste and smell a common and sometimes persistent feature. In severe cases, it damages the lungs. There it appears to be virus-triggered damage by host defenses and thrombosis that causes the pathology. It is likely that the amount of infecting virus, aka the viral load, is important.

Age, obesity, comorbid medical conditions, immunological compromise, and pregnancy complicate outcomes. Children are much less severely affected, whereas the elderly are at much greater risk of severe disease and death. Pregnancy, as discussed below, is associated with more-severe outcomes. A perplexing consequence of COVID-19 at any age, with profound health and social consequences, is the chronic condition known as long COVID (see below).

Pregnancy and newborns

Pregnancy is often a time of hopefulness—anticipation of the arrival of a child. But in the time of COVID, pregnancy became a time of apprehension—of anticipation of a possible bad clinical outcome for the mother and worry for the unborn child. Despite concerns generated by rubella, cytomegalovirus, and Zika virus, congenital abnormalities were not found. But a national U.S. CDC surveillance study did document worse outcomes for pregnant women with COVID-19 compared to infected but not pregnant peers (123). It turned out that these worse outcomes could be prevented by vaccination. Furthermore, it was later found that unvaccinated women infected with COVID-19 during pregnancy "were far more likely than the general pregnant population to have a stillborn infant or one that dies in the first month of life (124).

Regrettably, pregnancy myths grew up among those who wanted to discredit vaccination. Notable among these was the myth that vaccination would cause infertility. It was completely disproven (125). Vaccination protected pregnant women, and studies demonstrated that antibody is transferred to neonates via placenta and breast milk (126). Despite the CDC, the WHO, and the American College of Obstetrics and Gynecology giving their approval for the use of vaccines in pregnant women, adoption of vaccines in this population lagged. The CDC issued an urgent health advisory to increase vaccination among pregnant women due to low rates of vaccination and high rates of admission to ICUs and death (127).

COVID-19 in children

A crucial question sits at the heart of the disease in children. How does the infection in children differ biologically from the infection in adults (128)? For it surely must. First, the frequency of clinically apparent infection is much less than it is in adults. Second, the types of symptoms in children are more diverse. Third, the disease manifestations and outcomes in children are more benign. Nevertheless, a frightening complication arose, called multisystem inflammatory syndrome in children (MIS-C) in the United States and pediatric inflammatory multisystem syndrome temporally associated with SARS-CoV-2 (PIMS-TS) in the United Kingdom.

In mid-April of 2020, the South Thames Retrieval Service in London experienced "an unprecedented cluster of eight children with hyperinflammatory shock." It was a new syndrome with multiorgan involvement (129). The authors of the original report emphasized the need for multispecialty input. In a large U.S. study, fever, biomarkers of inflammation, and signs of multiple organ system impairment were found including gastrointestinal, cardiovascular, hematologic, mucocutaneous, and respiratory (130). A study of 46 children assessed 6 months following PIMS-TS (MIS-C) found them to be remarkably well (131). Frightening nonetheless.

Long COVID, aka long-haulers' disease or postacute COVID syndrome (PACS)

Long COVID is one of several confounding clinical COVID-19 conditions, and its personal and population effects have a long-term and costly impact. Long COVID joins a history of chronic or reemergent disease occurring in the wake of an acute infectious disease. The ones most often named as similar to long COVID are myalgic encephalomyelitis and, particularly, chronic fatigue syndrome. Much of the impetus to recognize and study the COVID-19-related prolonged illness, which was first called long-haulers' disease, was by self-organizing patient groups (132–134).

As much medical recognition and institutional concern as there is now for the great impairment of personal well-being of COVID survivors and the potentially immense economic loss to society, patients and health providers had a difficult time coming to grips with the cluster of symptoms. Profound changes in life as a result of disabling symptoms were described by a clinician testifying before the U.S. Congress: "A teacher experienced recurrent bouts of crushing chest pain mimicking a heart attack"; "A young mother described palpitations and dizziness while playing with her toddler"; "A local business owner could no longer remember the names of long-term customers and was unable to balance his books" (135). For the majority, less severe symptoms, such as fatigue, myalgia, and brain fog, constitute the condition.

Depending on the study, as many as a third of patients may show features of long COVID between 3 and 6 months after the diagnosis of the acute illness (136). If the percentage of recovered patients who go on to chronic long COVID is even in the range of 10%, the number of patients worldwide, the facilities necessary to care for them, and the costs to society stagger the mind. The number of sufferers will reach into the millions. In the United States, disability insurance mechanisms do not easily fit the circumstances of the long COVID patient population. There is no laboratory test or consolidated diagnostic symptom complex that defines the syndrome unambiguously. Nonetheless, in 2021 CDC established an ICD-10 diagnostic code for long COVID. To address the multiple dilemmas, the NIH launched a major initiative, well funded by Congress, to study the causes and to identify treatment and prevention of long COVID (137).

THERAPIES

While much of the focus and excitement centered on vaccine development to end the pandemic, antiviral therapies and anti-inflammatory medicines were intensively investigated. Throughout the pandemic, nonpharmaceutical interventions, such as social distancing, mask wearing, and avoidance of congested, poorly ventilated indoor spaces, were effective when thoroughly implemented. A massive randomized trial of mask promotion in rural Bangladesh demonstrated an increased use of masks and a reduction of SARS-CoV-2 symptomatic infections (138).

Proning of some hospitalized patients with compromised breathing was shown to be beneficial. Repurposed drugs were studied. For example, in hospitalized, critically ill patients, dexamethasone was found to enhance survival in patients requiring mechanical ventilation (139). Remdesivir, which failed to impact Ebolavirus outcomes, was found in randomized clinical trials to reduce time to recovery in hospitalized patients with severe COVID (140). In outpatients at high risk for COVID-19 progression, a three-day course of remdesivir treatment reduced the risk of hospitalization and death (141).

Monoclonal antibodies, either singly or in combination, were proven beneficial in the treatment of outpatients with early COVID-19 (142) and as prophylaxis. While the monoclonal antibodies proved efficacious, they were costly and cumbersome to administer, requiring intravenous delivery by specialized personnel in infusion centers. Additionally, some lost efficacy as subvariants mutated. The demonstration that monoclonal antibodies could be delivered subcutaneously in early asymptomatic COVID-19 was a therapeutic advance (143). Another advance came with Evusheld, a long-acting combination of tixagevimab and cilavimab monoclonal antibodies. It is delivered intramuscularly for preexposure prevention of infection in chronically immunosuppressed individuals who cannot respond to vaccination (144). As the field advances, the development of non-infusion treatment of all SARS-CoV-2 variants and preexposure prophylaxis are clinically attractive goals.

The advantages of oral medications were obvious. They are easier to store, take, and deliver, particularly in locations without good health infrastructure. Thus, they can relieve pressure on vaccine administration efforts in resource-poor countries (145). Several companies had drugs in clinical trials, including Merck and Pfizer. The Merck drug molnupiravir (named for Mjölnir, the Norse god Thor's hammer) works by inserting errors into the genetic code of the virus that prevent its replication (146). Among higher-risk adult patients with mild to moderate disease, in clinical studies it cut hospitalizations and deaths by about 30%. Britain was the first country to grant authorization (145). Pfizer announced that an independent monitoring committee had recommended stopping the clinical trial of its drug Paxlovid (nirmatrelvir tablets and ritonavir tablets, given together) because the outcomes in the interim analysis showed it reduced hospitalization or death by 89% (147). Its mechanism of action is as a protease inhibitor. It was granted an EUA by the FDA at the end of December 2021 (148). It emerged that a small percentage of patients treated with Paxlovid developed COVID-19 rebound of a mild nature (149).

In an ideal future, one can imagine the universal availability of rapid and accurate testing coupled with highly effective oral medications. Once the diagnosis of COVID-19 is made, antiviral treatment would begin immediately, halting the disease at an early stage. A foreshadowing of that future was seen in the NIH recommendations, released at the start of 2022, for the treatment of outpatients with mild to moderate COVID-19, but at high risk to progression (150).

CULTURAL IMPACT OF THE PANDEMIC

Societies varied in their capacity to manage the scourge. Tightly controlled societies such as China shut down with authoritative efficiency once the danger of the outbreak in Wuhan was recognized. Societies with looser control, such as the United States, suffered immeasurably (151).

The fate of populations, prior to vaccination, hinged on the nature of a nation's governance and on the strength of its leadership. There are examples of successful containment of the virus in authoritarian regimes and in democracies. In China, with a population of 1.4 billion people, after an initial period of attempting to deny the outbreak, authorities clamped down hard on Wuhan and other afflicted areas. Unfortunately, the virus had already escaped to other regions in China and had started enveloping the globe (152). Nonetheless, the shutdown was successful: new cases in China were curtailed. When flare-ups occurred, the Chinese authorities locked down cities and counties (153). When the highly infectious Omicron subvariants emerged in Shanghai, China's most populous city and financial center, authorities implemented a zero COVID policy, shutting down the city in 2022.

In the United States, the initial response was characterized throughout the first year of the pandemic by denials from leadership of its danger and a fumbled response to diagnosis and tracking of the infection, including political interference with public health recommendations. Shortages of crucial supplies and equipment complicated bedside care. For example, Scott Gottlieb, the former FDA chief, wrote that "there was a palpable fear that the United States would run out of ventilators to care for a rush of desperately ill COVID patients" (89).

The U.S. economy significantly faltered, with a drop in the gross domestic product (GDP) (154). There was a dramatic decrease in consumer spending: "this is the sharpest decline in consumer spending we've ever seen," said Luke Tilley, chief economist at Wilmington Trust, as quoted by Leatherby and Gelles (155). There was too a huge drop in employment (156). Essential workers, those necessary to maintaining the critical infrastructure, received plaudits as being heroes. They worked not only in health care, but also in food service industries, public transportation, and manufacturing (157). Frequently from Black and Hispanic communities, many were low-wage workers and were exposed to the threat of infection, often without PPE. They had little choice as to conditions of employment and could not work from home, as did many more privileged workers. Chen et al. reported that essential workers in several industries suffered significantly excess mortality, for example, a 39% increase among food/agricultural workers (158). People working in densely packed workspaces were also hit hard by the pandemic. The pandemic had a major impact on the meat processing industry through the high rate of infection of its workers. Using the basic interventions of universal mask wearing and installing physical barriers, Herstein et al. found a reduction of the incidence of COVID-19 in 8 of 13 locations tested (159).

Paradoxically tragic was the fate of homeless people during the pandemic. When brought into congregate care to reduce their exposure to the elements, their risk of becoming infected with SARS-CoV-2 increased (160). Thus, it was suggested to decongest congregate facilities into private or semiprivate housing arrangements. Compounding the stigma of homelessness, basic actions of public health are difficult to accomplish in this population. Good hygiene, social separation, consistent mask wearing, and following recommendations to "stay at home" are all but impossible to follow (161). Severe cold weather conditions, for example, might pose a greater risk of death than infection. "What's the option? Follow the health code for COVID or put them in the cold and let them die?" (162).

Life in the time of COVID-19

In March 2020, life as we knew it changed in the United States. The pandemic was growing and stay-at-home orders were issued in most states, mandating that people not leave home except to shop for essential needs or to go to a job providing health care or other critical services. Grocery store workers and truck drivers became heroes alongside doctors and nurses. For many, there were hardships of day-to-day living just to get by, especially during the early days of the pandemic when the nature of the virus and disease was not understood. As time passed and restrictions were implemented, rescinded, and reinstated, pandemic anxiety and depression were common.

Although masks were not initially recommended, handwashing and cleaning surfaces were advised. With the lockdown, trips to the grocery store were limited, encouraging people to stock up on items. Most notably, toilet paper was hoarded, creating shortages all over the country. Likewise, hand sanitizers, cleaning products, sugar, flour, and even yeast for baking disappeared from grocery shelves.

As more was learned about viral transmission, face masks were recommended by the CDC in April 2020, creating a critical shortage. Health care institutions were dramatically impacted due to overbuying by the general public and the huge needs of essential workers in the community. Some shoppers outfitted themselves not only with masks but also with helmets, face shields, gloves, and other protective garb. Masks were required for travel on all forms of public transportation.

Even going to the grocery store was different and required planning. The number of shoppers admitted to a store was limited at any one time, so people were forced to wait their turn in long lines while "social distancing" 6 feet apart. Many stores implemented early morning shopping hours for vulnerable seniors and one-way aisles to control foot traffic in the store. Ordering groceries online for home delivery also became a way of life, obviating a trip to the store and potential exposure to COVID.

Unpacking groceries at home became a ritual. Newscasts reported how long the virus survived on hard surfaces, on skin, and in the air, scaring people to take extreme measures. As instructed in tutorials by doctors on television news shows, grocery items were cleaned with bleach wipes at home decontamination stations or were quarantined in garages, and produce was disinfected. Mail and parcels were also quarantined in some households.

The general population was introduced to a new vocabulary. People learned about PPE, antigens, antibodies, and herd immunity. Washing hands meant soap and water for a minimum of 20 seconds or the time it takes to sing "Happy Birthday" two times. Dr. Fauci asked people to help "flatten the curve," meaning to take precautions to reduce new COVID cases so the steep upward trajectory of increasing numbers would gradually flatten and trend down.

When gyms closed, people walked, ran, or biked around neighborhoods and in local parks. They rediscovered their backyards. Home gyms and online yoga and exercise classes became popular. In the absence of in-person visits with friends, family, and colleagues, video calls or Zoom became popular for virtual cocktail hours and celebratory events that could only be shared remotely. Plays and concerts proliferated online. Likewise, there was an explosion of telemedicine visits, benefiting ongoing health care.

When vaccines were finally developed and offered at the end of 2020, cooped-up people hovered over computers or hung on phone lines, racing to secure coveted vaccination appointments. For a majority, the promise of immunity offered the hope of returning to life as it was prepandemic even though the end was not in sight.

Not only was there a profound effect on individuals, but fundamental institutions were altered and long-standing divisions in society were revealed and widened. We will start with a consideration of health care disparities, including poverty and racism. We will move to the vulnerability of those in prison and in nursing homes. Then we will discuss the central roles of schools in society.

Health care disparities

Data demonstrate that marginalized peoples have suffered most in the United States during the COVID-19 pandemic. After almost a half year of the pandemic, a large study found that hospitalization rates for Black and Hispanic populations were roughly 3 to 4 times the rates in White populations, and death rates were more than 2 times higher (163). Later in the first year of the pandemic a systematic review of the literature confirmed that Black and Hispanic populations "experience a disproportionate burden of SARS-CoV-2 infections and COVID-19-related mortality" (164).

The Agency for Healthcare Research and Quality has examined U.S. health care quality and disparities for 2016 to 2018. Black, American Indian, and Alaska Native populations received worse care than White populations on 40% of the quality measures. Hispanic groups received worse care for over one-third of quality measures. The results were mixed for Asian populations (165). Yet in the COVID-19 pandemic, Asian Americans have suffered compounded disparities. On top of high case fatality rates for health care workers of Asian descent and in Asian American immigrant communities, there has been scapegoating related to the initial outbreak being located in China (166), as well as subjection to racial slurs and incidents of physical harm.

Tightly linked to health care disparities has been poverty. It is the poor on whom the burden of COVID-19 has fallen most heavily (167, 168). It is in the poor that the social determinants of health are most readily seen (169). They often worked low-paying jobs deemed essential. Extensive contact with the public and coworkers exposed them to infection by aerosols. Public transportation to work or other travel needs further exposed them in close quarters to the risk of inhaling the virus. In cramped shared living quarters, sometimes in multigenerational families, social distancing became impossible. The frail and elderly among them became particular targets of the disease. For those who lost employment and with few skills to offer, living conditions worsened. Adequate nutrition decreased and weight gain resulted; chronic diseases such as diabetes, obesity, and hypertension increased further. These comorbidities led to more severe disease. The issues were compounded by low health literacy and in many cases by no or lost health insurance (163). Further complicating the distribution of health insurance is high cost and the lack of coverage for low wage earners. Later, vaccine hesitancy in these communities would contribute to the virus's toll.

Structural racism has contributed significantly to health care disparities. An example is "redlining," which is the practice of determining those areas deemed to be a poor financial risk and not worthy of mortgages (170–172). Snowden (170) and Ivey (171) have independently noted the correlation of historical redlining and the incidence of COVID-19. Thus, there is a direct link, historical and ongoing, between structural racism and the scourge of diseases such as COVID-19 in marginalized communities.

Prisons

Prisons have been the sites of some of the largest U.S. outbreaks. Because of excess imprisonment in the United States for nonviolent crime, Black and Hispanic populations are disproportionately affected (173). Conditions in prisons are ripe for generating outbreaks, such as confined living spaces, overcrowding, and high occupant turnover (174). Add to those factors uneven rates of vaccination among prisoners and corrections staff (175). Macmadu et al. recommend decarceration to mitigate infection and to reduce racial disparities (174).

Nursing homes

The lethality of COVID-19 in skilled nursing facilities burst on the scene near Seattle, WA, at the end of February 2020. The index patient was a 73-year-old woman with multiple comorbidities who was transferred to a local hospital with cough, fever, and shortness of breath (176). She was positive for SARS-CoV-2 by an rRT-PCR assay. There had been no known travel or contact with persons known to have COVID-19. She passed away 9 days after the onset of symptoms. A thorough public health workup found that by 18 March there were a total of 167 cases of COVID-19 linked to the skilled nursing facility. These were distributed among 101 residents, 16 visitors, and 50 health care personnel.

The Washington State nursing facility experience would presage the coming national experience. For much of the pandemic, nursing homes accounted for 4% of the nation's cases but 32% of the nation's fatalities from COVID-19 (177). By the last week of April 2021, more than 1,363,000 persons had been infected, with over 182,000 fatalities at about 32,000 facilities. Older age and multiple comorbidities in a congregate living setting contributed to high case fatality rates. But the high infection rate was also caused by health care workers moving from area to area and room to room, and by persons coming into the facilities from the community. Ultimately, infections and deaths would plummet as vaccination efforts prioritized the elderly and strict conditions were laid down to limit transmission of infection from the community. The latter prohibitions would cause great loneliness and sadness as elderly nursing home residents were precluded from seeing family members and subjected to dying alone. Poignant images appeared in the media of longing adult children looking through plate glass windows at their isolated loved ones in nursing homes (Fig. 9).

Internationally, nursing facilities have been vulnerable to the fatal and isolating effects of COVID-19. Regrettably, this sometimes came as a result of health care policy. It had been reported early in the pandemic that European leaders fixated on saving their hospitals, and in so doing "sometimes left nursing-home residents and staff to fend for themselves" (178). In Belgium, for example, nursing home

FIGURE 9 *Photo of adults looking at a family member through a nursing home glass door. (Photo credit: Don Knight,* The Herald Bulletin, *Anderson, IN.)*

deaths accounted for two-thirds of the country's deaths during a peak from March through mid-May 2020. Médecins Sans Frontières sent teams of experts to help. As the pandemic intensified, hospitals took fewer patients from nursing homes. A Médecins Sans Frontières report stated, "Paramedics had been instructed by their referral hospital not to take patients over a certain age, often 75 but sometimes as low as 65" (179).

Especially tragic COVID-19 losses were the deaths of retired religious sisters and nuns in convents and life centers. After they had devoted their lives to their beliefs in the service of others, life in congregate living centers at advanced ages made them especially vulnerable. For example, an outbreak was described from a Provincial House in Latham, NY, in which nearly half of the 100 residents became infected and 9 died (180). A report of other locations of infections and deaths among sisters explicitly noted that the virus "devastates religious congregate communities by infecting aging populations of sisters and nuns who had quietly devoted their lives to others" (181). In contrast, notable and uplifting was the report of Sister Andre, 117, in the Ste. Catherine Laboure Nursing Home in Toulon, France, who survived COVID-19 "with barely any complication." Of 88 residents, 81 became infected and 11 died. Sister Andre herself had lived through two world wars and the 1918 influenza pandemic before encountering COVID-19 (182).

We turn next to the other end of the age spectrum, children, and the effect of the COVID-19 pandemic on schooling and childhood psychosocial development.

Schools

Children, in general least affected by the illness but capable of transmitting the virus, were severely penalized as schools closed and recreation opportunities were suspended. At the start of the first wave of the pandemic, public school closures in the United States were held to be necessary to mitigate the spread of the virus. Closures were complete by the end of March 2020. With the shutdown of the economy, they would have far-reaching effects, revealing many of the chasms in American life. For children it meant not only a massive disruption in education, but also impaired socialization; loss or attenuation of school food programs; and the loss of medical, disability, and psychological support services. With the threat of increasing childhood food insecurity, the interruption of in-school food programs left school districts scrambling to find alternative mechanisms of food delivery (183). For many children it would spell the loss of a safe environment. For parents it would mean scrambling to find some arrangement for childcare or dropping out of the workforce. Women suffered a disproportionate loss of employment due to the need to remain at home.

Closures would affect close to 100 million children in the United States. This included 21 million children in preschool programs, 57 million children in the years from K to 12, and about 20 million students in colleges and universities. As Donohue and Miller pointed out, "the harms associated with school closures are profound." They further noted that "The pandemic has uncovered monumental inequalities in resources available to schools and families" (184). A crucial inequality when learning was resumed remotely was the unequal access to digital resources such as owning a computer and access to the internet. Children of impoverished parents might not have access to the internet, nor might children living in areas not served by broadband services.

International school closures in 192 countries by mid-April 2020 would affect almost 1.6 billion students (184). The effects for some children would be extreme. This was particularly true in low-income countries. For many young children it meant being pressed into the workforce (185). This included 10-year-olds mining sand in Kenya, 6- to 14-year-olds rummaging through garbage dumps in India, and children 8 years old and older begging while painted as statues in Indonesia. Girls were particularly affected, being forced into young marriages in Asia and Africa, and being sent out as sex workers in order for families to eat. For other girls it would mean being forced to undertake unpaid housework, widening the education gap between boys and girls (186).

Anthony Fauci

Throughout the pandemic, Anthony Fauci has been a consistent voice of medical and public health information and advice (Fig. 10). His steady presence has countered the pessimism and fatigue that accompanied the pandemic.

Of Italian extraction, Anthony S. Fauci was born on Christmas Eve, 1940, in Brooklyn, New York. Although Brooklyn was a bastion of the Brooklyn Dodgers baseball club and the home of the iconic Ebbets Field, Tony Fauci was a fan of the dreaded New York Yankees. Sports appeared to be an important element of his

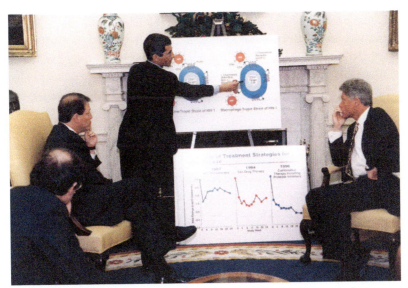

FIGURE 10 *Anthony Fauci (center), Director of the National Institute of Allergy and Infectious Diseases. He is shown with former President Clinton (right) and former Vice President Gore (left). (Courtesy of NIAID, under license CC BY 2.0.)*

growing-up years. He was the captain of his high school basketball team at the top-ranked NYC Jesuit high school, Regis High School. He was a quick, sharpshooting guard, but at 5'7" he realized that he needed another path in life (187, 188).

Scholarship would provide the path to his life's work, clinical and research medicine. He was first in his class at The College of the Holy Cross, in Worchester, MA, another Jesuit institution. He enrolled in the Bachelor of Arts-Greek Classics Premed program. He also ranked first in his class at the Cornell Medical School in Manhattan, from which he graduated in 1966. He had sensed that medicine would provide a "melding" of humanities and science (187). In those years, the Viet Nam War era, he secured an appointment to the Public Health Service, at the NIAID of the NIH. Following that work, he returned to New York City for residency training in internal medicine at the New York Hospital-Cornell Medical Center.

In 1971, he joined the staff of the NIAID as a senior investigator in the Laboratory of Clinical Investigation. It was with the HIV/AIDS global epidemic that he showed the greatest clinical foresight and courage prior to the COVID-19 pandemic. When the first reports of unusual infections in young gay men appeared in 1981, he recognized that something dramatic was occurring and changed the direction of his laboratory research. The 1980s were bleak and dark years for persons with AIDS and for AIDS clinicians and investigators. Much of the frustration and anger of the AIDS community was focused on the federal government. The initial contact with AIDS activists for Fauci was conflicted. He was seen as the face of the inadequate federal response. He changed; he transformed himself "from a conventional bench scientist into a public health activist who happened to work for the federal government" (187). Though he was originally castigated by Larry Kramer, a fierce and articulate AIDS activist, they came to have a mutually highly respectful relationship.

In the time of the COVID pandemic, Fauci has played two essential roles. First, and largely out of public view, has been his organizational expertise and his driving force behind COVID-19/SARS-CoV-2 research. His support of the work of Barney Graham and his colleagues in the Vaccine Research Center was based on respect. Of Graham, Fauci has said, "He understands vaccinology better than anyone I know" (55). The successful development of the structure-based immunogen used in many of the vaccines, including the mRNA vaccines, paved the way for possible control of the pandemic.

Second, that which has attracted constant public attention, is as his role as "America's top infectious diseases doctor." He has been the steady voice of science, communicating complex issues in readily understandable language with a directness born of his NYC heritage. He has countered the science deniers and the promoters of bogus claims, the objections to public health containment measures, and the vaccine resisters and deniers. Since the overwhelming surge of the Delta variant, he has actively pointed out the protective power of the vaccines and the risks to individuals and the country of those who refuse vaccination. These truths have exposed him to vilification and threats on his life, resulting in the Justice Department approving U.S. Marshals as bodyguards (189).

In this time of cultural division and animosity it is well to remember what Francis Collins said of Fauci. Collins is himself a distinguished physician-scientist, the former leader of the Human Genome Project and retired as the long-tenured Director of the NIH. "Tony is the complete model of what you want to see in a public servant. He is dedicated, he's incredibly smart and knowledgeable. He is both a fantastic scientist and a very experienced physician. . . . And he is incapable of doing anything except speaking truth" (190).

THE FUTURE

The fate of COVID-19; optimism fluctuates

After the first year of the pandemic in 2020, with the shutdown of much of society, the promise of the miracle vaccines caused hope to briefly emerge in the wealthy nations of the world. But the pandemic resurged with the dominance of the Delta variant and incomplete vaccination penetrance. In the United States, cultural conflict focused on actions such as vaccination, social distancing, and masking that could significantly control the infection. Despair resurfaced. Then hope emerged again with the availability of vaccine boosters and the announcements of effective oral antiviral medications for COVID-19 (124–126). Additionally, protective immunity was demonstrated for vaccination of children ages 5 to 11, and approval granted (191). Approval of vaccines for children down to six months of age would follow (192). In many areas, diagnostic testing was available at health centers, drive-through sites, and pharmacies, or at-home rapid tests were available for purchase (193). Yet many regions again reported frustrating testing delays as Delta surged, and once again testing demand outstripped supply (194). In December 2021, the highly infectious Omicron variant became predominant, further amplifying infection rates and testing shortages, and filling hospitals to capacity. As Omicron receded and mask requirements and other restrictions were being lifted, highly transmissible versions of Omicron appeared, with marked infectiousness. These subvariants, especially BA.5, overran the U.S. and other regions of the world.

In the face of scientific advances, in under 2.5 years' time the United States had suffered over one million deaths (195). Some commentators suggested that the pandemic in wealthy nations might be transitioning to an endemic disease (196). Others were more cautious, noting that the forces of waning immunity, hesitancy, and the possible emergence of more-virulent transmissible variants would dampen optimism. All of these changes produced a state of pandemic fatigue in many countries, notably in the United States.

Pandemic fatigue had settled like a depressing fog on whole sections of society. It was particularly hard on health care professionals and scientists. Through it all, health care workers, nurses and doctors, emergency medical technicians, respiratory therapists, laboratory technicians, and hospital staff pressed on while facing disease, death, and human loss on an hourly basis. Their own health was threatened, as was that of their families, to whom they might bring home the virus. They pressed on despite inadequate staffing, inadequate supplies of personal protective equipment, and overwhelming exhaustion in the face of massive human need. One physician was quoted as saying, "I just feel I'm empty" (197). Behavior among the general population often revealed the effects of pent-up frustrations. Particularly egregious were disruptive behaviors on airplanes and physical assaults on airline cabin personnel.

During COVID-19, depression and anxiety, which had already been rising globally, continued to rise. It was determined by a large meta-analysis that there was an urgent need to strengthen mental health systems (198). There may also be a multiplier effect, psychiatric illness compounded by complications of long COVID and responses to SARS-CoV-2 in the central nervous system (199).

Vaccination rates were low or barely measurable in many low-income regions and countries by the end of 2021. Vaccine distribution by COVAX, the international organization devoted to supplying vaccines to low-income countries, had fallen behind its goals for a multitude of reasons (200, 201). Many countries did not have sufficient funds to compete for vaccine purchase. Large populations of unvaccinated individuals are fertile ground for massive viral replication and the emergence of new viral mutants. Concern existed that variants might emerge that could evade vaccine- and disease-induced immunity. The Omicron subvariants proved the point, resulting in increased infections and reinfections. A further concern was that variants could emerge with greater virulence, sickening and killing more people than currently dominant variants. The global confirmed counts as of the start of July 2022 were over 560 million infected and over 6 million deaths (195). Concerns have been raised, particularly when excess deaths are examined, that the number of deaths is a significant undercount of the real total. In technologically less advanced countries in which diagnostic virological assays may not be widely employed, official data on the numbers of infected individuals and deaths are likely to be significantly lower than actual. As in industrialized nations, the future of the COVID-19 pandemic in low-income countries cannot be predicted with any certainty.

Future pandemics

The first point to be acknowledged is that COVID-19 is still with us. It has shown itself to be highly resourceful, throwing new variants at humanity in several waves of pandemics. Are there more variants to come, generating new waves of epidemics? Will COVID-19 adopt a pattern similar to the four seasonal coronavirus currently experienced? Some argue that both may come to pass, similar to influenza.

As described earlier, the development of potent oral antiviral medications and effective monoclonal antibodies, not requiring infusion, coupled with variant-specific adaptation of vaccines, could lead to a very different future in advanced societies. "Living with COVID" (202) might not mean balancing reduced vigilance with resumption of life's activities and the associated risk, but rather that we might have the antiviral treatments to effectively combat the infection. The transition of HIV/AIDS from a universally fatal disease to a chronic, treatable condition due to combination highly active antiretroviral therapy including a protease inhibitor in 1996, even without an effective vaccine, stands as an encouraging precedent.

There are at least two classes of evidence that more significant epidemics, and likely pandemics, are on their way. First is the frequency of epidemics in this century. These include the recognition of SARS-1 in 2003, H1N1 influenza in 2009, MERS in 2012, Ebola in 2014, Zika in 2015, and COVID-19 in 2019/2020. Add to these the continuing presence, albeit with effective treatment and preventative medication, of the global HIV/AIDS epidemic, which started in 1981. The second class of evidence encompasses the rapid changes in the environment that multiply the opportunities for spillover viral infections from other species to humans. Changes in human habitats also amplify the chances for rapid viral multiplication in human communities. Prominent forces include rapidly accelerating climate change, which shifts the habitats of animals, insects, and plants; encroachment and destruction of animal habitats by humans; development of huge cities without adequate sanitation, food and water supplies, and basic health care (203); and encampments of displaced persons as global conflicts proliferate. The demonstration of SARS-CoV-2 infection of white-tailed deer in several locations in the United States is a possible example of reverse spillover and the establishment of a nonhuman reservoir for the virus (204, 205). Hence, many locations at risk for emergent viral infections should be monitored prospectively. The importance of "One Health," the interwoven "health of humans, domestic and wild animals, plants, and the wider environment (including ecosystems) are closely interlinked and interdependent," becomes ever more apparent (206).

The role of diagnostic virology in international affairs

We stand at a transition point in the roles of virological diagnosis. The demonstration of the SARS-CoV-2 genetic sequence in the atypical pneumonia cases in Wuhan,

China, catapulted the use of next-generation sequencing (NGS) into international affairs. With its diagnostic role in the pandemic, its use for genetic epidemiology, and its definition of viral variants, NGS plays a major role in relations between nations and in geopolitical affairs. NGS technology has helped to demonstrate the need to strengthen international institutions and agreements (207). The U.S. Secretary of State, Antony J. Blinken, and the U.S. Secretary of the Department of Health and Human Services, Xavier Becerra, editorialized on the need to strengthen global health security in 2021 (208).

The laboratory diagnosis of viral infections in matters of international importance has been found throughout this history. We began our story of human viral disease with the first demonstration, yellow fever, in Cuba by Walter Reed and the Yellow Fever Commission after the Spanish-American War. A few years later, William Crawford Gorgas, using principles based on the Commission's findings of mosquito transmission, cleared the Canal Zone of yellow fever, opening the way for the construction of the Panama Canal (see chapter 1, and Fig. 1 therein). The first viral diagnostic lab anywhere in the world was established in 1941 at the Walter Reed Army Medical Center by Colonel Harry Plotz, MD. It was in preparation for the U.S. entry into World War II (see chapter 5, and Fig. 7 therein).

The 21st century has seen an increasing number of virulent and disruptive epidemics. At this writing (July 2022), monkeypox, which is endemic in west and central Africa, has emerged in non-endemic countries raising great concern (209). On 23 July 2022 WHO declared monkeypox a public health emergency of international concern (210). Hence, it is not surprising that diagnostic virology, using advanced molecular methods, should play a crucial role in global affairs. As the modern world becomes ever more contracted and interconnected, the role of diagnosing and tracking viral infections can only become even more central to our survival and well-being as a species.

REFERENCES

1. **Clarke R.** 2021. *Breathtaking: The UK's Human Story of Covid.* Little, Brown, London, United Kingdom.
2. **World Health Organization.** 26 November 2021. Classification of Omicron (B.1.1.529): SARS-Co-V-2 Variant of Concern. https://www.who.int/news/item/26-11-2021-classification-of-omicron-(b.1.1.529)-sars-cov-2-variant-of-concern.
3. **ProMED-mail.** 30 December 2019. Undiagnosed pneumonia - China (HU): RFI. Archive Number: 20191230.6864153. https://promedmail.org/promed-post/?id=6864153.
4. **Green A.** 2020. Li Wenliang. *Lancet* **395:**682 http://dx.doi.org/10.1016/S0140-6736(20)30382-2.
5. **Normile D, Huihui B, Hicks L, Langin K, Voosen P, O'Grady C, Kaiser J, Stone R, Ortega RP, Cho A, Mervis J.** 2020. Ones we've lost. *Science* **370:**1398–1401 http://dx.doi.org/10.1126/science.370.6523.1398.
6. **Cohen J.** 11 January 2020. Chinese researchers reveal draft genome of virus implicated in Wuhan pneumonia outbreak. ScienceInsider. https://www.science.org/content/article/chinese-researchers-reveal-draft-genome-virus-implicated-wuhan-pneumonia-outbreak.

7. **Huang C, Wang Y, Li X, Ren L, Zhao J, Hu Y, Zhang L, Fan G, Xu J, Gu X, Cheng Z, Yu T, Xia J, Wei Y, Wu W, Xie X, Yin W, Li H, Liu M, Xiao Y, Gao H, Guo L, Xie J, Wang G, Jiang R, Gao Z, Jin Q, Wang J, Cao B.** 2020. Clinical features of patients infected with 2019 novel coronavirus in Wuhan, China. *Lancet* **395**:497–506 http://dx.doi.org/10.1016/S0140-6736(20)30183-5.

8. **Wong E, Barnes JE, Kanno-Youngs Z.** 20 August 2020. Local officials in China hid coronavirus dangers from Beijing, U.S. agencies find, p A5. *The New York Times*, New York, NY.

9. **Horton R.** 2020. *The COVID-19 Catastrophe: What's Gone Wrong and How to Stop It Happening Again.* Polity Press, Cambridge, United Kingdom.

10. **Buckley C, Kirkpatrick DD, Qin A, Hernandez JC.** 31 December 2020. How coronavirus escaped from China's grasp, p A1. *The New York Times*, New York, NY.

11. **Cyranoski D.** 14 December 2020. Zhang Yongzhen: genome sharer. *In Nature's* 10: ten people who helped shape science in 2020. https://www.nature.com/immersive/d41586-020-03435-6/index.html.

12. **Campbell C.** 24 August 2020. Exclusive: The Chinese scientist who sequenced the first COVID-19 genome speaks out about the controversies surrounding his work. *Time.* https://time.com/5882918/zhang-yongzhen-interview-china-coronavirus-genome/.

13. **Qin A, Wang V, Hernandez JC, Li C, Chang Chien A.** 23 January 2021. The scientist, p A6–A7. *The New York Times*, New York, NY.

14. **Wright R.** 28 February 2020. How Iran became a new epicenter of the coronavirus outbreak. *The New Yorker.* https://www.newyorker.com/news/our-columnists/how-iran-became-a-new-epicenter-of-the-coronavirus-outbreak.

15. **Plucinski MM, Wallace M, Uehara A, Kurbatova EV, Tobolowsky FA, Schneider ZD, Ishizumi A, Bozio CH, Kobayashi M, Toda M, Stewart A, Wagner RL, Moriarty LF, Murray R, Queen K, Tao Y, Paden C, Mauldin MR, Zhang J, Li Y, Elkins CA, Lu X, Herzig CTA, Novak R, Bower W, Medley AM, Acosta AM, Knust B, Cantey PT, Pesik NT, Halsey ES, Cetron MS, Tong S, Marston BJ, Friedman CR.** 2021. Coronavirus disease 2019 (COVID-19) in Americans aboard the *Diamond Princess* cruise ship. *Clin Infect Dis* **72**:e448–e457 http://dx.doi.org/10.1093/cid/ciaa1180.

16. **Sakurai A, Sasaki T, Kato S, Hayashi M, Tsuzuki SI, Ishihara T, Iwata M, Morise Z, Doi Y.** 2020. Natural history of asymptomatic SARS-CoV-2 infection. *N Engl J Med* **383**:885–886 http://dx.doi.org/10.1056/NEJMc2013020.

17. **Azimi P, Keshavarz Z, Cedeno Laurent JG, Stephens B, Allen JG.** 2021. Mechanistic transmission modeling of COVID-19 on the *Diamond Princess* cruise ship demonstrates the importance of aerosol transmission. *Proc Natl Acad Sci USA* **118**:e2015482118 http://dx.doi.org/10.1073/pnas.2015482118.

18. **Horowitz J, Povoledo E.** 23 February 2020. Europe confronts coronavirus as Italy battles an eruption of cases. *The New York Times*, New York, NY. https://www.nytimes.com/2020/02/23/world/europe/italy-coronavirus.html.

19. **Winfield N.** 26 April 2020. Perfect storm: virus disaster in Italy's Lombardy region is a lesson for the world. *LA Times*, Los Angeles, CA. https://www.latimes.com/world-nation/story/2020-04-26/perfect-storm-virus-disaster-in-italys-lombardy-region-is-a-lesson-for-the-world.

20. **Alicandro G, Remuzzi G, La Vecchia C.** 2020. Italy's first wave of the COVID-19 pandemic has ended: no excess mortality in May, 2020. *Lancet* **396**:e27–e28 http://dx.doi.org/10.1016/S0140-6736(20)31865-1.

21. **Gonzalez-Reiche AS, Hernandez MM, Sullivan MJ, Ciferri B, Alshammary H, Obla A, Fabre S, Kleiner G, Polanco J, Khan Z, Alburquerque B, van de Guchte A, Dutta J, Francoeur N, Melo BS, Oussenko I, Deikus G, Soto J, Sridhar SH, Wang YC, Twyman K, Kasarskis A, Altman DR, Smith M, Sebra R, Aberg J, Krammer F, García-Sastre A, Luksza M, Patel G, Paniz-Mondolfi A, Gitman M, Sordillo EM, Simon V, van Bakel H.** 2020. Introductions and early spread of SARS-CoV-2 in the New York City area. *Science* **369**:297–301 http://dx.doi.org/10.1126/science.abc1917.

22. **Thompson CN, Baumgartner J, Pichardo C, Toro B, Li L, Arciuolo R, Chan PY, Chen J, Culp G, Davidson A, Devinney K, Dorsinville A, Eddy M, English M, Fireteanu AM, Graf L, Geevarughese A, Greene SK, Guerra K, Huynh M, Hwang C, Iqbal M, Jessup J, Knorr J, Latash J,**

Lee E, Lee K, Li W, Mathes R, McGibbon E, McIntosh N, Montesano M, Moore MS, Murray K, Ngai S, Paladini M, Paneth-Pollak R, Parton H, Peterson E, Pouchet R, Ramachandran J, Reilly K, Sanderson Slutsker J, Van Wye G, Wahnich A, Winters A, Layton M, Jones L, Reddy V, Fine A. 2020. COVID-19 Outbreak—New York City, February 29-June 1, 2020. *MMWR Morb Mortal Wkly Rep* **69:**1725–1729 http://dx.doi.org/10.15585/mmwr.mm6946a2.

23. **Gilbertson A, Melides R, Lieberman R, Schwartz ND.** 9 March 2021. A city ruptured. *The New York Times,* New York, NY. https://www.nytimes.com/interactive/2021/03/09/business/economy/covid-nyc-economy.html.

24. **Gettleman J, Kumar H, Singh KD, Yasir S.** 2 May 2021. Modi declared Covid defeated. India has paid a price ever since, p A1, A5. *The New York Times,* New York, NY.

25. **Chandra S.** 9 June 2021. Why India's second COVID surge is so much worse than the first. *Scientific American.* https://www.scientificamerican.com/article/why-indias-second-covid-surge-is-so-much-worse-than-the-first/.

26. **Laxminarayan R.** 20 April 2021. India's second Covid wave is completely out of control. *The New York Times,* New York, NY. https://www.nytimes.com/2021/04/20/opinion/india-covid-crisis.html.

27. **Slater J, Masih N.** 8 May 2021. In India's surge, a religious gathering attended by millions helped the virus spread. *Washington Post,* Washington, DC. https://www.washingtonpost.com/world/2021/05/08/india-coronavirus-kumbh-mela/.

28. **Mallapaty S.** 2021. India's massive COVID surge puzzles scientists. *Nature* **592:**667–668 http://dx.doi.org/10.1038/d41586-021-01059-y.

29. **Anand A, Sandefur J, Subramanian A.** 20 July 2021. Three new estimates of India's all-cause excess mortality during the COVID-19 pandemic. Center for Global Development Working Paper 589, Washington, DC. https://www.cgdev.org/publication/three-new-estimates-indias-all-cause-excess-mortality-during-covid-19-pandemic.

30. **Zimmer C.** 1 Feb 2022. How Omicron's mutations allow it to thrive, p D3. *The New York Times,* New York, NY. https://www.nytimes.com/2022/01/24/science/omicron-mutations-evolution.html.

31. **Maslo C, Friedland R, Toubkin M, Laubscher A, Akaloo T, Kama B.** 2022. Characteristics and outcomes of hospitalized patients in South Africa during the COVID-19 omicron wave compared with previous waves. *JAMA* **327:**583–584 http://dx.doi.org/10.1001/jama.2021.24868.

32. **Wolter N, Jassat W, Walaza S, Welch R, Moultrie H, Groome M, Amoako DG, Everatt J, Bhiman JN, Scheepers C, Tebeila N, Chiwandire N, du Plessis M, Govender N, Ismail A, Glass A, Mlisana K, Stevens W, Treurnicht FK, Makatini Z, Hsiao NY, Parboosing R, Wadula J, Hussey H, Davies MA, Boulle A, von Gottberg A, Cohen C.** 2022. Early assessment of the clinical severity of the SARS-CoV-2 omicron variant in South Africa: a data linkage study. *Lancet* **399:**437–446 http://dx.doi.org/10.1016/S0140-6736(22)00017-4.

33. **Kozlov M.** 2022. Omicron's feeble attack on the lungs could make it less dangerous. *Nature* **601:**177 http://dx.doi.org/10.1038/d41586-022-00007-8.

34. **Leonhardt D.** 5 Jan 2022. Not the same, p 4 (graph). *The New York Times,* New York, NY.

35. **Zhou P, Yang XL, Wang XG, Hu B, Zhang L, Zhang W, Si HR, Zhu Y, Li B, Huang CL, Chen HD, Chen J, Luo Y, Guo H, Jiang RD, Liu MQ, Chen Y, Shen XR, Wang X, Zheng XS, Zhao K, Chen QJ, Deng F, Liu LL, Yan B, Zhan FX, Wang YY, Xiao GF, Shi ZL.** 2020. A pneumonia outbreak associated with a new coronavirus of probable bat origin. *Nature* **579:**270–273 http://dx.doi.org/10.1038/s41586-020-2012-7.

36. **Temmam S, Vongphayloth K, Baquero Salazar E, Munier S, Bonomi M, Regnault B, Douangboubpha B, Karami Y, Chretien D, Sanamxay D, Xayaphet V, Paphaphanh P, Lacoste V, Somlor S, Lakeomany K, Phommavanh N, Perot P, Donati F, Bigot T, Nilges M, Rey F, van der Werf S, Brey P, Eloit M.** 2021. Coronaviruses with a SARS-CoV-2-like receptor-binding domain allowing ACE2-mediated entry into human cells isolated from bats of Indochinese peninsula. *In Review* http://dx.doi.org/10.21203/rs.3.rs-871965/v1.

37. **Dyer O.** 2021. Covid-19: China stymies investigation into pandemic's origins. *BMJ* **374:**n1890 http://dx.doi.org/10.1136/bmj.n1890.

38. **Dyer O.** 2021. Covid-19: China pressured WHO team to dismiss lab leak theory, claims chief investigator. *BMJ* **374:**n2023 http://dx.doi.org/10.1136/bmj.n2023.

39. **Mallapaty S.** 2021. The search for animals harbouring coronavirus - and why it matters. *Nature* **591:**26–28 http://dx.doi.org/10.1038/d41586-021-00531-z.

40. **Worobey M.** 2021. Dissecting the early COVID-19 cases in Wuhan. *Science* **374:**1202–1204 http://dx.doi.org/10.1126/science.abm4454.

41. **Cohen J.**28 Feb 2022. Do three new studies add up to proof of COVID-19's origin in a Wuhan animal market? ScienceInsider https://www.science.org/content/article/do-three-new-studies-add-proof-covid-19-s-origin-wuhan-animal-market.

42. **Lazarevic I, Pravica V, Miljanovic D, Cupic M.** 2021. Immune evasion of SARS-CoV-2 emerging variants: what have we learnt so far? *Viruses* **13:**1192 http://dx.doi.org/10.3390/v13071192.

43. **Mostafa HH, Pinsky BA.** 2021. Coronaviruses. *In ClinMicroNow: Manual of Clinical Microbiology.* American Society for Microbiology, Washington, DC. https://www.clinmicronow.org/doi/book/10.1128/9781683670438.MCM.ch92.

44. **Faulconbridge G.** 18 March 2021. UK's top COVID-19 virus hunter had a long and winding path to the top. Reuters. https://www.reuters.com/article/uk-health-coronavirus-britain-peacock-ne/uks-top-covid-19-virus-hunter-had-a-long-and-winding-path-to-the-top-idUSKBN2BA26G.

45. **Brierley C.** Hunting for COVID-10 variants. University of Cambridge. https://www.cam.ac.uk/stories/varianthunters. Accessed 20 December 2021.

46. **World Health Organization.** Tracking SARS-CoV-2 variants. https://www.who.int/en/activities/tracking-SARS-CoV-2-variants/. Accessed 10 March 2022.

47. **Harvey WT, Carabelli AM, Jackson B, Gupta RK, Thomson EC, Harrison EM, Ludden C, Reeve R, Rambaut A, Peacock SJ, Robertson DL, COVID-19 Genomics UK (COG-UK) Consortium.** 2021. SARS-CoV-2 variants, spike mutations and immune escape. *Nat Rev Microbiol* **19:**409–424 http://dx.doi.org/10.1038/s41579-021-00573-0.

48. **Stevens H, Berger M.** 23 December 2020. U.S. ranks 43rd worldwide in sequencing to check for coronavirus variants like the one found in the U.K. *Washington Post*, Washington, DC. https://www.washingtonpost.com/world/2020/12/23/us-leads-world-coronavirus-cases-ranks-43rd-sequencing-check-variants/.

49. **Kemp SA, Collier DA, Datir RP, Ferreira IATM, Gayed S, Jahun A, Hosmillo M, Rees-Spear C, Mlcochova P, Lumb IU, Roberts DJ, Chandra A, Temperton N, Sharrocks K, Blane E, Modis Y, Leigh KE, Briggs JAG, van Gils MJ, Smith KGC, Bradley JR, Smith C, Doffinger R, Ceron-Gutierrez L, Barcenas-Morales G, Pollock DD, Goldstein RA, Smielewska A, Skittrall JP, Gouliouris T, Goodfellow IG, Gkrania-Klotsas E, Illingworth CJR, McCoy LE, Gupta RK, CITIID-NIHR BioResource COVID-19 Collaboration, COVID-19 Genomics UK (COG-UK) Consortium.** 2021. SARS-CoV-2 evolution during treatment of chronic infection. *Nature* **592:**277–282 http://dx.doi.org/10.1038/s41586-021-03291-y.

50. **Choi PY.** 2021. Thrombotic thrombocytopenia after ChAdOx1 nCoV-19 vaccination. *N Engl J Med* **385:**e11 http://dx.doi.org/10.1056/NEJMc2107227.

51. **Tao K, Tzou PL, Nouhin J, Gupta RK, de Oliveira T, Kosakovsky Pond SL, Fera D, Shafer RW.** 2021. The biological and clinical significance of emerging SARS-CoV-2 variants. *Nat Rev Genet* **22:**757–773 http://dx.doi.org/10.1038/s41576-021-00408-x.

52. **Burki TK.** 2022. Omicron variant and booster COVID-19 vaccines. *Lancet Respir Med* **10:**e17.

53. **Oude Munnink BB, Worp N, Nieuwenhuijse DF, Sikkema RS, Haagmans B, Fouchier RAM, Koopmans M.** 2021. The next phase of SARS-CoV-2 surveillance: real-time molecular epidemiology. *Nat Med* **27:**1518–1524 http://dx.doi.org/10.1038/s41591-021-01472-w.

54. **Colmenares C.** Clear objectives—Barney Graham leaves Vanderbilt for NIH, but his feet stay planted. *VUMC Reporter* https://reporter.newsarchive.vumc.org/index.html?ID=1523. Accessed 20 December 2021.

55. **Wright L.** 28 December 2020. The plague year. *The New Yorker.* https://www.newyorker.com/magazine/2021/01/04/the-plague-year.

56. **Fauci AS.** 2021. The story behind COVID-19 vaccines. *Science* **372:**109 http://dx.doi.org/10.1126/science.abi8397.

57. **Demske KE.** 29 July 2021. St. Clair Shores native who helped develop technology for COVID-19 vaccine honored. *St. Clair Shores Sentinel*, Warren, MI. https://www.candgnews.com/news/st-clair-shores-native-who-helped-develop-technology-for-covid19-vaccine-honored-121219.

58. **Today@Wayne.** 1 March 2021. Alumnus plays role in vaccine development. Wayne State University. https://today.wayne.edu/news/2021/03/01/wsu-education-plays-role-in-vaccine-development-41623.

59. **Kirchdoerfer RN, Cottrell CA, Wang N, Pallesen J, Yassine HM, Turner HL, Corbett KS, Graham BS, McLellan JS, Ward AB.** 2016. Pre-fusion structure of a human coronavirus spike protein. *Nature* **531:**118–121 http://dx.doi.org/10.1038/nature17200.

60. **Pallesen J, Wang N, Corbett KS, Wrapp D, Kirchdoerfer RN, Turner HL, Cottrell CA, Becker MM, Wang L, Shi W, Kong WP, Andres EL, Kettenbach AN, Denison MR, Chappell JD, Graham BS, Ward AB, McLellan JS.** 2017. Immunogenicity and structures of a rationally designed prefusion MERS-CoV spike antigen. *Proc Natl Acad Sci USA* **114:**E7348–E7357 http://dx.doi.org/10.1073/pnas.1707304114.

61. **Kirchdoerfer RN, Wang N, Pallesen J, Wrapp D, Turner HL, Cottrell CA, Corbett KS, Graham BS, McLellan JS, Ward AB.** 2018. Stabilized coronavirus spikes are resistant to conformational changes induced by receptor recognition or proteolysis. *Sci Rep* **8:**15701 http://dx.doi.org/10.1038/s41598-018-34171-7.

62. **The Royal Swedish Academy of Sciences.** 4 October 2017. Press Release: The Nobel Prize in Chemistry 2017. https://www.nobelprize.org/prizes/chemistry/2017/press-release/.

63. **Cross R.** 29 September 2020. The tiny tweak behind COVID-19 vaccines. *Chemical & Engineering News.* https://cen.acs.org/pharmaceuticals/vaccines/tiny-tweak-behind-COVID-19/98/i38.

64. **Sheikh K, Thomas K.** 28 January 2020. Researchers are racing to make a coronavirus vaccine. Will it help? *The New York Times*, New York, NY. https://www.nytimes.com/2020/01/28/health/coronavirus-vaccine.html.

65. **Wrapp D, Wang N, Corbett KS, Goldsmith JA, Hsieh CL, Abiona O, Graham BS, McLellan JS.** 2020. Cryo-EM structure of the 2019-nCoV spike in the prefusion conformation. *Science* **367:**1260–1263 http://dx.doi.org/10.1126/science.abb2507.

66. **Heath D, Garcia-Roberts G.** 26 January 2021. Luck, foresight and science: how an unheralded team developed a COVID-19 vaccine in record time. *USA Today.* https://www.usatoday.com/in-depth/news/investigations/2021/01/26/moderna-covid-vaccine-science-fast/6555783002/.

67. **Trouillard S.** 18 December 2020. Katalin Kariko, the scientist behind the Pfizer Covid-19 vaccine. France24. https://www.france24.com/en/americas/20201218-katalin-kariko-the-scientist-behind-the-pfizer-covid-19-vaccine.

68. **Kolata G.** 9 April 2021. Her work laid foundation for effective Covid vaccines, p A6. *The New York Times*, New York, NY.

69. **Újszászi I.** 9 April 2020. Katalin Karikó, an alumna of the University of Szeged, is the founder of the most promising vaccine development against the coronavirus. University of Szeged, Szeged, Hungary. https://u-szeged.hu/news-and-events/2021/katalin-kariko-an-alumna.

70. **Cox D.** 2 December 2020. How mRNA went from a scientific backwater to a pandemic crusher. *Wired.* https://www.wired.co.uk/article/mrna-coronavirus-vaccine-pfizer-biontech.

71. **Simpson T.** 28 June 2021. Olympian Susan Francia on how her mother helped develop the COVID-19 vaccines and their American dream. *ESPN.* https://www.espn.com/olympics/story/_/id/31707795/olympian-susan-francia-how-mother-helped-develop-covid-19-vaccines-their-american-dream.

72. **Velasquez B.** Lexington native's mRNA research leads to coronavirus vaccine. *Lexington's Colonial Times Magazine*, Lexington, MA. https://colonialtimesmagazine.com/lexington-natives-mrna-research-leads-to-coronavirus-vaccine/. Accessed 24 August 2021.

73. **Karikó K, Buckstein M, Ni H, Weissman D.** 2005. Suppression of RNA recognition by Toll-like receptors: the impact of nucleoside modification and the evolutionary origin of RNA. *Immunity* **23:**165–175 http://dx.doi.org/10.1016/j.immuni.2005.06.008.

74. **Karikó K, Muramatsu H, Welsh FA, Ludwig J, Kato H, Akira S, Weissman D.** 2008. Incorporation of pseudouridine into mRNA yields superior nonimmunogenic vector with increased translational capacity and biological stability. *Mol Ther* **16:**1833–1840 http://dx.doi.org/10.1038/mt.2008.200.

75. **Garde D, Saltzman J.** 10 November 2020. The story of mRNA: how a once-dismissed idea became a leading technology in the Covid vaccine race. *STAT* and the *Boston Globe*, Boston, MA. https://www.statnews.com/2020/11/10/the-story-of-mrna-how-a-once-dismissed-idea-became-a-leading-technology-in-the-covid-vaccine-race/.

76. **Thomas K.** 19 November 2021. Pfizer says new results show vaccine is safe and 95% effective, p A7. *The New York Times*, New York, NY.

77. **Baden LR, El Sahly HM, Essink B, Kotloff K, Frey S, Novak R, Diemert D, Spector SA, Rouphael N, Creech CB, McGettigan J, Khetan S, Segall N, Solis J, Brosz A, Fierro C, Schwartz H, Neuzil K, Corey L, Gilbert P, Janes H, Follmann D, Marovich M, Mascola J, Polakowski L, Ledgerwood J, Graham BS, Bennett H, Pajon R, Knightly C, Leav B, Deng W, Zhou H, Han S, Ivarsson M, Miller J, Zaks T, COVE Study Group.** 2021. Efficacy and safety of the mRNA-1273 SARS-CoV-2 vaccine. *N Engl J Med* **384:**403–416 http://dx.doi.org/10.1056/NEJMoa2035389.

78. **Robertson C, Harmon A, Smith M.** 15 December 2020. "Healing is coming": US vaccinations begin. Dread persists as death toll tops 300,000, p A1, A6. *The New York Times*, New York, NY.

79. **Davis C, Tipton T, Sabir S, Aitken C, Bennett S, Becker S, Evans T, Fehling SK, Gunson R, Hall Y, Jackson C, Johanssen I, Kieny MP, Mcmenamin J, Spence E, Strecker T, Sykes C, Templeton K, Thorburn F, Peters E, Henao Restrepo AM, White B, Zambon M, Carroll MW, Thomson EC.** 2020. Postexposure prophylaxis with rVSV-ZEBOV following exposure to a patient with Ebola virus disease relapse in the United Kingdom: an operational, safety, and immunogenicity report. *Clin Infect Dis* **71:**2872–2879 http://dx.doi.org/10.1093/cid/ciz1165.

80. **Pollard AJ, et al, EBOVAC2 EBL2001 study group.** 2021. Safety and immunogenicity of a two-dose heterologous Ad26.ZEBOV and MVA-BN-Filo Ebola vaccine regimen in adults in Europe (EBOVAC2): a randomised, observer-blind, participant-blind, placebo-controlled, phase 2 trial. *Lancet Infect Dis* **21:**493–506 http://dx.doi.org/10.1016/S1473-3099(20)30476-X.

81. **Voysey M, et al, Oxford COVID Vaccine Trial Group.** 2021. Safety and efficacy of the ChAdOx1 nCoV-19 vaccine (AZD1222) against SARS-CoV-2: an interim analysis of four randomised controlled trials in Brazil, South Africa, and the UK. *Lancet* **397:**99–111 http://dx.doi.org/10.1016/S0140-6736(20)32661-1.

82. **Sadoff J, Le Gars M, Shukarev G, Heerwegh D, Truyers C, de Groot AM, Stoop J, Tete S, Van Damme W, Leroux-Roels I, Berghmans PJ, Kimmel M, Van Damme P, de Hoon J, Smith W, Stephenson KE, De Rosa SC, Cohen KW, McElrath MJ, Cormier E, Scheper G, Barouch DH, Hendriks J, Struyf F, Douoguih M, Van Hoof J, Schuitemaker H.** 2021. Interim results of a phase 1-2a trial of Ad26.COV2.S Covid-19 vaccine. *N Engl J Med* **384:**1824–1835 http://dx.doi.org/10.1056/NEJMoa2034201.

83. **Logunov DY, Dolzhikova IV, Shcheblyakov DV, Tukhvatulin AI, Zubkova OV, Dzharullaeva AS, Kovyrshina AV, Lubenets NL, Grousova DM, Erokhova AS, Botikov AG, Izhaeva FM, Popova O, Ozharovskaya TA, Esmagambetov IB, Favorskaya IA, Zrelkin DI, Voronina DV, Shcherbinin DN, Semikhin AS, Simakova YV, Tokarskaya EA, Egorova DA, Shmarov MM, Nikitenko NA, Gushchin VA, Smolyarchuk EA, Zyryanov SK, Borisevich SV, Naroditsky BS, Gintsburg AL, Gam-COVID-Vac Vaccine Trial Group.** 2021. Safety and efficacy of an rAd26 and rAd5 vector-based heterologous prime-boost COVID-19 vaccine: an interim analysis of a randomised controlled phase 3 trial in Russia. *Lancet* **397:**671–681 http://dx.doi.org/10.1016/S0140-6736(21)00234-8.

84. **Bucci EM, Berkhof J, Gillibert A, Gopalakrishna G, Calogero RA, Bouter LM, Andreev K, Naudet F, Vlassov V.** 2021. Data discrepancies and substandard reporting of interim data of Sputnik V phase 3 trial. *Lancet* **397:**1881–1883 http://dx.doi.org/10.1016/S0140-6736(21)00899-0.

85. **Keown A.** WHO delays authorization of Russia's Sputnik V vaccine indefinitely. https://www.biospace.com/article/who-delays-authorization-of-russian-made-covid-19-vaccine/. Accessed 17 March 2022.

86. **US Government Accountability Office.** 11 February 2021. Operation Warp Speed: accelerated COVID-19 vaccine development status and efforts to address manufacturing challenges. GAO-21-319. https://www.gao.gov/products/gao-21-319.
87. **Diamond D.** 17 January 2021. The crash landing of "Operation Warp Speed." *Politico.* https://www.politico.com/news/2021/01/17/crash-landing-of-operation-warp-speed-459892.
88. **Weintraub K.** 23 June 2021. Behind the historic US vaccine effort is FDA's Peter Marks. The job is "not for the faint of heart." *USA Today.* https://www.usatoday.com/story/news/health/2021/06/23/fda-peter-marks-behind-us-covid-vaccination-effort/7681024002/.
89. **Gottlieb S.** 2021. *Uncontrolled Spread: Why COVID-19 Crushed Us and How We Can Defeat the Next Pandemic.* Harper, New York, NY.
90. **Jacobson RM, St Sauver JL, Finney Rutten LJ.** 2015. Vaccine hesitancy. *Mayo Clin Proc* **90:**1562–1568 http://dx.doi.org/10.1016/j.mayocp.2015.09.006.
91. **Adams K.** 2 September 2021. Understanding the difference between vaccine hesitancy and resistance, Becker's Health IT. https://www.beckershospitalreview.com/digital-marketing/understanding-the-difference-between-vaccine-hesitancy-and-resistance.html.
92. **Kirzinger A.** 16 November 2021. The increasing importance of partisanship in predicting COVID-19 vaccination status. KFF Covid-19 Vaccine Monitor online. https://www.kff.org/coronavirus-covid-19/poll-finding/importance-of-partisanship-predicting-vaccination-status/.
93. **Graham DA.** 27 April 2021. It's not vaccine hesitancy. It's COVID-19 denialism. *The Atlantic.* https://www.theatlantic.com/ideas/archive/2021/04/its-not-vaccine-hesitancy-its-covid-denialism/618724/.
94. **Khazan O.** 6 December 2021. What's really behind global vaccine hesitancy. *The Atlantic.* https://www.theatlantic.com/politics/archive/2021/12/which-countries-have-most-anti-vaxxers/620901/.
95. **Sallam M.** 2021. COVID-19 vaccine hesitancy worldwide: a concise systematic review of vaccine acceptance rates. *Vaccines (Basel)* **9:**160 http://dx.doi.org/10.3390/vaccines9020160.
96. **Mallapaty S.** 2022. Researchers fear growing COVID vaccine hesitancy in developing nations. *Nature* **601:**174–175 http://dx.doi.org/10.1038/d41586-021-03830-7.
97. **Gostin LO, Parmet WE, Rosenbaum S.** 2022. The US Supreme Court's rulings on large business and health care worker vaccine mandates: ramifications for the COVID-19 response and the future of federal public health protection. *JAMA* **327:**713–714 http://dx.doi.org/10.1001/jama.2022.0852.
98. **Callaway E.** 2020. The unequal scramble for coronavirus vaccines - by the numbers. *Nature* **584:**506–507 http://dx.doi.org/10.1038/d41586-020-02450-x.
99. **Weiland S.** 1 July 2022. FDA wants Covid boosters updated to target subvariants, p A13. *The New York Times,* New York, NY.
100. **World Health Organization.** COVID-19 vaccine tracker and landscape. https://www.who.int/publications/m/item/draft-landscape-of-covid-19-candidate-vaccines. Accessed 20 December 2021.
101. **Cohen J.** 28 February 2020. The United States badly bungled coronavirus testing—but things may soon improve. ScienceInsider. https://www.science.org/content/article/united-states-badly-bungled-coronavirus-testing-things-may-soon-improve.
102. **Huff HV, Singh A.** 2020. Asymptomatic transmission during the coronavirus disease 2019 pandemic and implications for public health strategies. *Clin Infect Dis* **71:**2752–2756 http://dx.doi.org/10.1093/cid/ciaa654.
103. **Hanson KE, Caliendo AM, Arias CA, Hayden MK, Englund JA, Lee MJ, Loeb M, Patel R, El Alayli A, Altayar O, Patel P, Falck-Ytter Y, Lavergne V, Morgan RL, Murad MH, Sultan S, Bhimraj A, Mustafa RA.** 2021. The Infectious Diseases Society of America guidelines on the diagnosis of COVID-19: molecular diagnostic testing. *Clin Infect Dis* **22:**ciab048 http://dx.doi.org/10.1093/cid/ciab048.
104. **Woloshin S, Patel N, Kesselheim AS.** 2020. False negative tests for SARS-CoV-2 infection—challenges and implications. *N Engl J Med* **383:**e38 http://dx.doi.org/10.1056/NEJMp2015897.
105. **Krumholz HM.** 1 April 2020. If you have coronavirus symptoms, assume you have the illness, even if you test negative. *The New York Times,* New York, NY. https://www.nytimes.com/2020/04/01/well/live/coronavirus-symptoms-tests-false-negative.html.

106. **Mandavilli A.** 29 August 2020. Your coronavirus test is positive. Maybe it shouldn't be. *The New York Times*, New York, NY. https://www.nytimes.com/2020/08/29/health/coronavirus-testing.html.

107. **Hanson KE, Altayar O, Caliendo AM, Arias CA, Englund JA, Hayden MK, Lee MJ, Loeb M, Patel R, El Alayli A, Sultan S, Falck-Ytter Y, Lavergne V, Mansour R, Morgan RL, Murad MH, Patel P, Bhimraj A, Mustafa RA.** 2021. The Infectious Diseases Society of America guidelines on the diagnosis of COVID-19: antigen testing. *Clin Infect Dis* **23:**ciab557 http://dx.doi.org/10.1093/cid/ciab557.

108. **Tu YP, Jennings R, Hart B, Cangelosi GA, Wood RC, Wehber K, Verma P, Vojta D, Berke EM.** 2020. Swabs collected by patients or health care workers for SARS-CoV-2 testing. *N Engl J Med* **383:**494–496 http://dx.doi.org/10.1056/NEJMc2016321.

109. **Cox JL, Koepsell SA.** 2020. 3D-printing to address COVID-19 testing supply shortages. *Lab Med* **51:**e45–e46 http://dx.doi.org/10.1093/labmed/lmaa031.

110. **Butler-Laporte G, Lawandi A, Schiller I, Yao M, Dendukuri N, McDonald EG, Lee TC.** 2021. Comparison of saliva and nasopharyngeal swab nucleic acid amplification testing for detection of SARS-CoV-2: a systematic review and meta-analysis. *JAMA Intern Med* **181:**353–360 http://dx.doi.org/10.1001/jamainternmed.2020.8876.

111. **Food and Drug Administration.** 5 October 2021. Potential for false positive results with certain lots of Ellume COVID-19 home tests due to a manufacturing issue: FDA safety communication. https://www.fda.gov/medical-devices/safety-communications/potential-false-positive-results-certain-lots-ellume-covid-19-home-tests-due-manufacturing-issue-fda.

112. **Ganbaatar U, Liu C.** 2021. CRISPR-based COVID-19 testing: toward next-generation point-of-care diagnostics. *Front Cell Infect Microbiol* **11:**663949 http://dx.doi.org/10.3389/fcimb.2021.663949.

113. **Morales-Narváez E, Dincer C.** 2020. The impact of biosensing in a pandemic outbreak: COVID-19. *Biosens Bioelectron* **163:**112274 http://dx.doi.org/10.1016/j.bios.2020.112274.

114. **Vesga O, Agudelo M, Valencia-Jaramillo AF, Mira-Montoya A, Ossa-Ospina F, Ocampo E, Čiuoderis K, Pérez L, Cardona A, Aguilar Y, Agudelo Y, Hernández-Ortiz JP, Osorio JE.** 2021. Highly sensitive scent-detection of COVID-19 patients in vivo by trained dogs. *PLoS One* **16:**e0257474 http://dx.doi.org/10.1371/journal.pone.0257474.

115. **Barbieri D, Giuliani E, Del Prete A, Losi A, Villani M, Barbieri A.** 2021. How artificial intelligence and new technologies can help the management of the COVID-19 pandemic. *Int J Environ Res Public Health* **18:**7648 http://dx.doi.org/10.3390/ijerph18147648.

116. **Hanson KE, Caliendo AM, Arias CA, Englund JA, Hayden MK, Lee MJ, Loeb M, Patel R, Altayar O, El Alayli A, Sultan S, Falck-Ytter Y, Lavergne V, Morgan RL, Murad MH, Bhimraj A, Mustafa RA.** 2020. Infectious Diseases Society of America Guidelines on the diagnosis of COVID-19: serologic testing. *Clin Infect Dis* **12:**ciaa1343 http://dx.doi.org/10.1093/cid/ciaa1343.

117. **Centers for Medicare & Medicaid Services.** 15 October 2020. CMS changes Medicare payment to support faster COVID-19 diagnostic testing. https://www.cms.gov/newsroom/press-releases/cms-changes-medicare-payment-support-faster-covid-19-diagnostic-testing.

118. **Wu KJ.** 3 December 2020. "Nobody sees us": testing-lab workers strain under demand. *The New York Times*, New York, NY. https://www.nytimes.com/2020/12/03/health/coronavirus-testing-labs-workers.html.

119. **Greninger AL, Dien Bard J, Colgrove RC, Graf EH, Hanson KE, Hayden MK, Humphries RM, Lowe CF, Miller MB, Pillai DR, Rhoads DD, Yao JD, Lee FM.** 2022. Clinical and infection prevention applications of severe acute respiratory syndrome coronavirus 2 genotyping: an Infectious Diseases Society of America/American Society for Microbiology consensus review document. *J Clin Microbiol* **60:**e0165921 http://dx.doi.org/10.1128/JCM.01659-21.

120. **Daughton CG.** 2020. Wastewater surveillance for population-wide Covid-19: the present and future. *Sci Total Environ* **736:**139631 http://dx.doi.org/10.1016/j.scitotenv.2020.139631.

121. **Kirby AE, Walters MS, Jennings WC, Fugitt R, LaCross N, Mattioli M, Marsh ZA, Roberts VA, Mercante JW, Yoder J, Hill VR.** 2021. Using wastewater surveillance data to support the COVID-19 response—United States, 2020-2021. *MMWR Morb Mortal Wkly Rep* **70:**1242–1244 http://dx.doi.org/10.15585/mmwr.mm7036a2.

122. Ritchie H, Mathieu E, Rodés-Guirao L, Appel C, Giattino C, Ortiz-Ospina E, Hasell J, Macdonald B, Beltekian D, Roser M. 2020. Coronavirus pandemic (COVID-19). Published online at OurWorldInData.org. Retrieved from https://ourworldindata.org/coronavirus. Accessed 9 February 2022.

123. Zambrano LD, Ellington S, Strid P, Galang RR, Oduyebo T, Tong VT, Woodworth KR, Nahabedian JF III, Azziz-Baumgartner E, Gilboa SM, Meaney-Delman D, Akosa A, Bennett C, Burkel V, Chang D, Delaney A, Fox C, Griffin I, Hsia J, Krause K, Lewis E, Manning S, Mohamoud Y, Newton S, Neelam V, Olsen EOM, Perez M, Reynolds M, Riser A, Rivera M, Roth NM, Sancken C, Shinde N, Smoots A, Snead M, Wallace B, Whitehill F, Whitehouse E, Zapata L, CDC COVID-19 Response Pregnancy and Infant Linked Outcomes Team. 2020. Update: characteristics of symptomatic women of reproductive age with laboratory-confirmed SARS-CoV-2 infection by pregnancy status—United States, January 22-October 3, 2020. *MMWR Morb Mortal Wkly Rep* **69**:1641–1647 http://dx.doi.org/10.15585/mmwr.mm6944e3.

124. Wadman M. 2022. Studies reveal dangers of SARS-CoV-2 infection in pregnancy. *Science* **375**:253 http://dx.doi.org/10.1126/science.ada0233.

125. Lu-Culligan A, Iwasaki A. 26 January 2021. The false rumors about vaccines that are scaring women. *The New York Times*, New York, NY. https://www.nytimes.com/2021/01/26/opinion/covid-vaccine-rumors.html.

126. Gray KJ, Bordt EA, Atyeo C, Deriso E, Akinwunmi B, Young N, Baez AM, Shook LL, Cvrk D, James K, De Guzman R, Brigida S, Diouf K, Goldfarb I, Bebell LM, Yonker LM, Fasano A, Rabi SA, Elovitz MA, Alter G, Edlow AG. 2021. Coronavirus disease 2019 vaccine response in pregnant and lactating women: a cohort study. *Am J Obstet Gynecol* **225**:303.e1–303.e17 http://dx.doi.org/10.1016/j.ajog.2021.03.023.

127. Centers for Disease Control and Prevention. 29 September 2021. CDC statement on pregnancy health advisory. https://www.cdc.gov/media/releases/2021/s0929-pregnancy-health-advisory.html.

128. Davies NG, Klepac P, Liu Y, Prem K, Jit M, Eggo RM, CMMID COVID-19 working group. 2020. Age-dependent effects in the transmission and control of COVID-19 epidemics. *Nat Med* **26**:1205–1211 http://dx.doi.org/10.1038/s41591-020-0962-9.

129. Riphagen S, Gomez X, Gonzalez-Martinez C, Wilkinson N, Theocharis P. 2020. Hyperinflammatory shock in children during COVID-19 pandemic. *Lancet* **395**:1607–1608 http://dx.doi.org/10.1016/S0140-6736(20)31094-1.

130. Feldstein LR, Rose EB, Horwitz SM, Collins JP, Newhams MM, Son MBF, Newburger JW, Kleinman LC, Heidemann SM, Martin AA, Singh AR, Li S, Tarquinio KM, Jaggi P, Oster ME, Zackai SP, Gillen J, Ratner AJ, Walsh RF, Fitzgerald JC, Keenaghan MA, Alharash H, Doymaz S, Clouser KN, Giuliano JS Jr, Gupta A, Parker RM, Maddux AB, Havalad V, Ramsingh S, Bukulmez H, Bradford TT, Smith LS, Tenforde MW, Carroll CL, Riggs BJ, Gertz SJ, Daube A, Lansell A, Coronado Munoz A, Hobbs CV, Marohn KL, Halasa NB, Patel MM, Randolph AG, Overcoming COVID-19 Investigators, CDC COVID-19 Response Team. 2020. Multisystem inflammatory syndrome in U.S. children and adolescents. *N Engl J Med* **383**:334–346 http://dx.doi.org/10.1056/NEJMoa2021680.

131. Penner J, Abdel-Mannan O, Grant K, Maillard S, Kucera F, Hassell J, Eyre M, Berger Z, Hacohen Y, Moshal K, Wyatt M, Cavalli L, Mathias M, Bamford A, Shingadia D, Alders N, Grandjean L, Gaynor E, Brugha R, Stojanovic J, Johnson M, Whittaker E, Pressler R, Papadopoulou C, GOSH PIMS-TS MDT Group. 2021. 6-month multidisciplinary follow-up and outcomes of patients with paediatric inflammatory multisystem syndrome (PIMS-TS) at a UK tertiary paediatric hospital: a retrospective cohort study. *Lancet Child Adolesc Health* **5**:473–482 http://dx.doi.org/10.1016/S2352-4642(21)00138-3.

132. Velasquez-Manoff M. 25 January 2021. What if you never get better from Covid-19? *The New York Times*, New York, NY. https://www.nytimes.com/2021/01/25/world/what-if-you-never-get-better-from-covid-19.html.

133. **Yong E.** 19 August 2020. Long-haulers are redefining COVID-19. *The Atlantic.* https://www.theatlantic.com/health/archive/2020/08/long-haulers-covid-19-recognition-support-groups-symptoms/615382/.

134. **Lowenstein F, Davis H.** 17 March 2021. Long Covid is not rare. It's a health crisis. *The New York Times,* New York, NY. https://www.nytimes.com/2021/03/17/opinion/long-covid.html.

135. **Possick JD.** 28 April 2021. The long haul: forging a path through the lingering effects of COVID-19. Committee on Energy and Commerce, Subcommittee on Health: Testimony of Jennifer Possick, M.D. House Committee on Energy & Commerce, Washington, DC. https://energycommerce.house.gov/sites/democrats.energycommerce.house.gov/files/documents/Witness%20Testimony_Possick_HE_2021.04.28.pdf.

136. **Taquet M, Dercon Q, Luciano S, Geddes JR, Husain M, Harrison PJ.** 2021. Incidence, co-occurrence, and evolution of long-COVID features: a 6-month retrospective cohort study of 273,618 survivors of COVID-19. *PLoS Med* **18:**e1003773 http://dx.doi.org/10.1371/journal.pmed.1003773.

137. **Collins FS.** 23 February 2021. NIH launches new initiative to study "Long COVID." National Institutes of Health. https://www.nih.gov/about-nih/who-we-are/nih-director/statements/nih-launches-new-initiative-study-long-covid.

138. **Abaluck J, Kwong LH, Styczynski A, Haque A, Kabir MA, Bates-Jefferys E, Crawford E, Benjamin-Chung J, Raihan S, Rahman S, Benhachmi S, Bintee NZ, Winch PJ, Hossain M, Reza HM, Jaber AA, Momen SG, Rahman A, Banti FL, Huq TS, Luby SP, Mobarak AM.** 2022. Impact of community masking on COVID-19: a cluster-randomized trial in Bangladesh. *Science* **375:**eabi9069 http://dx.doi.org/10.1126/science.abi9069.

139. **Horby P, Lim WS, Emberson JR, Mafham M, Bell JL, Linsell L, Staplin N, Brightling C, Ustianowski A, Elmahi E, Prudon B, Green C, Felton T, Chadwick D, Rege K, Fegan C, Chappell LC, Faust SN, Jaki T, Jeffery K, Montgomery A, Rowan K, Juszczak E, Baillie JK, Haynes R, Landray MJ, RECOVERY Collaborative Group.** 2021. Dexamethasone in hospitalized patients with Covid-19. *N Engl J Med* **384:**693–704 http://dx.doi.org/10.1056/NEJMoa2021436.

140. **Beigel JH, Tomashek KM, Dodd LE, Mehta AK, Zingman BS, Kalil AC, Hohmann E, Chu HY, Luetkemeyer A, Kline S, Lopez de Castilla D, Finberg RW, Dierberg K, Tapson V, Hsieh L, Patterson TF, Paredes R, Sweeney DA, Short WR, Touloumi G, Lye DC, Ohmagari N, Oh MD, Ruiz-Palacios GM, Benfield T, Fätkenheuer G, Kortepeter MG, Atmar RL, Creech CB, Lundgren J, Babiker AG, Pett S, Neaton JD, Burgess TH, Bonnett T, Green M, Makowski M, Osinusi A, Nayak S, Lane HC, ACTT-1 Study Group Members.** 2020. Remdesivir for the treatment of Covid-19—final report. *N Engl J Med* **383:**1813–1826 http://dx.doi.org/10.1056/NEJMoa2007764.

141. **Gottlieb RL, Vaca CE, Paredes R, Mera J, Webb BJ, Perez G, Oguchi G, Ryan P, Nielsen BU, Brown M, Hidalgo A, Sachdeva Y, Mittal S, Osiyemi O, Skarbinski J, Juneja K, Hyland RH, Osinusi A, Chen S, Camus G, Abdelghany M, Davies S, Behenna-Renton N, Duff F, Marty FM, Katz MJ, Ginde AA, Brown SM, Schiffer JT, Hill JA, GS-US-540-9012 (PINETREE) Investigators.** 2022. Early remdesivir to prevent progression to severe COVID-19 in outpatients. *N Engl J Med* **386:**305–315 http://dx.doi.org/10.1056/NEJMoa2116846.

142. **Taylor PC, Adams AC, Hufford MM, de la Torre I, Winthrop K, Gottlieb RL.** 2021. Neutralizing monoclonal antibodies for treatment of COVID-19. *Nat Rev Immunol* **21:**382–393 http://dx.doi.org/10.1038/s41577-021-00542-x.

143. **O'Brien MP, et al, COVID-19 Phase 3 Prevention Trial Team.** 2022. Effect of subcutaneous casirivimab and imdevimab antibody combination vs placebo on development of symptomatic COVID-19 in early asymptomatic SARS-CoV-2 infection: a randomized clinical trial. *JAMA* **327:**432–441 http://dx.doi.org/10.1001/jama.2021.24939.

144. **FDA.** 8 December 2021. News release: coronavirus (COVID-19) update: FDA authorizes new long-acting monoclonal antibodies for pre-exposure prevention of COVID-19 in certain individuals. https://www.fda.gov/news-events/press-announcements/coronavirus-covid-19-update-fda-authorizes-new-long-acting-monoclonal-antibodies-pre-exposure.

145. **Francis E.** 4 November 2021. Britain authorizes Merck's molnupiravir, the world's first approval of oral covid-19 treatment pill. *Washington Post*, Washington, DC. https://www.washingtonpost.com/health/2021/11/04/covid19-pill-merck-molnupiravir-approval-uk/.

146. **Robbins R.** 2 October 2021. Merck says antiviral pill is first to effectively cut Covid danger, p A1, A14. *The New York Times*, New York, NY.

147. **Johnson CY.** 5 November 2021. Antiviral pills from Pfizer, Merck, show promise against worst Covid-19 outcomes. *Washington Post*, Washington, DC. https://www.washingtonpost.com/health/2021/11/05/pfizer-covid-pill/.

148. **FDA.** 22 December 2021. Coronavirus (COVID-19) update: FDA authorizes first oral antiviral for treatment of COVID-19. https://www.fda.gov/news-events/press-announcements/coronavirus-covid-19-update-fda-authorizes-first-oral-antiviral-treatment-covid-19.

149. **CDC.** 24 May 2022. COVID-19 rebound after Paxlovid treatment. https://emergency.cdc.gov/han/2022/pdf/CDC_HAN_467.pdf. Accessed 25 July 2022.

150. **NIH.** 19 January 2022. COVID-19 treatment guidelines, the COVID-19 treatment guidelines panel's statement on therapies for high-risk, nonhospitalized with mild to moderate COVID-19. https://www.covid19treatmentguidelines.nih.gov/

151. **Gelfand MJ, Jackson JC, Pan X, Nau D, Pieper D, Denison E, Dagher M, Van Lange PAM, Chiu CY, Wang M.** 2021. The relationship between cultural tightness-looseness and COVID-19 cases and deaths: a global analysis. *Lancet Planet Health* **5:**e135–e144 http://dx.doi.org/10.1016/S2542-5196(20)30301-6.

152. **Boghani P.** 2 February 2021. A timeline of China's response in the first days of COVID-19. Frontline. https://www.pbs.org/wgbh/frontline/article/a-timeline-of-chinas-response-in-the-first-days-of-covid-19/.

153. **Myers SL.** 13 January 2021. Facing fresh flare-ups, China puts more than 22 million on lockdown, p A6. *The New York Times*, New York, NY.

154. **Krugman P.** 6 August 2020. Coming next: the greater recession, p A26. *The New York Times*, New York, NY.

155. **Leatherby L, Gelles D.** 11 April 2020. How the virus transformed the way Americans spend their money. *The New York Times*, New York, NY. https://www.nytimes.com/interactive/2020/04/11/business/economy/coronavirus-us-economy-spending.html.

156. **Cohen P.** 7 May 2021. Job growth slowed in April, muddling expectations. *The New York Times*, New York, NY. https://www.nytimes.com/2021/05/07/business/economy/jobs-report-april-2021.html.

157. **McNicholas C, Poydock M.** 19 May 2020. Who are essential workers? Working Economics Blog. Economic Policy Institute, Washington, DC. https://www.epi.org/blog/who-are-essential-workers-a-comprehensive-look-at-their-wages-demographics-and-unionization-rates/.

158. **Chen YH, Glymour M, Riley A, Balmes J, Duchowny K, Harrison R, Matthay E, Bibbins-Domingo K.** 2021. Excess mortality associated with the COVID-19 pandemic among Californians 18-65 years of age, by occupational sector and occupation: March through November 2020. *PLoS One* **16:**e0252454 http://dx.doi.org/10.1371/journal.pone.0252454.

159. **Herstein JJ, Degarege A, Stover D, Austin C, Schwedhelm MM, Lawler JV, Lowe JJ, Ramos AK, Donahue M.** 2021. Characteristics of SARS-CoV-2 transmission among meat processing workers in Nebraska, USA, and effectiveness of risk mitigation measures. *Emerg Infect Dis* **27:**1032–1038 http://dx.doi.org/10.3201/eid2704.204800.

160. **Baggett TP, Gaeta JM.** 2021. COVID-19 and homelessness: when crises intersect. *Lancet Public Health* **6:**e193–e194 http://dx.doi.org/10.1016/S2468-2667(21)00022-0.

161. **Rubin R.** 2021. Helping people who are homeless stay healthy during the pandemic. *JAMA* **325:**517–519 http://dx.doi.org/10.1001/jama.2020.23436.

162. **Eligon J.** 14 February 2021. "Likely a death sentence": officials fear cold weather is greater risk for homeless than virus. *The New York Times*, New York, NY. https://www.nytimes.com/2021/02/14/us/coronavirus-homeless-cold-weather.html.

163. **Lopez L III, Hart LH III, Katz MH.** 2021. Racial and ethnic health disparities related to COVID-19. *JAMA* **325:**719–720 http://dx.doi.org/10.1001/jama.2020.26443.

164. **Mackey K, Ayers CK, Kondo KK, Saha S, Advani SM, Young S, Spencer H, Rusek M, Anderson J, Veazie S, Smith M, Kansagara D.** 2021. Racial and ethnic disparities in COVID-19-related infections, hospitalizations, and deaths: a systematic review. *Ann Intern Med* **174**:362–373 http://dx.doi.org/10.7326/M20-6306.

165. **Agency for Healthcare Research and Quality.** 2019. *2019 National Healthcare Quality and Disparities Report.* Content last reviewed June 2021. Agency for Healthcare Research and Quality, Rockville, MD. https://www.ahrq.gov/research/findings/nhqrdr/nhqdr19/index.html.

166. **Santos PMG, Dee EC, Deville C.** 2021. Confronting anti-Asian racism and health disparities in the era of COVID-19. *JAMA Health Forum* **2**(9):e212579.

167. **von Braun J, Zamagni S, Sorondo MS.** 2020. The moment to see the poor. *Science* **368**:214 http://dx.doi.org/10.1126/science.abc2255.

168. **Fair MA, Johnson SB.** 2021. Addressing racial inequities in medicine. *Science* **372**:348–349 http://dx.doi.org/10.1126/science.abf6738.

169. **Marmot M, Allen J, Bell R, Bloomer E, Goldblatt P, Consortium for the European Review of Social Determinants of Health and the Health Divide.** 2012. WHO European review of social determinants of health and the health divide. *Lancet* **380**:1011–1029 http://dx.doi.org/10.1016/S0140-6736(12)61228-8.

170. **Snowden F.** 2 December 2020. Maps and epidemiology: lessons for COVID-19: virtual mapping as knowing. Series talk by Professor Frank Snowden. Yale University. https://frankeprogram.yale.edu/event/maps-and-epidemiology-lessons-covid-19-virtual-mapping-knowing-series-talk-professor-frank.

171. **Ivey C.** 2020. Land use predicts pandemic disparities. *Nature* **588**:220 http://dx.doi.org/10.1038/d41586-020-03480-1.

172. **Bailey ZD, Feldman JM, Bassett MT.** 2021. How structural racism works—racist policies as a root cause of U.S. racial health inequities. *N Engl J Med* **384**:768–773 http://dx.doi.org/10.1056/NEJMms2025396.

173. **Rabin RC.** 30 November 2020. Prisons are Covid-19 hotbeds. When should inmates get the vaccine? *The New York Times*, New York, NY. https://www.nytimes.com/2020/11/30/health/coronavirus-vaccine-prisons.html.

174. **Macmadu A, Berk J, Kaplowitz E, Mercedes M, Rich JD, Brinkley-Rubinstein L.** 2020. COVID-19 and mass incarceration: a call for urgent action. *Lancet Public Health* **5**:e571–e572 http://dx.doi.org/10.1016/S2468-2667(20)30231-0.

175. **McEvoy J.** 30 September 2021. Prison inmates more vaccinated than corrections staff in at least 13 states. *Forbes*. https://www.forbes.com/sites/jemimamcevoy/2021/09/30/prison-inmates-more-vaccinated-than-corrections-staff-in-at-least-13-states/?sh=1224ddd14ebb.

176. **McMichael TM, Currie DW, Clark S, Pogosjans S, Kay M, Schwartz NG, Lewis J, Baer A, Kawakami V, Lukoff MD, Ferro J, Brostrom-Smith C, Rea TD, Sayre MR, Riedo FX, Russell D, Hiatt B, Montgomery P, Rao AK, Chow EJ, Tobolowsky F, Hughes MJ, Bardossy AC, Oakley LP, Jacobs JR, Stone ND, Reddy SC, Jernigan JA, Honein MA, Clark TA, Duchin JS, Public Health–Seattle and King County, EvergreenHealth, and CDC COVID-19 Investigation Team.** 2020. Epidemiology of Covid-19 in a long-term care facility in King County, Washington. *N Engl J Med* **382**:2005–2011 http://dx.doi.org/10.1056/NEJMoa2005412.

177. **The New York Times.** 27 June 2020. Nearly one-third of U.S. coronavirus deaths are linked to nursing homes. *The New York Times*, New York, NY. https://www.nytimes.com/interactive/2020/us/coronavirus-nursing-homes.html.

178. **Stevis-Gridneff M, Apuzzo M, Pronczuk M, Lima M.** 8 August 2020. As Covid hit, the world let its elderly die, p A1. *The New York Times*, New York, NY.

179. **Médecins Sans Frontières.** 2020. Left behind in the times of COVID-19. https://www.doctorswithoutborders.org/sites/default/files/documents/left-behind-in-the-time-of-covid-19_-a-report-into-nursing-care-homes-in-belgium.pdf.

180. **Zaveri M.** 3 January 2021. Sisters at upstate convent die as outbreak infects 47. *The New York Times*, New York, NY. https://www.nytimes.com/2021/01/03/nyregion/coronavirus-deaths-latham-convent.html.

181. **Hauser C, de Leon C.** 29 January 2021. "It's numbing": nine retired nuns in Michigan died of COVID-19. *The New York Times,* New York, NY. https://www.nytimes.com/2021/01/29/us/nuns-covid-deaths.html.

182. **Peltier E.** 10 February 2021. A French nun turns 117 after knocking down COVID-19. *The New York Times,* New York, NY. https://www.nytimes.com/2021/02/10/world/europe/sister-andre-covid19.html.

183. **Darville S.** 23 July 2020. Reopening school is harder than it should be, p SR6. *The New York Times,* New York, NY.

184. **Donohue JM, Miller E.** 2020. COVID-19 and school closures. *JAMA* **324:**845–847 http://dx.doi.org/10.1001/jama.2020.13092.

185. **Gettleman J, Raj S.** 27 September 2020. Virus closed schools, and world's poorest children went to work, p A1. *The New York Times,* New York, NY.

186. **Burzynska K, Contreras G.** 2020. Gendered effects of school closures during the COVID-19 pandemic. *Lancet* **395:**1968 http://dx.doi.org/10.1016/S0140-6736(20)31377-5.

187. **Specter M.** 10 April 2020. How Anthony Fauci became America's doctor. *The New Yorker.* https://www.newyorker.com/magazine/2020/04/20/how-anthony-fauci-became-americas-doctor.

188. **Gallin JI.** 2007. Introduction of Anthony S. Fauci, MD: 2007 Association of American Physicians George M. Kober Medal. *J Clin Invest* **117:**3131–3135 http://dx.doi.org/10.1172/JCI33692.

189. **Duncan C.** 2 April 2020. White House coronavirus expert Anthony Fauci now has bodyguards. Here's why. *McClatchy.* https://www.mcclatchydc.com/news/coronavirus/article241707131.html.

190. **Facher L.** 21 April 2020. NIH Director Francis Collins on Tony Fauci, the WHO, and running a $39 billion research agency from home. *STAT.* https://www.statnews.com/2020/04/21/francis-collins-q-and-a/.

191. **Centers for Disease Control and Prevention.** 2 November 2021. CDC recommends pediatric COVID-19 vaccine for children 5 to 11 years. https://www.cdc.gov/media/releases/2021/s1102-PediatricCOVID-19Vaccine.html.

192. **FDA.** 17 June 2022. Coronavirus (COVID-19) update: FDA authorizes Moderna and Pfizer-BioNTech COVID-19 vaccines for children down to 6 months of age. https://www.fda.gov/news-events/press-announcements/coronavirus-covid-19-update-fda-authorizes-moderna-and-pfizer-biontech-covid-19-vaccines-children. Accessed 25 July 2022.

193. **US Department of Health & Human Services.** Community-based testing sites for COVID-19. https://www.hhs.gov/coronavirus/community-based-testing-sites/index.html. Accessed 20 December 2021.

194. **Mervosh S, Fernandez M.** 4 August 2020. "It's like having no testing": coronavirus test results are still delayed. *The New York Times,* New York, NY. https://www.nytimes.com/2020/08/04/us/virus-testing-delays.html.

195. **Johns Hopkins University and Medicine.** Coronavirus Resource Center. Accessed 17 May 2022. https://coronavirus.jhu.edu/.

196. **Kupferschmidt K.** 2021. Pandemic enters transition phase-but to what? *Science* **374:**135–136 http://dx.doi.org/10.1126/science.acx9290.

197. **Wadman M.** 29 March 2021. "I'm empty." Pandemic scientists are burning out—and don't see an end in sight. *Science Careers.* https://www.science.org/content/article/i-m-empty-pandemic-scientists-are-burning-out-and-don-t-see-end-sight.

198. **Santomauro DF, et al, COVID-19 Mental Disorders Collaborators.** 2021. Global prevalence and burden of depressive and anxiety disorders in 204 countries and territories in 2020 due to the COVID-19 pandemic. *Lancet* **398:**1700–1712 http://dx.doi.org/10.1016/S0140-6736(21)02143-7.

199. **Nath A.** 2021. Neurologic manifestations of severe acute respiratory syndrome coronavirus 2 infection. *Continuum (Minneap Minn)* **27:**1051–1065 http://dx.doi.org/10.1212/CON.0000000000000992.

200. Editorial. 7 December 2021. The global response to Omicron is making things worse. *Nature* 600:190. https://www.nature.com/articles/d41586-021-03616-x.

201. **WHO.**16 January 2022. News release. COVAX delivers its 1 billionth COVID-19 vaccine dose. https://www.who.int/news/item/16-01-2022-covax-delivers-its-1-billionth-covid-19-vaccine-dose.

202. **Wolfe J.**16 February 2022. Preparing to live with Covid, p 2 sect A. *The New York Times*, New York, NY.

203. **Osterholm MT, Olshaker M.** July/August 2020. Chronicle of a pandemic foretold: learning from the COVID-19 failure—before the next outbreak arrives. *Foreign Affairs.* https://www.foreignaffairs.com/articles/united-states/2020-05-21/coronavirus-chronicle-pandemic-foretold.

204. **Chandler JC, Bevins SN, Ellis JW, Linder TJ, Tell RM, Jenkins-Moore M, Root JJ, Lenoch JB, Robbe-Austerman S, DeLiberto TJ, Gidlewski T, Kim Torchetti M, Shriner SA.** 2021. SARS-CoV-2 exposure in wild white-tailed deer (*Odocoileus virginianus*). *Proc Natl Acad Sci USA* **118:**e2114828118 http://dx.doi.org/10.1073/pnas.2114828118.

205. **Hale VL, Dennis PM, McBride DS, Nolting JM, Madden C, Huey D, Ehrlich M, Grieser J, Winston J, Lombardi D, Gibson S, Saif L, Killian ML, Lantz K, Tell RM, Torchetti M, Robbe-Austerman S, Nelson MI, Faith SA, Bowman AS.** 2022. SARS-CoV-2 infection in free-ranging white-tailed deer. *Nature* **602:**481–486 http://dx.doi.org/10.1038/s41586-021-04353-x.

206. **WHO.**1 December 2021. Tripartite and UNEP support OHHLEP's definition of "One Health" (Joint Tripartite (FAO, OIE, WHO) and UNEP Statement). https://www.who.int/news/item/01-12-2021-tripartite-and-unep-support-ohhlep-s-definition-of-one-health.

207. **Gostin LO, Halabi SF, Klock KA.** 2021. An international agreement on pandemic prevention and preparedness. *JAMA* **326:**1257–1258 http://dx.doi.org/10.1001/jama.2021.16104.

208. **Blinken AJ, Becerra X.** 2021. Strengthening global health security and reforming the international health regulations: making the world safer from future pandemics. *JAMA* **326:**1255–1256 http://dx.doi.org/10.1001/jama.2021.15611.

209. **Alakunle EF, Okeke MI.** 2022. Monkeypox virus: a neglected zoonotic pathogen spreads globally. *Nat Rev Microbiol* **Jul 20:**1–2 doi: 10.1038/s41579-022-00776-z.

210. **Hassan C, Sung C.** 23 July 2022. WHO declares monkeypox a public health emergency of international concern. https://www.cnn.com/2022/07/23/health/monkeypox-who-intl/index.html. Accessed 25 July 2022.

Appendix: Chapter timelines

CHAPTER 1

FEAR OR TERROR ON EVERY COUNTENANCE: YELLOW FEVER

Introduction
- Yellow fever outbreak, Philadelphia, 1793

Germ theory
- Precedents to germ theory
  - Fracastoro's *On Contagion*, 1546
  - Beginnings of microscopy, Antony van Leeuwenhoek and Robert Hooke, 1665
- Louis Pasteur (1822 to 1895)
  - Studies of fermentation, silk worm disease
  - Disproof of spontaneous generation
  - Rabies attenuation and vaccination
- Robert Koch (1843 to 1910)
  - Described the life cycle of anthrax
  - Developed photomicroscopy of bacteria
  - Developed semisolid media for pure cultures of bacteria
  - Isolated the agents of tuberculosis and cholera
  - Articulated Koch's Postulates
- Joseph Lister (1827 to 1912)
  - Aseptic surgical techniques, 1867
- Golden age of bacteriology, 1877 to 1906
  - Identification of the major causes of bacterial diseases

To Catch a Virus, Second Edition. Authored by John Booss and Marie Louise Landry.
© 2023 American Society for Microbiology. DOI: 10.1128/9781683673828.bapp01

Birth of virology

- Adolf Mayer, Dmitri Ivanowski, Martinus Beijerinck, 1885 to 1898
  - Transmission of tobacco mosaic disease by filtered sap
- Friedrich Loeffler and Paul Frosch, 1898
  - Transmission of foot-and-mouth disease with filtered pustular material
- Yellow Fever Commission, Walter Reed and colleagues in Cuba, 1898 to 1902
  - Demonstration that disease was transmitted by mosquitoes
  - Demonstration that filtered blood transmitted disease
  - Elimination of yellow fever from Havana by William Crawford Gorgas (1901) by mosquito eradication. Soon thereafter, using mosquito eradication, he eliminated malaria and Yellow Fever from the Panama Canal Zone, facilitating construction. NB cover image.

CHAPTER 2

OF MICE AND MEN: ANIMAL MODELS OF VIRAL INFECTION

Introduction

- Growing outcry limited further human studies of yellow fever in Cuba, 1901
- Animal models of human viral infection studied beginning in the early 19th century

Rabies

- Transmitted experimentally to several species by G. G. Zinke, 1804
- Transmitted to dogs from human saliva by F. Magendie, 1821
- Transmitted to experimental rabbits by M. Galtier, 1879
- Rabbit model perfected by Louis Pasteur and colleagues, 1880s
  - "Fixed virus" created a reproducibly fixed incubation time
  - Attenuation of rabies virulence by desiccation of rabbit spinal cords
- Treatment of Joseph Meister with a series of inoculations of rabies vaccine, 1885

Polio

- Description of first U.S. epidemic by Charles Solomon Caverly, 1894
- Ivar Wickman's epidemiological studies of Swedish epidemic of 1905
- Transmission to monkeys by Karl Landsteiner and Erwin Popper and by Simon Flexner and Paul A. Lewis, 1908–1909
- Filterability demonstrated by Landsteiner and Constantin Levaditi and by Flexner and Lewis, 1909

Yellow fever

- Transmission to monkeys by Adrian Stokes, Johannes H. Bauer, and N. Paul Hudson, 1928
- Adaptation to white mice by Max Theiler, facilitating a protection assay, 1930
- Development of 17D yellow fever vaccine by Theiler and H. G. Smith, 1937

Epidemic arboviral encephalitides

- Agent of St. Louis encephalitis isolated in monkeys by R. S. Muckenfuss et al. and in mice by L. T. Webster and G. L. Fite, 1933
- Agent of Japanese B encephalitis, named "B" in relation to encephalitis lethargica, isolated by several laboratories in the 1930s in monkeys and mice
- Several other arboviral encephalitis agents also isolated in the 1930s in animals

Influenza

- Pandemic of human influenza estimated to have killed 30 million people, 1918 to 1919
- Isolation of swine influenza and synergistic bacterium by Richard Shope and Paul Lewis, 1931
- Isolation of human influenza virus in ferrets by Wilson Smith, C. H. Andrewes, and Patrick Playfair Laidlaw, 1933
- Red blood cell agglutination for diagnosis demonstrated by George Hirst, 1941

Embryonated eggs

- Developed for use in biology by E. R. Clark, 1920
- Adapted to study of fowl plague virus by Alice Woodruff and Ernest William Goodpasture, 1931
- Used for isolation of influenza virus by Macfarlane Burnet, 1935

CHAPTER 3
FILLING THE CHURCHYARD WITH CORPSES: SMALLPOX AND THE IMMUNE RESPONSE

Introduction

- Antibodies are a crucial component of host defense
- Development of antibodies during viral infection as a means of diagnosis
- Immunological memory is the basis of protection against reinfection
- Persistent antibodies allow demonstration of patterns of disease in populations

Smallpox as a demonstration of protection

- Pocks on mummy show that Egyptian pharaoh Ramses V died of smallpox, 1157 BCE
- Rhazes, a Persian physician, wrote classic treatise on smallpox and measles, 10th century
- Scarring of smallpox a marker of protection against reinfection
- Variolation, inoculation of smallpox crusts, promoted by Lady Mary Wortley Montagu, 1717
- Vaccination, inoculation of cowpox, shown by Edward Jenner to prevent smallpox, 1798
- Global eradication program by the WHO rids the world of smallpox, 1979

Humoral immunity as serology

- Phagocytosis, the first host defense mechanism demonstrated by Elie Metchnikoff, 1884
- Lysis of bacteria by cell-free body fluids demonstrated by George Nuttall, 1888
- Emil Behring's demonstration of immunity to tetanus with Shibasaburo Kitasato and to diphtheria, 1890
- George Sternberg's demonstration of neutralization of vaccinia virus *in vivo*, 1892
- Paul Ehrlich's side chain theory of immunity, 1897
- The role of complement, called alexine, in bacteriolysis and hemolysis; the basis of the complement fixation reaction shown by Jules Bordet and Octave Gengou, 1901
- Opsonic effect demonstrated by A. E. Wright and S. R. Douglas, 1903
- Protective effect of convalescent-phase serum on experimental poliovirus infection demonstrated by A. Netter and C. Levaditi and by Simon Flexner and Paul A. Lewis, 1910
- Serological protection against yellow fever in mice demonstrated by Max Theiler, 1930
- Hemagglutination of red blood cells by influenza virus and its inhibition by convalescent-phase serum demonstrated by George Hirst, 1941 to 1942
- Hemadsorption and its inhibition as a measure of immunity demonstrated by Alexis Shelokov et al., 1958
- Observations that serological assays were the most frequently performed tests for viral diagnosis, 1950s, and the definitive evidence of infection, 1960s

CHAPTER 4
WHAT CAN BE SEEN: FROM VIRAL INCLUSION BODIES TO ELECTRON MICROSCOPY

Cell-based pathology of Rudolf Ludwig Virchow

- Cell theory and cell-based pathology
  - "Cell" first used in biology, Robert Hooke, 1665
  - Plant tissues composed of cells, M. J. Schleiden, 1838

- Animal tissues composed of cells, T. Schwann, 1839
- All cells derive from other cells, Virchow, 1855
- Establishment of the system of pathology based on cells: "cellular pathology," Virchow, 1858
- Improvement of microscopes
  - Compound microscope surpassed the simple microscope by 1830
    - First use of two lenses to improve magnification, Holland, 17th century
    - Correction of chromatic aberration, 1790s
    - Correction of spherical aberration, 1827 to 1830
    - Peak of resolving power reached, 1890s
- Improvement of tissue sections
  - Development of microtome, embedding materials, tissue fixation, cell and nuclear stains, and photomicroscopy by the end of the 19th century

Viral inclusion bodies and other cytopathology

- Smallpox and vaccinia
  - Elementary bodies, John Buist, 1887
  - Guarnieri body, 1892
- Rabies, Negri body, 1903
- Varicella, multinucleated cells, intranuclear inclusions, E. E. Tyzzer, 1906
- Cytomegalovirus
  - Inclusion-bearing cells, inclusions described as protozoa, 1904
  - "Cytomegalia" coined, 1921
  - Clinical description of cytomegalic inclusion disease, 1950
  - Human cytomegalovirus isolated in three laboratories, 1956 to 1957
  - Clinical description based on viral isolation, 1962

Beginnings of electron microscopy

- Theoretical basis established, 1920s
- First instrument, Ernst Ruska and Max Knoll, 1931
- Osmium impregnation of biological samples, Ladislaus Marton, 1934
- First electron micrographs of viruses, Helmut Ruska et al., 1939

CHAPTER 5
THE TURNING POINT: CYTOPATHIC EFFECT IN TISSUE CULTURE

Beginnings of tissue culture

- *In vitro* cultivation of neural tissue by Ross Granville Harrison, 1907
- Advances in technique, Alexis Carrel and Montrose Burrows, 1910 to 1923

First applications of tissue culture to virology

- Edna Steinhardt et al. grew vaccinia virus in corneas *in vitro*, 1913 to 1914
- Thomas Rivers et al. demonstrated cytoplasmic inclusions in vaccinia and intranuclear inclusions in herpetic infected corneas, 1928
- C. Hallauer reported morphological changes in cultures infected with fowl pox, a forerunner of cytopathic effect (CPE), 1932
- C. H. Wang found that infection with western equine encephalitis virus prevented cell growth in culture, 1942
- Thomas Weller and John Enders reported production of hemagglutinin by influenza and mumps viruses *in vitro*, 1948

Franklin Delano Roosevelt (FDR) and the fight against polio

- FDR developed polio as an adult at Campobello Island, summer, 1921. As President, he would later guide the U.S. out of the Great Depression and through WWII.
- With Basil O'Connor, established the National Foundation for Infantile Paralysis (NFIP) to fund the fight against polio, 1938
- Salk polio vaccine trial funded by the NFIP was successful, 1955

John Enders, Thomas Weller, and Frederick Robbins: studies of polio in cell culture

- Growth of poliovirus in nonneural cells *in vitro*, 1949
- Demonstration of CPE, allowing viral isolation and quantitation *in vitro*, 1950
- Demonstration of usefulness of addition of antibiotics to specimens, allowing direct culturing of clinical specimens, 1950
- Award of Nobel Prize, 1954

The first diagnostic virology laboratories in the United States

- Walter Reed Army Medical Center, Washington, DC, in anticipation of World War II; Harry Plotz appointed first director, January 1941, later directed by Joseph Smadel
- California Department of Health Viral and Rickettsial Disease Laboratory, Berkeley, CA; Edwin Herman Lennette ("father of diagnostic virology") appointed as chief, 1947, later joined by Nathalie Schmidt

CHAPTER 6
A TORRENT OF VIRAL ISOLATES: THE EARLY YEARS OF DIAGNOSTIC VIROLOGY

Beginnings of diagnostic virology in university laboratories; examples include the following:

- Children's Hospital of Pennsylvania
  - Clinical virology studies initiated by Werner and Gertrude Henle, 1940s
  - Associate Director of Virus Diagnostic Laboratory appointed, Klaus Hummeler, 1953; applied immunoelectron microscopy to viral identification
  - Relative sensitivities of various cell types for viral isolation studied by F. Deinhardt and G. Henle, 1957
  - Formal diagnostic laboratory set up by Stanley Plotkin and Harvey Friedman, 1972
  - Transition to molecular diagnostics by Richard Hodinka in the latter 1980s
- Yale-New Haven Hospital and West Haven Veterans Administration Medical Center (VAMC)
  - Gueh-Djen (Edith) Hsiung worked with J. L. Melnick to develop tissue culture systems, 1954 to 1960
  - Diagnostic laboratory established at Yale by Hsiung and Dorothy Horstmann, 1960
  - *Diagnostic Virology* published, 1964; ran to four editions to support Hsiung's diagnostic virology workshop, which ran annually or every other year from 1962
  - VA diagnostic virology laboratory at West Haven VAMC established by Hsiung, designated the Department of Veterans Affairs National Virology Reference Laboratory, 1984

Centers for Disease Control (CDC), beginnings of virological studies

- Beatrice Howitt, who had made the first isolations of western equine encephalitis virus, helped establish the laboratory in a former rabies facility
- Virus and rickettsial laboratories at Montgomery, AL, directed by Morris Schaeffer, virologist and infectious diseases specialist, 1949 to 1959
- CDC was designated the WHO Influenza Reference Laboratory for the Western Hemisphere, 1957

- Virology laboratory moved to Atlanta from Montgomery, 1960
- Virology was directed by Telford H. Work, an arbovirologist, from 1960 to 1967
- CDC became a WHO Collaborating Center for Arbovirus Reference and Research during Work's tenure
- The Influenza Coordinating Center led from 1968 to 1979 by Walter Dowdle, who had joined CDC in 1961

Laboratory of Infectious Disease (LID) at the National Institutes of Health (NIH)

- The LID was established in 1948 with Charles Armstrong as the first chief. He had discovered lymphocytic choriomeningitis virus while working on St. Louis encephalitis. He established the philosophy of fully working out an infectious disease.
- Robert J. Huebner, hired by Armstrong, had a distinguished record of accomplishments in discovering infectious disease associations
  - Isolation and characterization of rickettsialpox
  - Coxsackie A viruses and herpangina
  - Coxsackie B viruses and pleurodynia
  - With Joseph Bell, longitudinal studies of infection in a nursery
  - Developed the concept of the oncogene and moved to the National Cancer Institute (NCI) in 1968
- Huebner hired several outstanding investigators to the LID, including the following:
  - Wallace Rowe, who with Huebner isolated adenovirus and human cytomegalovirus (CMV), and studied tumor viruses with Huebner and Janet Hartley
  - Robert Chanock, who had isolated a human paramyxovirus and respiratory syncytial virus (RSV) before coming to the LID. He collaborated with Robert Parrott of Children's Hospital in Washington, DC, and played a major role in elucidating respiratory pathogens at the LID.
  - Albert Kapikian, who defined agents of acute viral gastroenteritis in adults and children and, with others, hepatitis A virus and coronavirus by immunoelectron microscopy

Advances in viral isolation and characterization techniques, and proliferation of isolates

- Arboviruses
  - "Arthropod borne" coined by W. M. Hammon
  - Use of mice as the principal means of isolation from the 1930s
  - Highly sensitive serological assays used by Jordi Casals, with colleagues L. V. Brown and D. H. Clarke, to establish the taxonomy of the arboviruses, 1954 and 1955
  - Sonja M. Buckley, who later with Casals and another colleague would isolate Lassa virus, adapted tissue culture for the growth and characterization of arboviruses, 1959
- Enteric viruses
  - Large numbers of viral isolates were obtained from the human gastrointestinal tract in the absence of disease, prompting the term enteric cytopathogenic human orphan (ECHO) viruses, 1950s
- Common cold viruses
  - The technique of tracheal and nasal pharyngeal organ culture was applied to nasal washings from patients with the common cold. David Arthur John Tyrrell and colleagues, including June Almeida, isolated a new class of viruses, the coronaviruses, 1965 to 1968
- Chronic viral infections
  - Cocultivation of cells from biopsied brain in subacute sclerosing panencephalitis, a chronic neurological disease, with susceptible cells resulted in the isolation of free measles virus, 1969
- CMV
  - Shell vial technology, utilizing inoculation of cultures with centrifugation, introduced by C. A. Gleaves et al. It greatly accelerated the time to diagnosis of CMV and was later applied to other viruses, 1984.

CHAPTER 7
IMAGING VIRUSES AND TAGGING THEIR ANTIGENS

Introduction
- Illnesses yet undefined
  - Viral gastroenteritis
  - Viral hepatitis
- Timeliness of laboratory viral diagnosis

Electron microscopy (EM) advances
- Thin-sectioning improvements, 1940s to 1950s
- Negative staining, Sydney Brenner and Robert Horne, 1959
- Immunoelectron microscopy (IEM)
  - Basic studies, 1941
  - Adaptation to clinical diagnosis, June Almeida and colleagues, 1960s to 1980s
  - Winter vomiting disease, Norwalk agent imaged, Albert Kapikian et al., 1972
  - Acute infantile diarrhea, rotavirus visualized, Ruth Bishop et al., 1973, Thomas Flewett et al., 1973, and Albert Kapikian, et al., 1974.
  - Hepatitis A agent visualized, Stephen Feinstone et al., 1973

Fluorescent-antibody (FA) studies, timeliness of diagnosis
- Albert Hewlett Coons developed the technique
  - First work on tagging an antibody, 1941 to 1942
  - Demonstrated usefulness in studies of infectious diseases, 1950s
- Initial adaptations to diagnostic virology
  - Diagnosis of influenza, Ch'ien Liu, 1955
  - Diagnosis of rabies, R. A. Goldwasser and R. E. Kissling, 1958
- Rapid laboratory viral diagnosis
  - Phillip S. Gardner
    - Studies of acute respiratory disease in children from 1959
    - Developed rapid FA assay for respiratory syncytial virus (RSV) with Joyce McQuillin for children with acute respiratory disease, 1968
    - Published text with McQuillin, *Rapid Virus Diagnosis*, 1974
    - Helped establish European and Pan American groups for rapid viral diagnosis, 1975 and 1977

CHAPTER 8
IMMUNOLOGICAL MEMORY: INGENUITY AND SERENDIPITY

Introduction
- Hepatitis viruses had not been isolated in experimental animals or tissue culture by the 1960s
- Immunologically based techniques would uncover agents of infectious and serum hepatitis

Forms of hepatitis
- Infectious hepatitis, originally called catarrhal jaundice by Rudolf Ludwig Virchow, based on the flawed notion of a mucous plug in the bile duct, 1865
- Serum hepatitis, first termed homologous serum jaundice, was associated with human serum in vaccines and transfusions, 1930s and 1940s
- The terms hepatitis A and B offered for infectious hepatitis and serum hepatitis, respectively, by F. O. MacCallum, 1947

Australia antigen and serum hepatitis, Baruch Blumberg

- Australia antigen discovered incidentally by Baruch Blumberg in the search for serum protein polymorphisms, 1964
- Association of Australia antigen and posttransfusion serum hepatitis, 1969
- Discovery of a "virus-like" particle, the Dane particle, in serum in patients with Australia antigen-positive hepatitis, 1970
- Agreement that the original means to detect Australia antigen or related antigens, double diffusion in agar (Ouchterlony method), was insufficiently sensitive to protect the blood supply, 1970
- Blumberg awarded the Nobel Prize, 1976

Radioimmunoassay (RIA), Rosalyn Yalow and Solomon Berson

- Rosalyn Yalow, a physicist, and Solomon Berson, a physician, teamed up to apply radioisotopes to problems in endocrinology, 1950
- Yalow and Berson developed radioisotopic methodology to measure minute amounts of plasma insulin immunologically, transforming endocrinology and immunology, 1959
- John Walsh developed a highly sensitive and specific RIA for the Australia antigen and the anti-Australia antigen antibody in the Yalow-Berson lab, 1970
- Rosalyn Yalow awarded the Nobel Prize, 1977, Berson having passed away

Enzyme-linked immunosorbent assay (ELISA) and related assays

- RIA, while highly sensitive and specific, had problems associated with radioactivity. It would be supplanted for many assays with a methodology in which antibodies and enzymes were coupled.
- Stratis Avrameas and B. Guilbert's "enzyme-immunological method," 1971
- Bauke K. van Weemen and Anton Schuurs's enzyme immunoassay (EIA), 1971
- Eva Engvall and Peter Perlmann's ELISA, 1971, based on the following:
  - RIA format
  - Coupling of antibodies and enzymes without steric hindrance
  - Adsorption of antibodies to a solid surface
  - Enzymatic production of colored product, allowing measurement

Immunoblotting, also called Western blotting

- Separation of proteins by polyacrylamide gel electrophoresis, transferred electrophoretically to nitrocellulose sheets and measured by EIA, described by Harry Towbin et al., 1979
- Used to demonstrate viral proteins, such as in HIV diagnosis

Immunoglobulin classes to diagnose acute or convalescent viral infections

- Changes in specificity and mobility of antibodies during recovery from foot-and-mouth disease in cattle, F. Brown and J. H. Graves, 1959
- Change in sedimentation of antibodies, 19S to 7S, found to be associated with recovery from human viral infections, A. Schluederberg, 1965
- Use of immunoglobulin M (IgM) (19S) antibody measurement, such as in IgM capture assays, to facilitate diagnosis of acute viral infections, 1979

Monoclonal antibodies, Georges Kohler and Cesar Milstein

- Production of polyclonal antisera inconsistent, labor-intensive, and costly
- Fusion of myeloma cell lines achieved by Richard Cotton and Milstein, 1973
- Fusion of specific antibody-secreting cell line with myeloma cell line to confer immortality achieved by Kohler and Milstein, 1975
- Kohler and Milstein awarded the Nobel Prize, 1984

CHAPTER 9
TO THE BARRICADES: THE MOLECULAR REVOLUTION

Inheritance, DNA, the double helix, and the genetic code
- Linus Pauling's description of the nature of chemical bonds and molecular stability, 1939
- Oswald Avery et al.'s demonstration of DNA as the molecule of inheritance, 1944
- Erwin Chargaff's demonstration of the molar ratios of bases in DNA, Chargaff's Rules, 1950
- Rosalind Franklin and R. G. Gosling's X-ray crystallographic studies of DNA, 1953
- James D. Watson and Francis Crick's solution: an antiparallel double-helical structure, 1953
- Arthur Kornberg's discovery of DNA polymerase I, 1958
- Messenger RNA (mRNA) shown as the unstable intermediate between DNA and ribosomes, 1961
- Genetic code, codons of three bases coding for amino acids, established, 1966
- Demonstration of reverse transcriptase by Howard Temin and Satoshi Mizutani and by David Baltimore, 1970

HIV/AIDS epidemic as a driver of molecular diagnostic virology
- *Pneumocystis carinii* pneumonia and Kaposi's sarcoma (KS) described in young homosexual males, heralding the AIDS epidemic, 1981
- Other risk groups soon defined: individuals abusing intravenous drugs, babies born to infected mothers, blood and blood product recipients, sex workers, heterosexual partners, notably in Africa
- Isolation of a human immunosuppressant retrovirus, human immunodeficiency virus (HIV), 1983
- An antibody assay licensed to diagnose the presence of HIV, 1985
- Control of infection demonstrated following use of combination antiretroviral drugs which included protease inhibitors, 1996
- Measurement of HIV viral load in blood established as a crucial tool in disease management, 1997

Early application of molecular methods to viral diagnosis
- Restriction endonuclease mapping, 1978
- *In situ* nucleic acid hybridization, 1983

Polymerase chain reaction (PCR) and other nucleic acid amplification tests
- Conception of the PCR technique by Kary Mullis, 1983
- First publication of use of PCR for prenatal diagnosis of sickle cell anemia, 1985
- PCR applied to clinical viral diagnosis, 1991
- Commercial systems development
  - Systems based on gene amplification: reverse transcriptase PCR (RT-PCR), nucleic acid sequence-based amplification (NASBA)
  - Systems based on signal amplification: branched-chain DNA
  - Systems based on antibody capture of RNA-DNA hybrids
  - Real-time quantitative PCR
  - Multiplex PCR assays to seek multiple agents simultaneously
  - "Walk-away" technology for flexibility and rapid turnaround

Future of diagnostic virology, as envisioned in 2013 in the first edition
- Expanded role in patient management, not only diagnosis
- Point-of-care (POC) testing and applications to rapid viral diagnosis and health care delivery driven by advances in technology

- Miniaturization of instrumentation and sample volume requirements
- Development of testing for discovery of newly emergent viral diseases
- Ever-increasing role of bioinformatics for virus recognition
- Establishment of syndromic and viral surveillance systems
- Anticipation of future pandemics

CHAPTER 10
THE WORLD CHANGED: THE COVID-19 PANDEMIC

Most notable viral epidemics/pandemics of the 21st century

- Human immunodeficiency virus (HIV)/AIDS, ongoing
- Severe acute respiratory syndrome (SARS), 2003
- H1N1 influenza, 2009
- Middle East respiratory syndrome (MERS), 2012
- Ebola, 2014
- Zika, 2015
- COVID-19, 2019

Crucial scientific advances

- Next-generation sequencing (NGS)
  - Sanger sequencing, the most widely used nucleic acid sequence method for 4 decades, was developed by Fred Sanger and colleagues in 1977, led to a Nobel Prize in 1980, and remains widely used for small-scale projects
  - Next-generation sequencing (NGS) greatly reduces the time and cost of nucleic acid sequencing by using multiple parallel sequence determinations, which are then integrated using bioinformatics programs
  - Allowed the rapid identification and sequencing of SARS-CoV-2 (SARS coronavirus 2)
  - Provided the sequences for construction of COVID-19 diagnostic PCRs
  - Permitted rapid tracking of global spread and generation of new variants
  - Facilitated rapid production of vaccines, especially messenger RNA (mRNA) and viral vector platforms
- Cryo-electron microscopy (cryo-EM)
  - Succeeded X-ray crystallography for the imaging of biomolecules, with more rapid processing and higher resolution
  - Three scientists awarded the Nobel Prize for its development in 2017: Jacques Dubochet, Joachim Frank, and Richard Henderson
  - With NGS, cryo-EM established the design for the structure-based immunogen critical to both the mRNA and viral vector COVID-19 vaccines

Events in first year of the COVID-19 pandemic

- Atypical pneumonia cases recognized in Wuhan, China, December 2019
- Dr. Li Wenliang alerted friends of the outbreak, was punished by authorities, died of the disease, and became a symbol of open communication
- A SARS-related virus was identified and its genetic sequence posted to a public website on 10 January 2020, by Dr Yong-Zhen Zang, aided by a colleague, Dr Edward Holmes. Work on vaccines, diagnostic PCR, and genetic tracking of viral spread began immediately.
- Origin of the virus was unclear: whether a species jump from bats to humans via an intermediate animal or a laboratory leak
- Despite lockdown of Wuhan on 23 January 2020, virus escaped to multiple locations within and outside of China

- Disease was called COVID-19 and the virus was named SARS-CoV-2
- WHO declared a pandemic, 11 March 2020
- A severe outbreak in Lombardy, Italy, shattered Europe's sense of safety, February 2020
- New York City became the epicenter of the disease in the United States, March 2020. Subsequently, the United States became the worldwide leader in COVID-19 deaths.
- After succeeding in "flattening the curve" from May 2020 to January 2021, India suffered a huge outbreak from March to May 2021, associated with large political and religious gatherings. The Delta variant emerged during this time.

The virus, viral variants, diagnosis, and vaccines

- The virus and its variants
  - SARS-CoV-2, a betacoronavirus, resembles solar corona when visualized by EM
  - Viral spike proteins attach to angiotensin-converting enzyme-2 (ACE-2) receptors on the target cell as the initial step in infection
  - COVID-19 Genomics UK Consortium (COG-UK) was organized by Professor Sharon Peacock at Cambridge University to track the spread of the virus
  - As of the end of 2021, variants of concern included Alpha, which emerged in the United Kingdom in September 2020; Delta, in India in October 2020; and Omicron, in southern Africa in November 2021. Of these, Omicron was most transmissible and more likely to evade natural, vaccine-induced, and therapeutic neutralizing antibodies. Omicron subvariants BA 2.12.1, BA.4 and BA.5 emerged, demonstrating great infectiousness and immune-evasion. By mid-2022, BA.4 and especially BA.5 predominated.
  - The need to expand viral sequencing capacity and rapidly identify phenotypic traits of new variants recognized
- Vaccines
  - Barney Graham, with Jason McLellan and colleagues, developed the structure-based immunogen that served as the basis for the mRNA and viral vector COVID-19 vaccines
  - Katalin Karikó and Drew Weissman in their 2005 and 2008 publications solved the riddle of making mRNA vaccines feasible
  - BioNTech with Pfizer, and Moderna with the National Institutes of Health (NIH), produced, tested, and gained Emergency Use Authorization (EUA) from the U.S. Food and Drug Administration (FDA) for their mRNA vaccines within 10 months of the posting of the SARS-CoV-2 genome
  - Virus vector vaccines, successful against Ebola, also proved effective against COVID-19
  - Operation Warp Speed, a multipronged strategy, including manufacturing at risk, to accelerate vaccine development, was underwritten by the U.S. government
  - Vaccine hesitancy and resistance had multiple causes. In the United States, roughly 15% of adults failed to be vaccinated as of the end of 2021
  - Vaccine hesitancy coupled with resistance to nonpharmaceutical interventions compromised pandemic control. It was part of a growing cultural divide in America.
  - Vaccination rates in low- and medium-income countries were low, despite the efforts of COVAX, an organization devoted to equitable access globally
  - Vaccine boosters approved and shown to be clinically effective but their uptake has been variable
- Diagnosis
  - Testing delays in United States impaired the national response to the pandemic
  - PCR is the diagnostic standard, but access and time to result were limitations
  - Rapid antigen tests are less sensitive but can provide results within minutes. Home testing was facilitated to improve access to testing and reduce time to results.
  - Lack of data on protective antibody levels limited the utility of serology
  - Wastewater surveillance recognized as an important tool to monitor transmission

Clinical dilemmas and treatments

- Great variability in outcomes observed, from asymptomatic infection to severe respiratory compromise and death
- Primarily a respiratory illness often associated with loss of taste and smell, progressing in some to severe lower respiratory tract disease, especially if risk factors present such as older age, obesity, diabetes, hypertension, or immunocompromise.
- Long-term symptoms called post-COVID syndrome, long COVID, or post-acute COVID syndrome (PAC-S) range from mild to severely disabling. People with long COVID were originally referred to as "long-haulers."
- Unvaccinated pregnant women found at increased risk for severe disease, preterm birth, and stillbirth
- Children, who in general have mild or no symptoms, rarely experienced a condition known as multisystem inflammatory syndrome in children (MIS-C). In the UK it is called pediatric inflammatory multisystem syndrome temporally associated with SARS-CoV-2 (PIMS-Ts).
- Systemic steroid treatment improved outcomes in hospitalized patients who required supplemental oxygen
- Monoclonal antibodies found useful in treating early infection in patients at risk for progressing to severe disease. However, spike protein mutations reduced efficacy of specific monoclonal antibodies.
- At the end of 2021, two antiviral pills received EUA for use within the first 5 days of symptoms, molnupiravir and nirmatrelvir. Molnupiravir (Lagevrio) introduces mistakes into the viral genome. Paxlovid is a combination of nirmatrelvir, a protease inhibitor, boosted with ritonavir. It had significantly higher clinical efficacy than molnupiravir. In a number of patients a COVID-19 rebound occurred after conclusion of therapy.

Cultural impacts

- Health care disparities in the United States were manifest in greater illness and mortality among Black and Hispanic populations than among White populations
- Poverty was associated with a higher disease incidence and mortality
- "Essential workers," such as those in health care, public safety, grocery stores, and transportation, faced constant exposure to infection, in contrast to those who could work from home
- Businesses suffered under lockdown conditions. Trade and the production and distribution of goods were impacted. Gross domestic product fell and unemployment rose.
- Persons in congregate living facilities such as nursing homes and prisons suffered high infection rates and their consequences
- School closures had serious consequences beyond impaired education, such as loss of socialization, meals, and health care; disproportionate loss of jobs for women; and in some countries, increased child labor
- Anthony Fauci, Director of the National Institute of Allergy and Infectious Diseases, became the face of science and explained the pandemic and recommended public health measures to the American people. Astoundingly, he and many other public health officials were vilified by a portion of the population.

The future

- Vaccines, rapid diagnostic testing, monoclonal antibodies, and antiviral medications offer hope of control.
- Evolution of COVID-19 from a pandemic to an endemic infection with periodic epidemics speculated by some
- Certainty of continuing epidemics in the 21st century
- Need for global surveillance systems for emerging infections
- Need for enforceable global health treaties to rapidly communicate health threats
- Rapid diagnosis and international tracking of newly emerging viral infections will continue to play a crucial role in global affairs

Glossary

Acquired immunodeficiency syndrome (AIDS) A chronic disease caused by the human immunodeficiency virus from the *Retroviridae* family. HIV is transmitted through contact with infected blood, semen, or vaginal fluid and infects CD4 T cells, a key component of the human immune system. CD4 T cells are eventually depleted by the HIV infection, leading to an acquired immunodeficiency characterized by the development of opportunistic infections. AIDS was universally fatal until the implementation of multidrug regimens including protease inhibitors in 1996.

Agar A gelatin substance composed of polysaccharides isolated from algae that can be used as a growth medium on which many microbes can be grown, such as bacteria and fungi.

Agglutination The clumping together of particles (e.g., blood cells, bacteria) as a result of the presence and binding of an antibody or complement.

Angiotensin-converting enzyme-2 receptor (ACE-2 receptor) The host cell ligand used by SARS-CoV and SARS-CoV-2 to enter cells.

Antibodies Immune system proteins produced by plasma cells in response to an antigen. They circulate in the blood and recognize and bind specific epitopes (locations) on antigen proteins. *See also immunoglobulin.*

Antigen A molecule that triggers an immune response. Antigens may be proteins, polysaccharides, lipids, nucleic acids, or other biomolecules.

Antiretroviral therapy (ART) An HIV treatment regimen that involves taking multiple drugs (a "cocktail") aimed at different steps in the viral infection pathway (e.g., reverse transcriptase inhibitors, protease inhibitors, integrase inhibitors, fusion inhibitors). *See also AIDS.*

Arbovirus A name derived from "arthropod-borne virus" indicating a virus transmitted by arthropods (e.g., mosquitos, flies, ticks).

Bacteriophage A virus that infects bacteria.

Betacoronavirus One of the four genera of coronaviruses; includes SARS-CoV, SARS-CoV-2, MERS-CoV, and the common cold coronaviruses OC43 and HKU1.

To Catch a Virus, Second Edition. Authored by John Booss and Marie Louise Landry.
© 2023 American Society for Microbiology. DOI: 10.1128/9781683673828.bgloss

Biotechnology An applied science combining disciplines of the natural sciences and engineering to use biological organisms (or their components) to develop new technologies, products, and services.

Capsid The protein shell surrounding the genetic material of a virus. May or may not be surrounded by an envelope.

Central dogma In molecular biology, the concept that the conversion of genetic information flows from DNA to RNA to protein.

ChAdOx1 An adenoviral vector developed at University of Oxford from a chimpanzee adenovirus and used for the Oxford-AstraZeneca COVID-19 vaccine.

Chargaff's rules The rule establishing base pair complementarity, that is, DNA always features an equal amount of the base adenine compared with thymine and an equal amount of the base guanine compared with cytosine.

Chickenpox Also known as varicella, a highly transmissible disease spread by respiratory aerosols or by direct contact with infected blisters, mucus, or saliva and caused by the varicella-zoster virus of the *Herpesviridae* family. The disease features a characteristic itchy skin rash consisting of blisters that progress to pustules and eventually scab over. Other symptoms include fever, fatigue, and headache with the potential for severe complications including pneumonia and brain inflammation. Following chickenpox, the varicella-zoster virus becomes latent in neurons and may reactivate many years later, causing the disease known as shingles or zoster.

Chromatic aberration In light microscopy, the failure of the microscope lens to focus all colors to the same point and resulting in the appearance of "fringes" of color along edges.

Clade In virology, a group of viruses determined to be similar based on their genetic sequences.

Complement An immunological system made up of a large number of proteins that promotes inflammation and aids antibodies and phagocytic cells in the clearance of pathogens.

Continuous cell line Also known as an immortalized cell line, cultured cells of a single type that can be grown and passaged in culture either for a high known number of cell divisions or indefinitely (as opposed to primary cell lines, which can only be passaged a limited number of times).

Cowpox A zoonotic disease that causes skin pustules in humans and confers immunity to smallpox. It is caused by the cowpox virus from the *Poxviridae* family. Observations that milkmaids were immune to smallpox led Jenner to inoculate pus from cowpox lesions as a vaccination against smallpox. The word "vaccination" is derived from the Latin word *vacca* for cow. The vaccinia virus used in modern smallpox vaccines was originally thought to be cowpox, but instead appears to be most closely related to horsepox.

Coxsackievirus Viruses within the *Enterovirus* genus in the *Picornaviridae* family whose members, designated as group A or group B, cause diseases in humans including aseptic meningitis, hand, foot, and mouth disease, and Bornholm disease. The name derived from a small village in New York state where cases were recognized.

Cryoelectron microscopy (cryo-EM) A fixative-free type of electron microscopy using cryogenically frozen samples that allows for high-resolution imaging of materials in their native states.

Cycle threshold (Ct) In qPCR, the number of cycles needed for the fluorescent signal indicating amplification to cross a designated threshold that is above background level.

Cytomegalovirus A virus in the *Herpesviridae* family. Named for the presence of enlarged infected cells with characteristic inclusions. In humans, often asymptomatic or associated with mononucleosis, hepatitis, or pneumonia. May be life-threatening for newborns and the immunocompromised.

Cytopathic effect Also known as cytopathogenic effect (CPE), structural changes in a cell as a result of viral infection. Includes cell destruction, degeneration, swelling or shrinking, clumping, and formation of syncytia and inclusion bodies.

Cytopathology The microscopic study and diagnosis of diseased cells, either in free cells or fragments of tissue.

Dendritic cells Immune cells that process and present antigen to T cells and are key to initiating the adaptive immune response to a specific pathogen.

Diagnostic virology The laboratory diagnosis of viral infections. Also refers to the field of study and all encompassed methods for the laboratory diagnosis of viral infections.

Disease vector An organism responsible for pathogen transmission from an infected host to a new host.

Echoviruses A contraction of enteric cytopathogenic human orphan viruses, originally discovered in children without symptoms, but later associated with many illnesses typical of the *Picornaviridae* family to which they belong. Most are in the *Enterovirus* genus, but some strains were later separated into the *Parechovirus* genus.

Electron microscope A microscope capable of producing very high-resolution images by using a beam of electrons as the illuminating radiation (as opposed to a beam of visible light used in light microscopy). Subtypes include transmission electron microscopy and scanning electron microscopy.

Embryonated eggs In virology, fertilized eggs containing an embryo and its supporting membranes that can be used for virus propagation.

Endemic disease A disease that is consistently maintained at some level in a particular population or geographic region and is expected to remain so.

Envelope In virology, the outermost layer of some viruses, surrounding the capsid, composed of a lipid bilayer that is derived from host cellular membranes.

Enzyme-linked immunosorbent assay (ELISA) A test which uses the basic properties of antibody-antigen binding to determine whether a substance of interest is in a sample. Depending on the type of ELISA, an antigen (or antibody) from a sample is attached to a surface so that an antibody (or antigen) linked to an enzyme can be applied. Unbound antibody (or antigen) is removed, leaving only that which bound the antigen (or antibody). The substrate for the enzyme is then applied, leading to a detectable signal if the antigen (or antibody) of interest is present in the sample.

Epidemic An outbreak of an infectious disease affecting a large number of individuals within a population or area.

Epidemiology The scientific study of the factors (e.g., incidence, distribution, spread, control) influencing disease and other health-related conditions in a population.

Epstein-Barr virus A member of the *Herpesviridae* family known for causing infectious mononucleosis in humans as well as being associated with a number of lymphoproliferative diseases.

Etiology The cause(s) of a disease or other health-related condition.

Fluorescent antibody staining (FA) Direct staining of cells in clinical specimens with fluorescent-labeled antibodies against specific viruses to rapidly detect viral proteins without the need for culture.

Fomite A nonliving object by which a pathogen may be transmitted (e.g., doorknob, light switch, article of clothing).

Gastroenteritis An inflammation of the digestive system characterized by diarrhea, abdominal cramping, nausea, and vomiting. It may be caused by viruses, bacteria, parasites, or certain chemicals.

Guarnieri bodies Also called B type inclusions, microscopic inclusions in the cytoplasm of epithelial cells infected with a poxvirus (e.g., smallpox).

Hemagglutination The clumping (agglutination) of red blood cells in response to antibodies or certain viruses. In virology, may be used to determine relative concentrations of virus across dilutions or to detect antibodies, such as heterophile antibodies.

Hemolysis The destruction of red blood cells and release of hemoglobin from the cell interior.

Histopathology The microscopic study and diagnosis of diseased tissue in biopsy, surgical, or autopsy specimens.

Human papillomavirus (HPV) A large group of viruses in the *Papillomaviridae* family that cause both skin and genital warts or papillomas. HPV is the most common sexually transmitted infection in humans. While usually benign, some types are high risk for the development of cancer of the cervix, anus, or oropharynx. Vaccines are available to protect against cancer-causing types.

Hyperimmune serum Blood plasma that contains a high amount of antibody to a particular pathogen.

Immunity Innate immunity refers to the nonspecific immune defenses an individual is born with that react quickly to broad categories of pathogens. Adaptive or acquired immunity refers to immunity that develops after exposure to a specific pathogen, involving specialized white cells and antibodies.

Immunoelectron microscopy A technique used in electron microscopy in which samples are first stained with antibodies conjugated with markers that are electron-opaque (e.g., colloidal gold particles) that can be viewed with the electron microscope, allowing for the identification of specific components within the sample.

Immunofluorescence staining A method of detecting antigens in a cell or tissue via the binding of antibodies that have been conjugated with fluorescent dyes that may be visualized with a fluorescent microscope.

Immunoglobulins Also called antibodies, part of the immune system. Glycoproteins are produced by plasma cells (a type of white blood cells) that specifically recognize and bind specific antigens for destruction. Human classes include immunoglobulin G (IgG), immunoglobulin M (IgM), immunoglobulin A (IgA), immunoglobulin D (IgD), and immunoglobulin (IgE). *See also antibodies.*

Immunological memory The ability of the immune system to quickly recognize and immunologically respond to antigens that have been previously encountered (adaptive immunity).

Influenza A disease spread via the respiratory route and caused by influenza viruses from the *Orthomyxoviridae* family. Types A and B cause seasonal epidemics and are targeted by vaccinations annually. The disease may manifest in a population in an endemic, epidemic, or pandemic form. Symptoms can range from mild to severe depending on prior immunity and an individual's age and underlying conditions, and include fever, headache, sore throat, cough, runny nose, muscle aches, and fatigue.

in vitro A Latin term meaning "within glass," such as in a test tube or culture dish, and encompassing a very wide range of scientific studies performed using microbes, cells, or biological molecules outside of a living organism.

in vivo A Latin term meaning "within the living" and encompassing a very wide range of scientific studies performed on whole living organisms.

Koch's postulates The criteria by which a causative relationship between a pathogen and a disease may be determined. These include isolation of the etiological agent from diseased tissue with absence from healthy tissue, growth of the etiological agent in pure culture, re-creation of the disease in a susceptible host, and reisolation of the agent from the infected experimental host. Developed by Robert Koch in the 19th century.

Kyasanur forest disease A disease transmitted by ticks and caused by the Kyasanur forest virus of the *Flaviviridae* family. Recognized in India in 1957. Symptoms include fever, headache, chills, muscle pain, vomiting, bleeding, low blood pressure, and low platelet and blood cell count. Patients may recover or move to a second wave with symptoms including mental disturbance, tremor, and vision loss.

Lineage In virology, a group of viruses that have been identified via sequencing to be closely related to each other and derived from a common predecessor.

Lymphocytic choriomeningitis virus A rodent-borne member of the *Arenaviridae* family responsible for lymphocytic choriomeningitis, a disease of the membranes surrounding the brain, spinal cord, and cerebrospinal fluid.

Middle East respiratory syndrome (MERS) A respiratory disease caused by the MERS coronavirus, first identified in 2012.

Monoclonal antibody Antibodies made from cloning a single white blood cell and that bind a single epitope of an antigen. They have been employed in the treatment of early COVID-19 infection.

mRNA vaccine A vaccine that delivers messenger RNA into a cell to use as instructions for the cell to produce an immunogen.

Multiplex testing In diagnostic virology, the ability to simultaneously test for multiple viruses in a single sample (e.g., testing for influenza A and B, RSV, and SARS-CoV-2 in one test.)

Myxoviruses A term used to refer to any members of the former single-stranded RNA family of viruses *Myxoviridae* which included orthomyxoviruses (e.g., influenza virus) and paramyxoviruses (e.g., parainfluenza viruses, measles virus). "Myxo" refers to the mucus associated with respiratory virus infections.

Negative staining A microscopy method in which the background of a sample is stained leaving the specimen unstained and in relief (contrast) to the background to allow for imaging and the visualization of enhanced morphological detail.

Negri bodies Also called Negri inclusions, are eosinophilic inclusions found in the cytoplasm of some neurons upon infection with the rabies virus. The inclusions contain the rabies virus within the cell and are supportive of a diagnosis of rabies when found by histopathology. They were described by Dr. Adelchi Negri at the start of the 20th century.

Neutralization In virology, the loss of the ability of a virus to cause infection due to the binding of antibodies to the viral surface protein(s), thus preventing the virus from attaching to a host cell.

New World monkey A common name used to refer to monkeys native to Mexico, Central America, and South America, including marmosets, capuchins, and spider monkeys.

Next-generation high-throughput sequencing Sequencing methods that are capable of being done rapidly and massively in parallel. The first generation method, Sanger sequencing, was slower and generated more limited data.

Norwalk virus A species in the genus *Norovirus* in the family *Caliciviridae* that is responsible for "winter vomiting disease" gastroenteritis. Named for an outbreak of gastroenteritis in a school in Norwalk, Ohio in 1968.

Nucleic acid amplification test (NAAT) An assay that makes use of DNA or RNA amplification to aid in the diagnosis of infectious diseases. The first and most commonly used NAAT is PCR (*see Polymerase chain reaction*).

Old World monkey A common name used to refer to monkeys native to Africa and Asia, including baboons and macaques.

Open reading frame (ORF) A section of DNA that, when transcribed, has no stop codon.

Organelle Specialized substructure within a cell with defined functions (e.g., mitochondria, nucleus, lysosomes).

Pandemic An epidemic on a very large (often global) scale. COVID-19 was declared a pandemic by the WHO on 11 March 2020.

Pathogenesis The process by which a disease or other health-related disorder develops. In virology, includes viral transmission, entry, replication, release, and other related interactions in the host at various levels (e.g., molecular, cellular, systemic).

Phylogenetics In virology, the study of the evolutionary history of viral taxa.

Plaque assay A method used to determine the concentration, or titer, of virus in a sample. Virus is serially diluted and introduced to a cell monolayer with a semisolid medium overlay that restricts movement of virus to neighboring cells, causing circular regions of infected or lysed cells which may become large enough to see without a microscope. Fixatives and stains are used to visualize and count these circular regions, or plaques. A calculation is done using the number of plaques and serial dilution to determine the concentration in plaque-forming units (PFU) per milliliter.

Point-of-care (POC) testing Diagnostic testing that is able to be done at the time and place of patient care (e.g., doctor's office, clinic, hospital, or at home).

Polio Also called poliomyelitis, a disease caused by the poliovirus of the *Picornaviridae* family. Infections are transmitted mainly via the fecal-oral route and are often asymptomatic or result in a mild illness. In a small number of patients, the virus infects the central nervous system, leading to aseptic meningitis or to paralysis from infection of the spinal cord (myelitis).

Polyclonal antibodies A collection of antibodies that recognize different epitopes of a specific antigen.

Polymerase chain reaction (PCR) A method of amplifying segments of DNA. Subtypes include qPCR (also called real-time PCR), which provides quantitative and semiquantitative readouts regarding the amplification process as it is happening; and reverse transcription-PCR (RT-PCR), which uses RNA as its starting material and reverse transcribes the RNA into DNA before amplification.

Postexposure prophylaxis Also called PEP, the use of a vaccine, therapeutic, antibodies, or other treatment to either prevent infection or lessen the likelihood of severe disease or death after a known or suspected exposure to a pathogen such as rabies or HIV.

Primary cell line A cell line derived from a specific organ that can only divide and be passaged a limited number of times (as opposed to continuous cell lines, which can divide and be passaged for a long time or indefinitely).

Protozoa An informal term used to refer to single-celled eukaryotes (containing a clearly defined nucleus), which may be either free-living or parasitic (e.g., amoeba, diatoms, paramecia).

Rabies A fatal encephalitis transmitted by exposure to saliva from the bite of an infected animal and caused by the *Rabies lyssavirus* of the *Rhabdoviridae* family. Initial symptoms may include fever, headache, tingling at the site of the bite, and muscle weakness which progress to more severe symptoms including cerebral dysfunction, delirium, excess saliva production, hydrophobia, and insomnia. Postexposure prophylaxis with immune globulin and vaccination can prevent a fatal outcome.

Radioimmunoassay A method in which a known amount of radioactively labeled antigen is mixed with a known amount of antibody for specific binding. Sample serum suspected of containing the same antigen is then added. The unlabeled antigen in the sample serum competes for binding of the antibody with the radiolabeled antigen. The bound antigen is separated, and the radioactivity of the remaining supernatant can be measured to determine the amount of unbound radiolabeled antigen remaining.

Receptor binding domain The portion of a viral surface protein that interfaces with a host-cell protein for viral entry.

Receptor binding motif (RBM) The amino acid residues within the receptor binding domain that actually make contact with the host cell ligand.

Respiratory syncytial virus (RSV) A member of the *Pneumoviridae* family responsible for infection of the respiratory tract and spread via respiratory droplets. The disease is the most common cause of bronchiolitis and pneumonia in children under the age of 1 and can cause serious disease in the elderly or immunocompromised. Disease can range from mild to severe and manifests in both the upper and lower respiratory tracts.

RNA-dependent RNA polymerase (RdRp) An enzyme that replicates RNA from an RNA template; also called an RNA replicase.

RNA replicase *See RdRp.*

Rotavirus A genus within the *Reoviridae* family that is a common cause of diarrheal disease in infants and children under 5 years old.

Sarbecovirus The subgenus of *Betacoronavirus* into which SARS-CoV and SARS-CoV-2 are classified, along with a number of related bat coronaviruses.

Serial passage The process of serially growing a pathogen. In virology, this is growth of a virus (either in an *in vitro* cell culture or *in vivo* animal) followed by a removal and transfer of a small portion of progeny virus to a new environment. Often used to study viral evolution under specific conditions and/or produce altered virulence, for purposes of study or vaccine creation.

Serology The scientific study of blood serum. In diagnostic virology, usually refers to the diagnostic study of antigen or antibodies in the blood for determination or confirmation of disease etiology.

Shell vial culture A culture-based technique used for rapid, sensitive detection of virus *in vitro*. A clinical sample is centrifuged at a low speed onto a single layer of cells on a coverslip in a small shell vial culture tube, incubated for 1 to 2 days, and then viral proteins are detected in the cells on the coverslip via an antigen staining assay.

Smallpox A disease with a mortality rate of 30% spread by respiratory droplets or other bodily fluids and caused by variola viruses of the *Poxviridae* family. Eliminated by a global eradication program by 1980, the major symptoms of the disease were mucosal and skin lesions (pox), fever, back and muscle pain, headache, and fatigue. Survivors often had extensive scarring and some had blindness.

Spherical aberration In microscopy, failure of light rays to all focus to the same plane when passed through a spherical lens, resulting in blurriness along the edges.

Spontaneous generation The idea, disproven by Louis Pasteur, that living organisms could be produced from nonliving material.

St. Louis encephalitis A disease transmitted by mosquitos and caused by the St. Louis encephalitis virus from the *Flaviviridae* family. Symptoms include fever, headache, nausea, dizziness, and weakness. The disease may progress to central nervous system infection and coma with a ~ 5–20% mortality rate. First recognized in and around St Louis, Missouri in a 1933 epidemic.

Structural proteins Viral proteins that make up the composition of the virus particle, e.g., nucleoprotein (N), matrix (M), envelope (E), and spike (S) for coronaviruses.

Structure-based vaccine design A method for creating vaccine immunogens based on data from structural biology techniques (e.g., X-ray crystallography, electron microscopy, nuclear magnetic resonance).

Syncytia A type of cytopathology in which cells fuse or nuclei divide, resulting in a single cell mass containing multiple nuclei.

Tissue culture The growth and maintenance of cells or tissue in a nutrient medium in the absence of the source organism.

Trephination The intentional surgical creation of a hole in the skull, allowing removal of blood or other fluid, relieving pressure on the brain.

Tzanck smear A test used to diagnose a variety of cutaneous and blistering skin infections, including herpes, which involves the staining of sampled cells and microscopic examination to identify the presence of multinucleated Tzanck cells. Arnault Tzanck described the diagnostic tool in the 20th century.

Ultracentrifuge A centrifuge that is capable of accelerations as high as 1,000,000 g that allows for separation of very small particles, including viruses and viral proteins, within a sample.

Vaccination Treatment with a preparation (i.e., vaccine) that stimulates the immune system and confers acquired immunity against a specific pathogen.

Variolation Inoculation against smallpox using material taken from a smallpox pustule.

Viral vector vaccine A vaccine that uses a modified version of a virus to deliver the instructions for a cell to produce an immunogen.

Viruria The presence of virus in the urine.

Western blotting Also called immunoblotting, a technique using antibody to detect specific proteins in a sample that has been separated via gel electrophoresis and transferred to a solid membrane. Described by W. N. Burnette and named in contrast to Southern blotting to detect DNA described by E. M. Southern.

Whole-genome sequencing (WGS) A method for determining the order of nucleotides in an entire genome at once.

X-ray crystallography A method for determining the atomic and molecular structure of a crystallized material using X-ray diffraction.

Yellow fever A disease spread by *Aedes* mosquitos and caused by the yellow fever virus from the *Flaviviridae* family. First human disease proven to be due to a virus. Called yellow fever because of the jaundice which may occur. Initial symptoms may include fever, headache, chills, back and muscle pains, fatigue, appetite loss, nausea, and vomiting. Patients may progress to a more severe disease characterized by liver damage and internal bleeding with a fatality rate of approximately 20–50%. Vaccine preventable.

Zoonosis A disease that is transmitted from animals to humans.

Index

To Catch a Virus, Second Edition. Authored by John Booss and Marie Louise Landry.
© 2023 American Society for Microbiology. DOI: 10.1128/9781683673828.bindex